AF794088

MODIFIED ASPHALT

Edited by **Jose Luis Rivera-Armenta**
and **Beatriz Adriana Salazar-Cruz**

Modified Asphalt
http://dx.doi.org/10.5772/intechopen.71157
Edited by Jose Luis Rivera-Armenta and Beatriz Adriana Salazar-Cruz

Contributors

Farzaneh Tahmoorian, Bijan Samali, John Yeaman, Ragab Abd Eltawab, Ahmet S. Karakas, Faruk Ortes, Vladislav Podolski, Dmitry Chernousov, Alexandr Lukashuk, Bagdat Teltayev, Russell Chianelli, Eva Deemer

© The Editor(s) and the Author(s) 2018
The rights of the editor(s) and the author(s) have been asserted in accordance with the Copyright, Designs and Patents Act 1988. All rights to the book as a whole are reserved by INTECHOPEN LIMITED. The book as a whole (compilation) cannot be reproduced, distributed or used for commercial or non-commercial purposes without INTECHOPEN LIMITED's written permission. Enquiries concerning the use of the book should be directed to INTECHOPEN LIMITED rights and permissions department (permissions@intechopen.com).
Violations are liable to prosecution under the governing Copyright Law.

(cc) BY

Individual chapters of this publication are distributed under the terms of the Creative Commons Attribution 3.0 Unported License which permits commercial use, distribution and reproduction of the individual chapters, provided the original author(s) and source publication are appropriately acknowledged. If so indicated, certain images may not be included under the Creative Commons license. In such cases users will need to obtain permission from the license holder to reproduce the material. More details and guidelines concerning content reuse and adaptation can be found at http://www.intechopen.com/copyright-policy.html.

Notice

Statements and opinions expressed in the chapters are these of the individual contributors and not necessarily those of the editors or publisher. No responsibility is accepted for the accuracy of information contained in the published chapters. The publisher assumes no responsibility for any damage or injury to persons or property arising out of the use of any materials, instructions, methods or ideas contained in the book.

First published in London, United Kingdom, 2018 by IntechOpen
IntechOpen is the global imprint of INTECHOPEN LIMITED, registered in England and Wales, registration number: 11086078, The Shard, 25th floor, 32 London Bridge Street
London, SE19SG – United Kingdom
Printed in Croatia

British Library Cataloguing-in-Publication Data
A catalogue record for this book is available from the British Library

Additional hard copies can be obtained from orders@intechopen.com

Modified Asphalt, Edited by Jose Luis Rivera-Armenta and Beatriz Adriana Salazar-Cruz
p. cm.
Print ISBN 978-1-78923-726-9
Online ISBN 978-1-78923-727-6

We are IntechOpen, the world's leading publisher of Open Access books

Built by scientists, for scientists

3,750+
Open access books available

115,000+
International authors and editors

119M+
Downloads

151
Countries delivered to

Our authors are among the
Top 1%
most cited scientists

12.2%
Contributors from top 500 universities

CLARIVATE ANALYTICS BOOK CITATION INDEX INDEXED

WEB OF SCIENCE™

Selection of our books indexed in the Book Citation Index in Web of Science™ Core Collection (BKCI)

Interested in publishing with us?
Contact book.department@intechopen.com

Numbers displayed above are based on latest data collected.
For more information visit www.intechopen.com

For EU product safety concerns:
IN TECH d.o.o., Prolaz Marije Krucifikse Kozulić 3, 51000 Rijeka, Croatia,
info@intechopen.com or visit our website at intechopen.com.

Meet the editors

José Luis Rivera-Armenta was born in Tampico, Mexico, in 1971. He has a BSc in chemical engineering (1994), an MSc in Petroleum Technology and Petrochemicals (1998), and a PhD in Chemical Engineering (2002) at the Technological Instituto of Madero City (ITCM). Since 2003 he has been a full-time professor in postgraduate programs at ITCM and project manager of several projects sponsored by the National Technologic of Mexico and CONACYT. Since 2005 he has been a member of the National Research System in CONACYT level 1. He is responsible for the injection and extrusion, and thermal analysis laboratory at the Petrochemical Research Center at ITCM. He has advisored eight PhD, 15 master's degree and three bachelor theses. He also supervised three posdoctorate students. He has had 48 articles and five book chapters published.

Beatriz A. Salazar Cruz, PhD since 2014, has been an associate member of MATCO since 2016, and an associate professor since 2012 of the Technological Institute of Madero City. She has had experience in the chemical process industry from 1995 to 2008 (Dynasol Elastomers) developing projects with SBS, SBR and SEBS polymers, and characterizing and innovating the quality of products in a wide range of applications: asphalts, adhesives and compounds. She is the author or coauthor of several scientific publications in English and Spanish, and author or coauthor of several book chapters. She has taught several rheology courses and is a collaborator of several projects supported by the National Technologic of Mexico and CONACYT. She has advisored a master's degree thesis and also supervised engineering students.

Contents

Preface

This book presents specific information on asphalt modifications, which has become an interesting area in recent years, due to the relevance of these kinds of materials for applications in roads and highways. The current requirements in terms of the capacity to withstand greater loads on roads and highways make it necessary to develop more resistance asphalt with better properties, for instance deformation resistance at high temperatures, low-temperature cracking, and fatigue cracking, with the aim of meeting the needs of different geographical areas.

Asphalt is a byproduct of oil refinery and is a complex mix of hydrocarbons. Because it has a varied composition that mainly consists of heavy oil fractions, such as saturated asphaltenes, resins, and aromatics, and depending on its origin, the proportions of these fractions can vary to make asphalt heavier.

The main problems with modified asphalts are phase separation and thermal stability, which are responsible for failures such as ruts, fractures, and cracks, among others. Depending of the environmental conditions of use, these failures can increase in some cases, so it becomes necessary to use modifier agents that improve the characteristics of asphalts.

Asphalt modifiers by excellence are elastomers, mainly styrene-butadiene; however, even modified asphalt with these kinds of elastomers have fallen short in fulfilling their main purpose, which is why it is necessary to look for new modifiers and especially new developments to adapt and evaluate the performance of modified asphalts.

Among the main properties that a modified asphalt must have are mechanical properties, since final performance depends on their supporting the loads and conditions to which they are subjected during their application.

In recent years, environmental issues have become an important aspect of caring, so not only should modified asphalts allow for lower fuel consumption, but also the modifying agents such as polymers or natural products can be recycled, because in this way natural resources would be used less and less pollution would be generated.

This book is divided into two sections: the first "Asphalts with Environmental Focus" and the second "Evaluation of Asphalt Performance." The objective of presenting in this way is to analyze the options to generate modified asphalts using environmentally friendly materials and, on the other hand, to analyze the performance of modified asphalts with conventional agents such as rubbers, polymers, or clays, or to evaluate properties such as fatigue failures or thermal properties.

This book is a compilation of work that studies how modified asphalts are developed using unconventional materials that can be environmentally friendly and, on the other hand, work where the performance of modified asphalts is evaluated mainly for their mechanical properties, due to the relevance of these areas in the applications of modified asphalts.

Dr. José Luis Rivera-Armenta and Dr. Beatriz Adriana Salazar-Cruz
National Technologic of Mexico (TecNM)
Technological Institute of Madero City, Mexico

Section 1

Asphalts with Environmental Focus

Chapter 1

Asphalt Modified with Biomaterials as Eco-Friendly and Sustainable Modifiers

Ragab Abd Eltawab Abd El-latief

Additional information is available at the end of the chapter

http://dx.doi.org/10.5772/intechopen.76832

Abstract

High construction costs, when combined with awareness regarding environmental stewardship have encouraged the use of waste and renewable resources in asphalt modification. Increasing energy costs and the strong worldwide demand for petroleum have encouraged the development of alternative binders to modify or replace asphalt binders. The benefits of using alternative binders are that they can help save natural resources and reduce energy consumption while maintaining and in some cases improving asphalt performance. Common alternative binders include engine oil residue, bio-binder, soybean oil, palm oil, fossil fuel, swine waste, and materials from pyrolysis. Chemical compositions of the majority of these alternative binders are similar to those of unmodified asphalt binders (e.g. Resin, saturates, aromatics, and asphaltene). On the other hand, tests indicate the wide variability in the properties of alternative binders. Also, the chemical modification mechanism for asphalt with alternative binders depends clearly on the unmodified asphalt and is consequently not well understood. For energy sustainability, environment-friendly materials and an urgent need for infrastructure rehabilitation that more research is needed to evaluate the alternative binders for use in asphalt modification. The alternative binders should have moisture resistance and good aging characteristics.

Keywords: asphalt, sustainable, alternative binders, biomaterials, eco-friendly, bio-binder

IntechOpen

© 2018 The Author(s). Licensee IntechOpen. This chapter is distributed under the terms of the Creative Commons Attribution License (http://creativecommons.org/licenses/by/3.0), which permits unrestricted use, distribution, and reproduction in any medium, provided the original work is properly cited.

1. Introduction

1.1. General introduction

Bio-materials that are used as binders in asphalt mixtures are termed "bio binders" [1]. The interest on using bio-binders in pavement engineering has significantly grown over the last decades due to the increasing scarcity of raw materials and environmental concerns about the use of non-recoverable natural resources.

Common alternative binders include engine oil residue, bio-binder, soybean oil, palm oil, fossil fuel, swine waste, and materials from pyrolysis [2]. Chemical compositions of the majority of these alternative binders are similar to those of unmodified asphalt binders (e.g. Resin, saturates, aromatics, and asphaltene) [3]. On the other hand, tests indicate the wide variability in the properties of alternative binders. Also, the chemical modification mechanism for asphalt with alternative binders depends clearly on the unmodified asphalt and is consequently not well understood [4, 5]. For energy sustainability, environment-friendly materials and an urgent need for infrastructure rehabilitation that more research is needed to evaluate the alternative binders for use in asphalt modification. The alternative binders should have moisture resistance and good aging characteristics [6].

2. Bio binder's definition and resources

2.1. Definition

Bio-binder is an eco-friendly asphalt binder alternative obtained from non-petroleum-based renewable resources, which should not rival any food material. From the definition, bio-binder can be described as dark brown, high flow, organic liquids that are comprised mainly of highly oxygenated compounds [7–10].

2.2. Bio-binder material sources

A range of different vegetable oils has been investigated in recent times, through the application of scientific research and development, to determine their physical and chemical properties to study their applicability to be used as bio-binders in the pavement industry [11–13]. Bio-oils are produced from plant matter and residues, such as municipal wastes, agricultural crops, and byproducts from agricultural and forestry [8]. Other biomass sources include molasses and rice, sugar, potato starches and corn, gum resins and natural tree, vegetable oils and natural latex rubber, cellulose, lignin, waste oil of palm, peanut oil waste, coconut waste, potato starch, canola oil waste, dried sewerage effluent, and others.

Utilize this bio-binder can be a great potential as a modifier for asphalt binder because of similar chemical properties when compared with crude petroleum as shown in **Table 1**.

Currently, bio-binder is the second leading renewable energy in the nation after hydropower [14, 15].

Physical property	Value
Percent of moisture (wt %)	15–30
pH	2.5
Specific gravity	1.2
Elemental composition (H, C, O, N) (wt %)	(5.5–7.0, 54–58, 35–40, 0–0.2)
Distillation residual (wt %)	Up to 50
Viscosity @500°C (pa)	40–100

Table 1. Typical characteristics of wood-based bio-oils.

3. Materials used for the production of bio-binder

3.1. Bio-oil

Bio-Oil is the liquid produced from the rapid heating of biomass in a vacuum condition [17]. There are many advantages for bio-oils over asphalt from crude oil as they are environment-friendly, renewable, present a great economic opportunity, and provide energy security.

3.1.1. Bio-oil sources and production

3.1.1.1. Thermochemical liquefaction

To obtain more gasoline and other liquid fuels, the quality of the bio-oil is improved by using processes such as thermal cracking and hydrogenation [16, 17]. Upon fractionation, the light hydrocarbon fraction can be used as fuel. The remaining heavy residue called bio-binder can be used as an asphalt binder modifier.

3.1.1.2. Pyrolysis process

Fast pyrolysis is a thermal decomposition process that requires a high heat transfer rate to the biomass particles and a short vapor residence time in the reaction zone [8], which is a high-temperature process for the production of vapors, aerosols and some coal-like char where biomass is quickly heated in the vacuum and then decomposes. The dark brown mobile fluid (bio-oil) is formed after cooling and condensation of these vapors and aerosols. When organic matter is biomass, consisting of biopolymers (such as cellulose, hemicelluloses, and lignin), the oils produced are called bio-oils. Generally, fast pyrolysis is used to obtain high-grade bio-oil. Fast pyrolysis processes produce 60–75 wt% of liquid bio-oil, 15–25 wt% of solid char, and 10–20 wt% of non-condensable gases. Fast pyrolysis initially starts with slow heating rates, and then involves a rapid heating rate of the biomass, that can reach up to 300°C/min, but not as fast as flash pyrolysis.

Fast pyrolysis design variables include, but are not limited to the following ones reported by [8]: feed moisture content, particle size, pretreatment, reactor configuration, heat supply, heat transfer, heating rates, reaction temperature, vapor residence time, secondary cracking, char separation, ash separation, and liquid collection.

3.1.2. Importance of using bio-oil in the modification of asphalt

The bio-oil obtained from waste biomass at low cost is an environmentally friendly material containing the natural antioxidant lignin, bio-renewable asphalt modifier and asphalt substitutes potential to be successfully applied as an antioxidant additive in asphalt pavements. The other various chemicals that using as antioxidant additives are not environmentally and economically preferable [18].

3.2. Biopolymers

Bio-plastics or organic plastics are a form of plastics derived from renewable biomass sources, such as vegetable oil, corn starch, pea starch, or microbiota. Some of its advantages are often biodegradable, not toxic to produce and alternative to traditional plastics this is included in the definition of biodegradable polymers mentioned in [19]. Biopolymers are an alternative to petroleum based polymers produced by living organisms. The field of biopolymers is still in its early stage but is growing in popularity every day. Biodegradable polymers are produced by using micro-organisms, plants and animals (biological systems), or biological starting materials, which have been synthesized chemically from (e.g. starch, sugars, oils or natural fats, etc.). The biopolymers are biodegradable most of them and water-soluble some of them. Most of the biopolymer are compostable or will decompose in landfills, but the time can vary from a few days to even years, and will eventually decompose [20]. The natural rubber is the best example classified as a biopolymer. The natural rubber is the most common elastomer in almost all human activities due to its unique characteristics, raw or combined with synthetic elastomers. This elastomer has the ability to improve pavements performance and durability for being applied in road pavements as a recycled material from tires [21].

3.2.1. Importance of using the biopolymers

The bitumen degree controls the performance of the pavement mixture during the service temperature. In many cases, bitumen properties need to change to enhance their flexible characteristics at low temperatures to withstand adequate cracking and increase shear resistance through high-temperatures and continuous loads to resist corrosion. With the addition of SBS polymers, the physical properties of polymers are usually modified to produce an improved asphalt grade that improves the performance of hot mix asphalt. In 2008, there was a shortage of type spherical polymers for the asphalt industry forcing asphalt mixing producers and owner/agencies to search for different products that could be used as bitumen [21]. With expectations of increased demand from soft asphalt in the next time, the need for viable asphalt is cost-effective that can be used instead of typical spike-type rates will still strong. Of asphalt-modified polymer mixture, nearly 80% of the polymer modified asphalt uses the SB type polymers. Thus there is a great market opportunity to create new polymers that can complement and/or replace the type SBS polymers used in asphalt paving. The researchers of chemistry and engineering are currently working to obtain chemicals and polymers from renewable raw materials, composite materials and manufactured products known as "biological materials." When deliberating biomaterials, it is noteworthy that there are no direct negative effects caused by plants on the ecosystem, they can recycle carbon dioxide for the Earth's crust, grow in different climatic zones, and much work on soil fertility improvement [22].

3.2.2. Bio-elastomers

Elastomers materials, or rubber materials, have a cross-linked structure. Natural and synthetic rubbers are both common examples of Elastomers [23]. Elastomer is a polymer with viscoelasticity property in general, especially at low Yung's modulus, and high yield strain compared to other materials. The plastics are very flexible and flexible, which means they can undergo large elastic deformities without rupture and greatly restore the shape and size after removal has been removed. Plastics are usually resistant to oil and fuel, not permeable to liquids and gases, but tend to deteriorate by oxidation [23]. Furthermore, plastics are either thermoplastic (can be thawed) or thermoplastic (which cannot be melted). The use of plant-derived triglycerides is a substance used in the production of plastics because it provides two interactive sites, the double bond in the series of unsaturated fatty acids, and the ester group [24].

3.2.3. Using of bio-elastomers in asphalt modification

The introduction of crumb rubber in the production of asphalt-rubber mixtures for road pavement should be considered a sustainable technology that turns unwanted residues into a new mixture with high resistance to fatigue and breakage. According to ASTM D 6814–02, rubber is a synthetic or natural synthetic rubber that can be chemically cross-linked/vulcanized to enhance its useful properties. Rubber or elastic joints across the link are three-dimensional molecular networks, with long molecules held together by chemical bonds. It absorbs solvents and swells but does not melt. Moreover, it cannot be reprocessed simply by heating [25–27].

3.3. Waste cooking oil

The amount of waste cooking oil collected every year from food shops and restaurants is found to be 3 billion gallons; this is according to the U.S. Environmental Protection Agency (2011) [28]. This can be treated by polymerization and used as an alternative to bitumen [29].

The researchers investigated the potential use of cooking oil waste as an approved bitumen rejuvenation agent [30, 31]. The result was very promising because of the successful application of waste cooking oil with bitumen as an activated agent for used bitumen or older leads to an economical and environmentally friendly solution. More modification and research is needed to obtain more efficient and effective results.

3.3.1. Problems caused by waste cooking oil and its treatment methods

In general, the dumping of untreated waste into a land fill or river leads to a negative environmental impact. One of the main environmental issues arises the enrichment process that occurs when there is an obstacle to sunlight to penetrate the surface of the river caused by the blocking of a thin layer of oil. Ultimately, oxygen supply to aquatic life is disturbed when nutrient enrichment occurs in the river [32, 33]. The balance of the water ecosystem balance in the lake or river has also affected water quality. Engine oil from vehicles or cooking oil waste from residential areas contributes to the main source of river pollution. The responsibility to overcome high construction costs and reduce waste disposal issues has begun to practice the recycling of waste as an alternative and an alternative way to prevent these problems [32]. According to Hamad et al. [33] and Singhbando and Tezuka [34], only the designated licensed companies responsible for collecting cooking waste World Customs Organization (WCO) from the food industry and especially the fast food restaurant before being sent to the recycling centers.

3.3.2. Production of waste cooking oil

Waste cooking oil originates from frying activity at high-temperatures during food preparation, usually in the food industry, restaurants, hotels, and residences. The application of global customs organizations is diverse, for example, the use of yellow grease [35, 36], potential as a fuel source in biodiesel production [37], animal food [38], and making soap production [37]. Despite efforts to collect up to 15 million tons of WCO annually, only a small amount of cooking waste is properly managed through the recycling process [38]. Regular monitoring and management of the WCO creates a challenge to be addressed and becomes the primary consideration for overcoming serious dumping problems that eventually lead to serious water pollution [39]. **Figure 1** shows the total production of the WCO by States. The production of 10 million tons of WCO per year, accounting for 55%, makes the United States the highest producer of the WCO.

A large amount of vegetable oil consumption about 17 million tons causes a huge amount of waste that has reduced the production of oil resources, and this amount increases about 2%

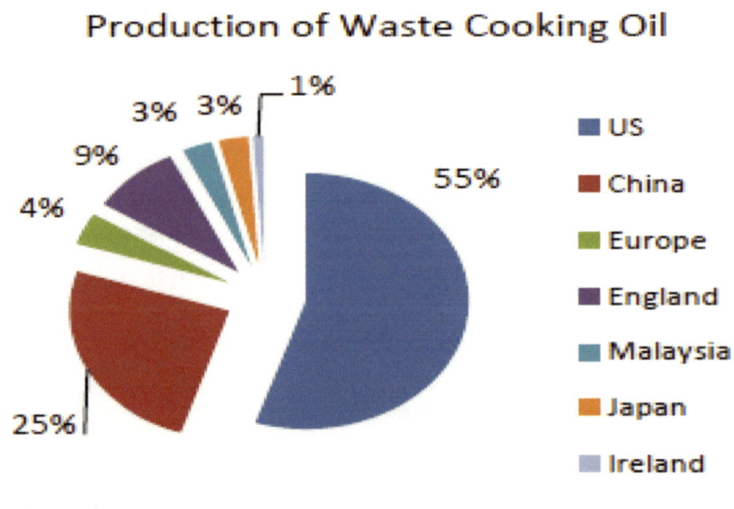

Figure 1. Production of cooking oil based on the country [36, 40, 41].

every passing year [42]. According to the food and drug administration (FDA) report [42, 43], 40% of the sewage system is blocked with frying oil which is cast in the kitchen sink.

3.3.3. Chemical properties and characterization of waste cooking oil

Chemical properties gas chromatography–mass spectrometry test (GCMS) is one of the chemical tests that are used for the purpose of the identification of chemical compound in unknown waste cooking oil. Based on the observation, the major chemical compound identified in the WCO is oleic acid, which represents 43.67% of the entire compound. Meanwhile, palmitic acid represents for 38.35% whereas 11.39% is recorded linoleic acid as shown in **Table 2**.

Due to long chain of palmitic acid and oleic acid [44], the potential of being cracked is high by thermal cracking or catalytic cracking process. Based on the research conducted by Zahoor et al. [45], the properties of unused oil is differ from used cooking oil especially in term of density, kinematic viscosity, and moisture content which presented in **Table 3**.

3.3.4. Experimental application of bio-binders

The main field of application of bio-binders is in the paving industry. However, market opportunities exist in housing products via roofing shingles and sealants [46]. There are three ways that a bio-binder reduces the use of bitumen from fossil fuels (as shown in the percentage) [8].

1. Directly alternative (75–100% bitumen substitute.
2. Bitumen extender (10–75% bitumen substitute)
3. Bitumen modifier (<10% bitumen substitute)

Free fatty acid's type	Waste cooking oil %
Heneicosanoic acid	0.08
Cis-11-Eicosenoic acid	0.16
Linolenic acid	0.29
Palmitic acid	38.35
Linoleic acid	11.39
Stearic acid	4.33
Myristic acid	1.03
γ-Linolenic acid	0.37
Lauric acid	0.34
Oleic acid	43.67
Total	100

Table 2. Chemical properties of waste cooking oil [44].

Properties	Used cooking oil values	Unused cooking oil values
Acid value (mg KOH/gm)	4.03	0.3
Calorific value (J/gm)	39,658	—
Saponification value (mg KOH/gm)	177.97	194
Peroxide value (meq/kg)	10	< 10
Density (gm/cm^3)	0.9013	0.898
Kinematic viscosity (mm^2/s)	44.956	39.994
Dynamic viscosity (mpa.s)	40.519	35.920
Flash point (°C)	222–224	161–164
Moisture content (wt. %)	0.140	0.101

Table 3. Comparison between properties of unused oil and used cooking oil.

In almost, all the road construction has been used modifiers in conventional bitumen binder. To decrease the demand for petroleum-based bitumen research is being done on bitumen extender as 100% replacement of bitumen. Additives are the resin, emulsions, crumb rubber, polymer and so on. Waste materials like waste cooking oil, waste engine oil and so on, may be the promising alternatives [47, 48].

4. Overview of the results of using of bio-modified asphalt

4.1. Predictable overall specifications

Asphalt is a mixture of moderately molecular weight hydrocarbons, aliphatic, and aromatic hydrocarbons containing moderate amounts of sulfur, small amounts of oxygen and nitrogen, and other rare elements including transitional metals. The physical and chemical properties of asphalt are results of their chemical composition. Large changes in chemical composition may lead to changes in unexpected physical properties. Bio-binders are at best linked to performance, so important chemical changes may give a false test. There are many noticeable, tacit assumptions about asphalt to consider:

- Expected aging characteristics.
- It is expected to have rheological properties.
- Will have the characteristics of the expected adhesion to aggregates.
- Has the expected coating behavior in the mixing plant.
- Have predictable flow characteristics throughout construction.

There are many other less obvious predictions:

- Predictable outflow characteristics.
- Expectable water solubility.
- Expected interactions with fuels, oils, and so on.
- Predictable environmental characteristics.
- It will have a predictable smell.
- It will have a predictable mixing with a virgin binder as a reclaimed asphalt pavement
- It will have a predictable interaction with contiguous mixtures.
- Will be available in large quantities upon request.

4.2. Aging

Rolling thin film oven test (RTFO) and pressure aging vessel (PAV) may not sufficiently represent plant and field aging because an alternative binder may have significantly different aging characteristics. Start by identifying an aging index to compare with unmodified asphalt is suggested in experimental respects. RTFO It should be operated over a variety of times and temperatures to see if normal temperature correspondence is still present. The same is true for PAV. Use PAV at 60°C for excessive times and compare output performance with results in standard conditions.

4.3. Rheological properties of bio-modified asphalt

Bio-binder may it may completely break down or give suggestively different time-temperature relationships. A clear example is an alternative bond with a fine melting point. A binder with a melting point at 73°C may be well suited for use in heavy road traffic in the PG 64 environment, but the dynamic shear rheometer (DSR) test at a specific grade of 76°C will give inappropriate results.

4.4. Strength properties and cracking

While strength characteristics and cracking are precarious to pavement performance, they are controlled by the total and the gradient and the effect of the alternative binder properties is not easily expected. Unlike low-temperature characteristics, it may be necessary to reduce the mixture test to explain the effect of a new material on cracking and strength properties.

4.5. Rutting performance at high-temperature

As illustrious above, bio-binder may have a different shape to the master curve. They may also have varied stress tolerance. For more than a few reasons, the creep compliance is unrecoverable, Jnr, from the multiple creep and recovery creep (MSCR), is a better choice than the G */sin δ of the oscillating DSR to describe the performance of high-temperatures. MSCR is run at the expected high pavement temperature and therefore, does not depend on the time-overlay overlap. Jnr has been shown to be well correlated with the actual progressive performance of a larger group of substances from G */sin δ. Finally, stress tolerance is already a unit for MSCR testing. However, it would be practical to check the range of temperatures and pressures to look for unusual behavior that may affect progressive performance. The mixture test is also strongly recommended.

4.6. Cracking in cold climates

Cold climates tensile and crack the characteristics of a bio-binder may differ from conventional asphalt, so testing the bending beam rheometer (BBR) may not be sufficient. A fracture test such as a direct tensile test or asphalt cracking will provide information on low-temperature break characteristics. If there is a significant difference in expected temperatures, the fracture test results can still be used to re-evaluate the BBR results to simplify the test over the long term. One more caution, these still only characterize one single thermal cracker. May not adequately address fatigue at low-temperatures.

4.7. Fatigue performance

There is currently a no good way to characterize a binder for fatigue performance, although there are some talented tests in development. The G * sin δ from DSR has been included in the superpave specification to control the shape of the main curve, and as mentioned above, it may or may not be suitable for a substance different from asphalt. There are differences in mixing tests that are related to fatigue performance, and at least today, this is the only correct position to handle fatigue performance.

5. Comparison between bio-modified asphalt binders and unmodified ones

The temperature range of viscous behavior for bio-oils may be lower than the virgin asphalt rate, by about 30–40°C. The rheological properties of the original bio-binder differ from those of the asphalt, but the rheological properties of these modified biological bonds vary greatly when polymer rates are added. For the developed bio-binders the high-temperature performance grade may not vary significantly from that of the asphalt binders, nevertheless, the low-temperature performance grade may vary significantly [6].

5.1. Structure of bio-oil compared with asphalt

The chemical properties indicated that the amount of furfural and phenols were varying due to the different aging processes and intervals.

The chemistry of bio-oils is complex, similar to asphalt; thus, a complete chemical characterization is difficult or practically impossible. The complication of chemical characterization or analysis resulted from the attendance of high molecular weight of phenolic species [4]. In addition, the fragmented oligomeric products occur with different numbers of phenolic and carboxylic acids, and hydroxyl groups as well as aldehyde, alcohol, and ether purposes. Thus, phenolic species occur as different hydrogen-bonded aggregates, micelles, droplets, and gels.

5.2. Temperature of performance

In general, the mixing and compaction the temperature range for bio-oils may be lower than bitumen inhibitors at about 30–40°C.

- The flow characteristics, that is, the temperature and shear, of the biomedical biomaterials of the bio-oils differ from those of the asphalt, but when adding the polymer hosts, the rheological properties of these modified biomarkers change dramatically.
- Polymers should be carefully chosen because the temperature range of the bio-groups established differs from the polymers used extensively in the asphalt industry.
- The high-temperature performance of biologically determined groups may not change significantly from sediment deposits. However, the low-temperature performance level may change significantly because of the high oxygen content in biomaterials compared to typical asphalt volumes.

5.3. Comparison from viewpoints of environment, economy, and energy

Significantly, The United States is working to create a biobased economy that generates energy from renewable organic matter rather than fossil fuels. Because of access to large amounts of vital sources such as triglycerides, proteins, starch, and other carbohydrates from various plant sources, there are interesting technical and economic forecasts for their use to produce vital bonding materials. At present, research on the application of bio-oils has focused on their use as bio-defense fuel to replace fossil fuels. Based on the findings of these surveys, the use of bio-oils as asphalt metal is very promising.

On the other hand, no research has yet been conducted on the feasibility of using bio-oils as an alternative to asphalt (replacing 100%) for use in the paving industry. As a result, there is a lack of data demonstrating the development of biomaterials from essential oils. Biomass boxes (artificial bonding materials) can be used in three different ways to reduce the demand for fossil-based asphalt compounds.

6. Conclusion

High construction costs, when combined with awareness regarding environmental stewardship have encouraged the use of waste and renewable resources in asphalt modification.

Increased energy costs and strong global demand for oil have encouraged the development of alternative bonding materials to modify or replace asphalt bonding materials. The benefits of using alternative bonding materials are their ability to help conserve natural resources and reduce energy consumption while at the same time improving asphalt performance.

The use of alternative (or secondary) materials in asphalt mixtures may be one of the most complex methods of highway use. It is not a matter of throwing alternative materials into the mixture and coating them with cement. The best use should be designed and used, including the design of the mixture itself, and the effects of alternative materials on the behavior of the asphalt binder, and the pavement to be incorporated into it. It is necessary to know how to test the resulting mixture in order to conform to the specifications; in some cases, this knowledge is lacking. Also, the expected finding from this chapter is as follows:

- Bio-Binder modified asphalt is capable to change rheological properties of the asphalt binder which will improve the performance against pavement distress
- The bio-binder will increase viscosity of the asphalt binder at high service temperature
- The binder test will show that the addition of bio-oils is expected to improve the rutting performance
- Most of the bio-oil modified asphalt mixture has higher fatigue lives than the control asphalt mixture.
- The addition of bio-modified asphalt with asphalt binder reduced the PG of the base binder, representing an increased resistance to thermal cracking, nevertheless reduced resistance to rutting.
- The adding of bio-modified asphalt reduced the crack energy at middle temperatures when compared to unoriginal asphalt under monotonic loading, representing reduced resistance to fatigue cracking.
- The dynamic modulus of hot mix asphalt (HMA) reduced with addition of bio-modified asphalt.
- The flow numbers of the bio-modified asphalt HMA mixes were lower than those of unoriginal mixes. This finding indicated that bio-modified asphalt HMA mixes are more disposed to rutting than unoriginal mixes.
- The cracking energy reduced with addition of bio-modified asphalt in the HMA and crack work obtained from IDT testing at intermediate temperatures, indicating that the bio-modified asphalt mixes are disposed to fatigue failure as soon as compared to control HMA.
- Thermal cracking energy increased with addition of bio-modified asphalt in the HMA, which indicating that an improved resistance to low-temperature thermal cracking. Bio-modified asphalt mixes are highly elastic even at lower temperatures than unoriginal mixes.
- Overall bio-modified asphalt can be a promising applicant for low-temperature behavior. However, further modification of bio-modified asphalt for higher temperature is essential to increase its performance against rutting.

http://dx.doi.org/10.5772/intechopen.76832

It will be a positive step in the direction of achieving mixture modified with bio-binder that has similar or improve performance when compare to conventional mixtures.

Author details

Ragab Abd Eltawab Abd El-latief

Address all correspondence to: chemragab83@yahoo.com

Asphalt Lab, Petroleum Applications Department, Egyptian Petroleum Research Institute (EPRI), Cairo, Egypt

References

[1] Jiménez del Barco-Carrión A, Pérez-Martínez M, Themeli A, Lo Presti D, Marsacb P, Pouget S, Hammoum F, Chailleux E, Airey GD. Evaluation of bio-materials' rejuvenating effect on binders for high-reclaimed asphalt content mixtures. Materiales de Construcción. 2017;**67**:327

[2] Yang X, You X, Dai Q. Performance evaluation of asphalt binder modified by bio-oil generated from waste wood resources. International Journal of Pavement Research and Technology. 2009;**6**(4):431-439

[3] Walters RC, Fini EH, Abu-Lebdeh T. Enhancing asphalt rheological behaviour and aging susceptibility using bio-char and nano-clay. American Journal of Engineering & Applied Sciences. 2014;**7**(1): 66-76

[4] Mohammad LN, Elseifi MA, Cooper SB, Challa H, Naidoo P. Laboratory evaluation of asphalt mixtures containing bio-binder technologies. 92nd Transportation Research Board Annual Meeting; 2013

[5] Huang SC, Salomon D, Haddock J. Alternative binders for sustainable asphalt pavements. In: Laboratory of Waste Cooking oil Based as sustainable binder for hot-mix asphalt. Washington, D.C: Papers from a Workshop; January 22, 2012. p. 625

[6] Peralta J, Raouf MA, Tang S, Williams RC. Bio-renewable asphalt modifiers and asphalt substitutes. In: Sustainable Bioenergy and Bioproducts, Green Energy and Technology. London: Springer-Verlag London Limited; 2012. DOI: 10.1007/978-1-4471-2324-8_6

[7] Asokan P, Firdoous M, Sonal W. Properties and potential of bio fibres, bio binders, and bio composites. Review on Advanced Materials Science. 2012;**30**:254-261

[8] Mohan D, Pittman CU, Steele PH. Pyrolysis of wood/biomass for bio-oil: A critical review. Energy & Fuels. 2006;**20**(3):848-889

[9] Oasmaa A, Czernik S, Johnson DK, Black S. Stability of wood fast pyrolysis oil. Biomass Bioenergy. 1999;**7**:187-192

[10] Oasmaa A, Sipil K, Solantausta Y, Kuoppala E. Quality improvement of pyrolysis liquid: Effect of light volatiles on the stability of pyrolysis liquids. Energy & Fuels. 2005;**19**(6) 2556-2561

[11] Airey GD, Mohammed MH. Rheological properties of polyacrylates used as synthetic road binders. Rheologica Acta. 2008;**47**:751-763

[12] Shields J. Adhesives Handbook. London: Butterworth; 1976

[13] Tan CP, Che Man YB. Comparative differential scanning calorimetric analysis of vegetable oils: Effects of heating rate variation. Phytochemical Analysis. 2002;**13**:129-141

[14] Urbanchuk JM. Biomass energy resources. US Department of Energy LECG LLC. February 19, 2007

[15] Xiu SN, Zhang Y, Shahbazi A. Swine manure solids separation and thermochemical conversion to heavy oil. BioResources. 2009A;**4**(2):458-470

[16] Ancheyta J, Speigh JG. Hydroprocessing of Heavy Oils and Residua. CRC Press; 2007

[17] Gevert BS, Otterstedt JE. Upgrading of directly liquefied biomass to transportation fuels-hydroprocessing. Biomass. 1987;**13**(2):105-115

[18] Williams RC, Peralta J, Ng Puga KL. Development of non-petroleum-based binders for use in flexible pavements—Phase II. Final Report. 2015

[19] Goyal HB, Seal D, Saxena RC. Bio-fuels from thermochemical conversion of renewable resources: A review. Renewable and Sustainable Energy Reviews. 2006;**12**:504-517

[20] Liu M, Ferry MA, Davidson RR, Glover CJ, Bullin JA. Oxygen uptake as correlated to carbonyl growth in aged asphalt and corbett fractions. Industrial & Engineering Chemistry Research. 1998;**37**:4669-4694

[21] Ouyang C, Wang S, Zhang Y, Zhang Y. Improving the aging resistance of styrenebutadiene-styrene tri-block copolymer modified asphalt by addition of antioxidants. Polymer Degradation and Stability. 2006;**91**:795-804

[22] Lukkassen D, Meidell A. Advanced Materials and Structures and their Fabrication Processes. Book manuscript, Narvik University College, HiN; 2007

[23] Rus AZM. Polymers from renewable materials. Science Progress. 2010;**93**(3):285-300

[24] Hamed GR. Materials and Compounds. Engineering with Rubber: How to Design Rubber Components from Alan N. Gent. Germany: Hanser Publishers; 1992

[25] Rahman MM. Characterisation of dry process crumb rubber modified asphalt mixtures [Dissertation]. University of Nottingham; 2004

[26] Wan Azahar WNA, Bujang M, Jaya RP, Hainin MR, Mohamed A, Ngadi N, Sri Jayanti D. The potential of waste cooking oil as bio-asphalt for alternative binder—An overview. Jurnal Teknologi (Sciences & Engineering). 2016;**78**(4):111-116

[27] U.S Environmental Protection Agency. 2011. [cited 10.7.2011]

[28] Wen H, Bhusal S, Wen B. Laboratory evaluation of waste cooking oil-based bioasphalt as an alternative binder for hot mix asphalt. Journal of Materials in Civil Engineering. 2012;**25**(10):1432-1437

[29] Asli H, Ahmadinia E, Zargar M, Karim MR. Investigation on physical properties of waste cooking oil—Rejuvenated bitumen binder. Construction and Building Materials. 2012;**37**:398-405

[30] El-Fadel M, Khoury R. Strategies for vehicle waste-oil management: A case Study. Resources, Conservation and Recycling. 2001;**33**:75-91

[31] Hamad BS, Rteil AA, El-fadel M. Effect of used engine oil on properties of fresh and hardened concrete. Construction and Building Materials. 2003;**17**:311-318

[32] Singhabhandhu A, Tezuka T. The waste-to-energy framework for integrated multi-waste utilization: Waste cooking oil, waste lubricating oil, and waste plastics'. Energy. 2003;**35**(6):2544-2551

[33] EPA. 2011. Available from: http://www.epa.gov/region9/waste/biodiesel/questions.html [Accessed: August 2011]

[34] Gui MM, Lee KT, Bhatia S. Feasibility of edible oil vs. non-edible oil vs. waste edible oil as biodiesel feedstock. Energy. 2008;**33**(11):1646-1653

[35] Peiro LT, Lombardi L, Méndez GV, Durany XG. Life cycle assessment (LCA) and exergetic life cycle assessment (ELCA) of the production of biodiesel from used cooking oil (UCO). Energy. 2010;**35**:2:889-893

[36] Castellanelli CA. Analyzes of the used oil under environmental perspective and its possibilities for production of biodiesel. Courtesy of UFSM—Federal University of Santa Maria; 2007

[37] Math MC, Kumar SP, Chetty SV. Technologies for biodiesel production from used cooking oil—A review. Energy for Sustainable Development. 2010;**14**:4:339-345

[38] Kulkarni MG, Dalai AK. Waste cooking oil—An economical source for biodiesel: A review. Industrial & Engineering Chemistry Research. 2006;**45** 9:2901-2913

[39] Moghaddam TB, Karim MR, Abdelaziz M. A review on fatigue and rutting performance of asphalt mixes. Scientific Research and Essays. 2011;**6** 4:670-682

[40] Agriculture and Food Development Authority, Waste Oils and Fats as Biodiesel Feedstocks. 2000. An assessment of their potential in the EU, ALTENER program NTB-NETT phase IV, Task 4, Final Report. March 2000

[41] Sanli H, Canakci M, Alptekin E. Characterization of waste frying oil obtained from different facilities. Linkoping, Sweden: World Renewable Energy Congress 2011-Sweden; 8-13 May 2011

[42] Khalisanni K, Khalizani K, Rohani MS, Khalid PO. Analysis of waste cooking oil as raw material for biofuel production. Global Journal of Environmental Research. 2008;**2**(2):81-83

[43] Zahoor U, Mohamad AB, Zakaria M. Characterization of waste palm cooking oil for biodiesel production. International Journal of Chemical Engineering and Application. 2014;**5**(2):134-137

[44] Md Maniruzzaman AA, Md Tareq R, Mohd. Rosli H, WAWA Bakar. An overview on alternative binders for flexible pavement Construction and Building Materials. 2015;**84**:315-319

[45] Raouf MA, Williams RC. Temperature susceptibility of non-petroleum binders derived from bio-oils. In: The 7th Asia Pacific Conference on Transportation and the Environment Semarang, Indonesia. 2010

[46] Demirbas M, Balat M. Recent advances on the production and utilization trends of bio-fuels: A global perspective. Energy Conversion and Management. 2006;**47**(15):2371-2381

[47] Rahman MT, Aziz MMA, Hainin MR, Bakar WAWA. Impact of bitumen binder: Scope of bio-based binder for construction of flexible pavement. Jurnal Teknologi. 2014;**70**(7): 105-109

[48] Kluttz R. Considerations for use of alternative binders in asphalt pavements: Material characteristics transportation research circular E-C165: Alternative Binders; 2012

Section 2

Evaluation of Asphalt Performance

Chapter 2

Aging Effects on Mechanical Characteristics of Multi-Layer Asphalt Structure

Ahmet Sertac Karakas

Additional information is available at the end of the chapter

http://dx.doi.org/10.5772/intechopen.75698

Abstract

This chapter focuses on Asphalt ages during construction, transportation, application phases and service life, as well. Exposure of environmental conditions such as traffic and climate is one of the prominent reasons aging in asphalt. The most common mechanism of aging is the degradation in the chemical structure of the binder by oxidation. Asphalt aging could cause several serious issues on the pavement such as stiffening, stripping that accelerates fatigue cracking and different moisture-induced problems such as raveling and potholes. Therefore, various additives are used as modifiers to improve the mechanical properties of asphalt. The most commonly used modifiers are polymers. Styrene butadiene styrene (SBS) polymers are utilized to prevent from deteriorating against external factors during its service life and to delay its aging in asphalt pavement. In the scope of this study, it was aimed to provide the general perspectives on mechanical characteristics of multi-layer asphalt structure under the aging effects besides traffic conditions.

Keywords: asphalt, SBS, aging, polymer additive, numerical approach

1. Introduction

Asphalt ages during construction, transportation, application phases and during service life. One of the biggest causes of aging in asphalt is caused by exposure to environmental conditions such as traffic and climate. The most common type of degradation is the degradation of the chemical structure of the binder by oxidation. Some serious problems can be seen in the asphalt coating as a result of the aging of the asphalt. Oxygen, moisture, ultraviolet irradiation, heat radiation and traffic action ultimately lead to the changes in the molecular structures and chemical functional groups of asphalt. The consequence of asphalt aging can be several serious issues in the pavement, such as stiffening, stripping; that accelerates fatigue cracking, and

IntechOpen

© 2018 The Author(s). Licensee IntechOpen. This chapter is distributed under the terms of the Creative Commons Attribution License (http://creativecommons.org/licenses/by/3.0), which permits unrestricted use, distribution, and reproduction in any medium, provided the original work is properly cited.

different moisture induced distresses such as raveling, potholes, respectively. Therefore, various additives are used as modifiers to improve the engineering properties of the bitumen material. The most commonly used modifiers are polymers.

The investigations in asphalt pavement technology and its application fields have enabled the development and practical use of polymer modifiers. A number of researches have been carried out to investigate in asphalt pavement [1–15].

Various additives which are called modifiers have been contributed binder to increase the performance of asphalt. Modifiers are provided long term service life of asphalt and prevented them long term aging [17–20].

Polymer has been used commonly with asphalt for nearly 50 years as of additives. Many tests are being applied for increasing the properties of asphalt combination. [3].

The traffic load and temperature cause the asphalt coating to lose strength over time. SBS Block Copolymer is used commonly with binders in order to ensure the high quality of materials, increase the endurance of binder and hot mix asphalt pavement. Many studies have been found that Modified asphalt with SBS provided to endurance rutting in high temperature, fatigue behavior and low temperature cracking [17–20].

Hamid et al. determined different types of HMAs which are prepared with grinded and non-grinded SBS polymers. It was seen better result with grinded SBS polymer modifier. According to the test result, it was determined that there was slight increase in the air gap and aggregate volume, and endurance to stresses increased and the density decreased.

Qadi et al. probed the effect of multiple additives and modifiers on asphalt pavement. Polyphosphoric acid (PPA), liquid anti strip (LAS), and hydrated lime were selected for using of laboratory study. It was seen that the moisture sensitivity in asphalt mixtures decreased when used of Liquid Anti Stripe and hydrated lime [21].

Styrene butadiene styrene copolymers (SBSC) are classified as thermoplastic elastomers because of their elastic and thermoplastic characteristics, and are used as modifiers for the bitumen because of their important features contributing to the mechanical properties of the asphalt [22–24].

In numerical studies, it is performed Abaqus and Ansys program are widely used with non-linear viscoplastic finite element model (FEM) analysis. Furthermore; Due to the longitudinal dimension, 2- and 3-dimensional models are generally approved for the design of modeling [25, 26].

2. Aging behavior of asphalt

Most of the coating systems are located at moderate temperatures. The asphalt cement has both a fluid and an elastic solid character. Since the asphalt is an organic material, it reacts with the surrounding oxygen. Oxidation changes the structure and complexity of asphalt molecules. Oxidation causes to oxidation on the asphalt, aging or hardening, leading to further fractures.

Oxidation occurs more rapidly at higher temperatures. When the asphalt cement is heated, mixed more easily and compressed, the amount of breakage is considerable. The characteristics of asphalt cement under temperature changes and loading rates and aging stages are determined by their ability to perform as a binder in coating systems.

Asphalt cement within the coating; it is hardened under the influence of air, environmental condition and heat during the service life. It is occurred aging hardening and viscosity increase in the bitumen over the time. In the bitumen subjected to aging hardening, lower penetration and higher viscosity are seen, and it is stated that aging hardening results in lower adhesion and brittle fracture.

The aging index of asphalt is shown the following **Figure 1** in term of years.

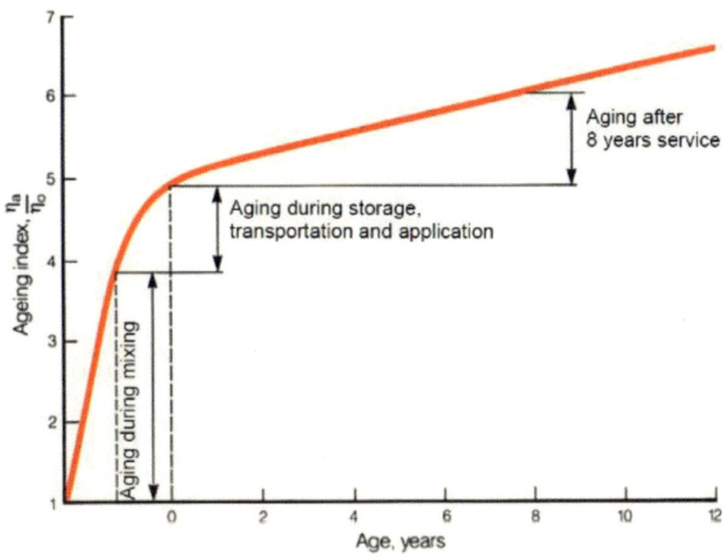

Figure 1. Asphalt aging index by years [27].

3. Factors effect on asphalt aging

As indicated before, asphalt in service condition interacts with many environmental and mechanical factors such as, atmospheric oxygen, dissolved oxygen, moisture, ambient and pavement layer temperatures, UV irradiation, traffic. Therefore, all these factors have minor to very severe effects in physical & chemical characteristics of asphalt, and subsequently these effects are reflected in either increase in viscosity, loosening adhesibility, or changes in moduli.

Stimulate asphalt aging has been shown in **Table 1** and the Petersen's findings have been presented in **Figure 2** respectively.

Factors	Influenced by					Occurring	
	Time	Heat	Oxygen	Sun-light	Beta & gamma rays	At the surface	In mass
Oxidation (in dark)	√	√	√			√	
Photo-oxidation (direct light)	√	√	√	√		√	
Volatilisation	√	√				√	
Photo-oxidation (reflect light)	√	√	√	√		√	
Photo-chemical (direct light)	√	√		√		√	
Photo-chemical (reflected light)	√	√		√		√	√
Polymerization	√	√				√	√
Steric or physical	√					√	√
Exudation of oils	√	√				√	
Changes by nuclear energy	√	√			√	√	√
Action by water	√	√	√	√		√	
Absorption by solid	√	√				√	√
Absorption of components at a solid surface	√	√				√	
Chemical reactions	√	√				√	√
Microbiological deterioration	√	√	√			√	√

Table 1. List of individual factors and conjoint factors that affect asphalt aging [28].

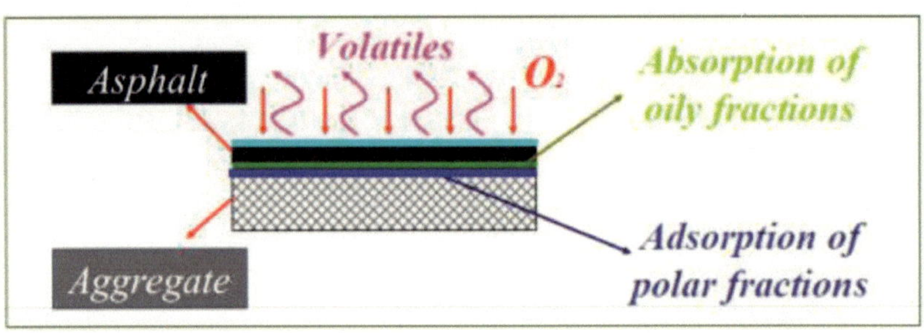

Figure 2. Illustration of asphalt aging process due to volatilization, oxidation, absorption and adsorption [28].

4. Bitumen modification

Bitumen exhibits viscoelastic properties as rheological structure. Bitumen, which plays a major role in many parameters of road performance, mainly cracking and permanent deformation resistance, also causes viscoelastic properties of asphalt mixtures. Generally, the amount of deformation in asphalt coating varies depending on the loading time and the temperature value.

The behavior of bitumen affected by static and dynamic loads is shown in **Figure 3**.

Permanent deformation (creeping), cracking (thermal or fatigue), moisture damage are the most common types of degradation in flexible coatings. Shortening the service life of road coverings under increasing traffic load has necessitated modification of bitumen. In recent years, based on polymers, the use of highly modified bituminous blends and mixtures has been increasing. As polymer additives and non-polymer additives as shown in **Table 2**.

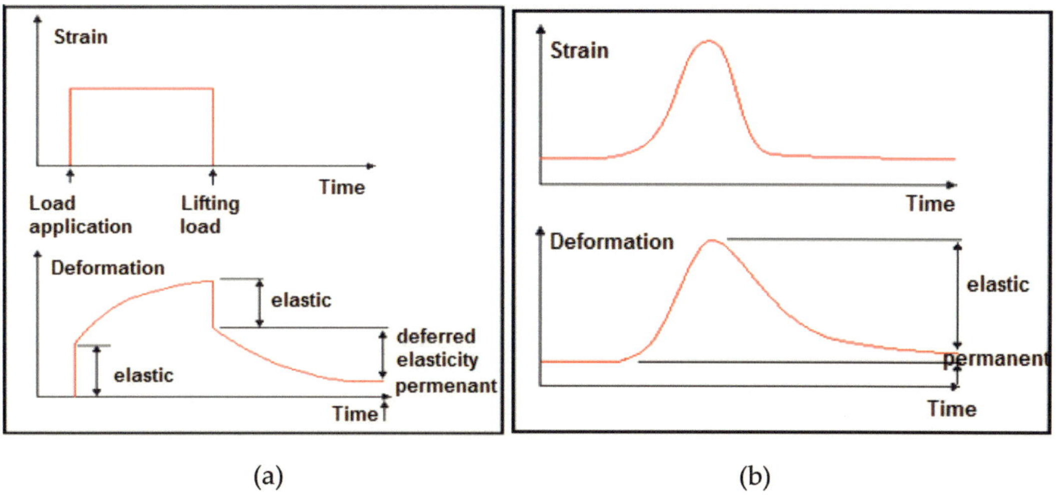

Figure 3. The behavior of bitumen affected by static load (a) and dynamic load (b) [27].

Modification types	Specimens
I. Modification with non-polymer additive	
a. Filler	Clay, black carbon, fly ash
b. Anti-peel additives	Organic Amines and Amids
c. Expander (Extendents)	Lignin and sulfur
d. Antioxidants	Zinc and lead antioxidants, phenolics
e. Organo-metal compounds	Amines
f. Others	Organo manganese compositions
	Organo carbon compositions
II. Modification with polymer additives	
a. Plastics	Polyethylene (PE), polypropylene (PP), polyvinyl chloride (PVC),
1. Thermoplastics	Polystyrene (PS), ethylene vinyl acetate (EVA)
2. Thermosets	Epoxy resins
b. Elastomers	Synthetic butadiene copolymer (SBR), Styrene-butadiene styrene copolymer (SBS),
1. Natural Rubbers	
2. Artificial Elastomers	Ethylene prokplendien harmoliper (EPDM), Isobutene isoprene copolymer (IIR)
3. Processed Rubbers	
4. Fibers	Polyester, fibers, polypropylene fibers
III. Chemical reaction modification	Additive reaction (bitumen + monomer) Vulcanization (bitumen + sulfur) Nitrogen reaction (bitumen + nitric acid)

Table 2. Bitumen modification types [29].

Usage purposes of modified bitumen applications:

- Control of fatigue cracks,
- Water impermeability,

- Increase of adhesion,
- Reduction of noise,
- Reduction of groove marks on wheel load can be summarized [37, 38].

Additives used in the modification of bituminous and bituminous mixtures and the benefits provided according to their shape of deterioration are shown in **Table 3**.

Modifier	Permanent deformation	Thermal cracking	Fatigue crack	Moisture damage	Aging
Elastomers	+	+	+		+
Plastomers	+				
Processed rubber		+	+		
Black carbon	+				+
Lime					+
Sulfur	+				
Chemical modifier	+				
Antioxidants					+
Hydrated lime				+	+

Table 3. Benefits of different types of modifications [27].

5. SBS polymer modified bitumen and properties

Styrene-butadiene styrene (SBS) polymer additives used as additives in bitumen are one of the most preferred additive materials since they give positive results in physical and mechanical properties of HMA blends.

SBS additive and its three-dimensional structure is indicated in **Figure 4**.

SBS polymers are randomly formed as bond shapes in different forms such as block, styrene, butadiene, and linear.

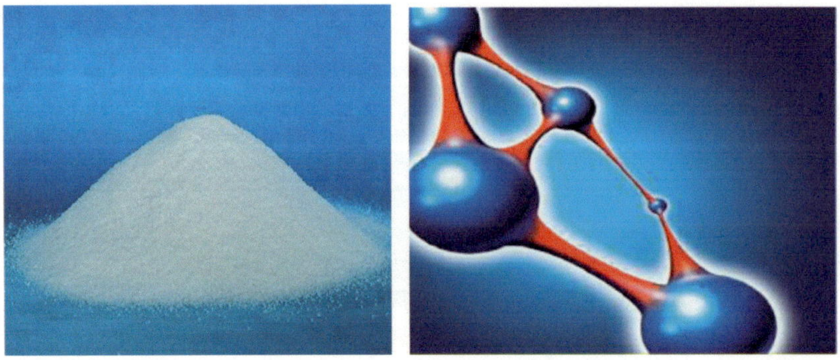

Figure 4. General view of SBS-additive and three-dimensional SBS structure [30].

Molecular structure of SBS additives is illustrated in **Figure 5**.

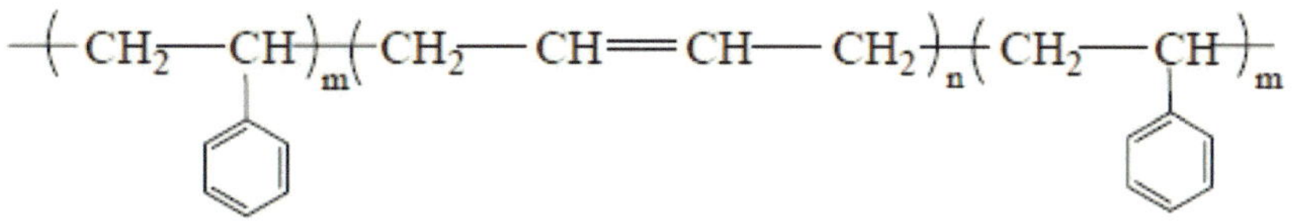

Figure 5. SBS molecular structure [31].

Figure 6 is indicated a three-dimensional view of asphalt-coated SBS molecules and the SBS modified bitumen structure consisting of asphaltic cell form and styrene butadiene bonds.

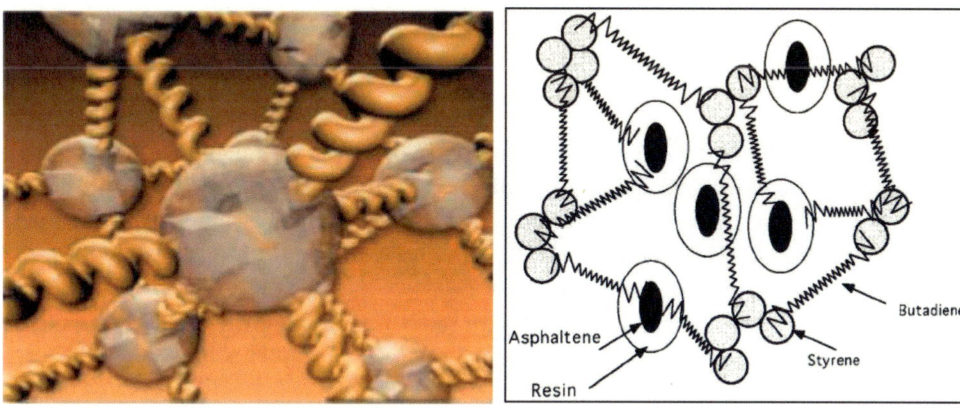

Figure 6. Three-dimensional appearance of asphalt-coated SBS molecules bond structure of SBS coated asphalt film [30, 32].

When the structure of SBS is examined, the following conclusions can be drawn:

- The polystyrene crowlings gives strength by forming physical cross-links.
- The polybutadiene bridges provide elasticity and flexibility.
- At 100°C, the polymer becomes fluid and the three-dimensional network structure recurs. Since the material is a thermoplastic elastomer, it does not lose anything in its heating and cooling properties.
- It maintains its characteristic between −40°C and + 80°C.

6. Materials and methodology

In this study, initially, some conventional tests were applied. The properties of bitumen and aggregates to be used in bituminous hot blends have been examined. In the performance tests; core samples were taken different time in 1 year period (1st, 4th, 8th and 12th months) and tested in **Figure 7**. Core samples achieved from the implementation of asphalt pavement which are prepared with SBS modified and neat bitumen of the asphalt pavement. When taking the

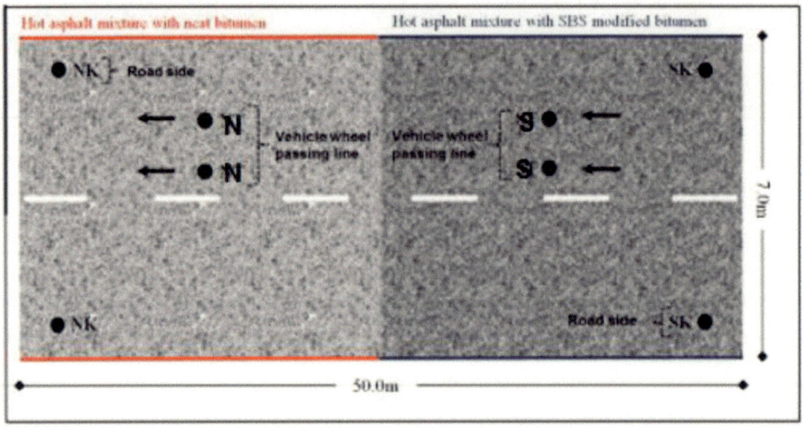

Figure 7. Core samples of hot mix asphalt (HMA) [33].

core samples, the edge of the road and the wheel passage areas are taken into consideration in **Figure 8**. In a one-year time period, 7432 equivalent axle loads were evaluated on trial path. Small axle tracks are not considered and are not taken into consideration because of using a secondary road. The effects of temperature on stability, stiffness, indirect tensile strength and fatigue resistance were investigated. Furthermore, it was examined Von Mises stress and vertical deformation of asphalt pavement on different time period with numerical analysis.

The location of the samples in asphalt and identification of the sample types are given in **Table 4**.

As a neat binder, asphalt binder of B 50/70 type taken from TUPRAS refinery was used. In modification, KRATON D 1101 contains styrene-butadiene styrene (SBS) block copolymer produced by Shell Bitumen was used.

The materials used in the layers that constituted the super structure of the road are; crushed aggregate chosen in the lower base layer, granular crushed aggregate chosen in the base layer, and crushed limestone aggregate chosen in road coating layer.

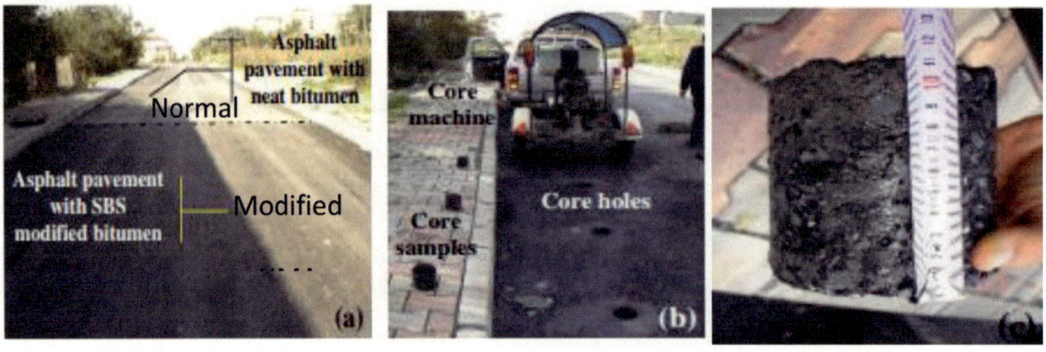

Figure 8. Asphalt pavement type (a), the core holes (b) and core sample (c) [33, 34].

Identification of the samples	State and location of samples
SK	SBS modified samples on road side
S	SBS modified samples on vehicle wheel passing line
NK	Neat samples on road side
N	Neat samples on vehicle wheel passing line

Table 4. Classification of drilling core samples of the HMA mixtures [33–35].

A solid model consisting of superstructure layers was created with namely Ansys using the finite element program. The solid model is designed by using the field parameters obtained by considering the environmental conditions such as traffic and climate. HMA design has been applied in the solid model, and only by doing so could real conditions were simulated with the analyses.

6.1. Numeric analyses

Numerical analyses assure significant advantages in using parameters derived from empirical studies to determine the stress and deformation characteristics of the superstructure of roads, corrosion, binder, base and sublayers, and to make future estimates. The Finite Element Method is often favored among other numerical approaches since it allows to make various changes and provides precise measurements in creating numerical models of physical problems. The Finite Elements Program used in numerical analysis is a numerical solution method developed for the analysis of problems expressed by differential equations. In the program, a continuous medium is branched by finite elements, and the equations are enrolled for one element, and integrated to derived system equations. As a result, complex differential equations considered for continuous media transformed into matrix format and it is reduced to a set of linear equations [36].

6.2. The finite elements method

The Finite Elements Method is a numerical solution method used frequently in engineering practices for the purpose of examining the continuous medium problems by examining them after dividing into a certain number of elements. In the scope of this method, the solution is obtained by dividing a medium or an object by finite elements and writing the rigidity matrix for each element, and then integrating the solutions for all elements. The numerical solutions of the differential equations that express the mechanical behavior of the system in question are written in matrix format. Generally, bigger matrices appear for the geometrical applications that require multiple elements. In solving such problems, the necessary linear algebraic operations are performed via computers.

6.2.1. The advantages of the finite elements programs

- The Finite Elements Formulation may be applied to many problems.

- It ensures the opportunity of speed and optimization that can be analyzed via computer.
- Any types of complex geometrical, material status and loading limit conditions may be defined.
- It facilitates the solution of integrated problems in ways such as tension, shape-shifting (statics) and consolidation (dynamics).
- The primary independent variables like place-shift, flow potential, which are selected, and the secondary unknown factors depending on these such as tension, shape-shifting, speed, and the amount of flow, etc., which are depending on these, are assessed together.
- Results that are close to reality are obtained with adequate element definitions [36].

In this chapter, a finite elements model was developed by using the ANSYS Finite Elements Program for the super-structure of the road. The models, which consist of asphalt coating, the base and sub-base include 79,045 elements and 558,224 nodal point in average.

Different models were prepared for various time periods, which were 1st, 4th, 8th and 12th Months, by considering the multi-layer structure of asphalt, the traffic and the environmental conditions [35].

The Finite element model (FEM) of road is given in **Figure 9**.

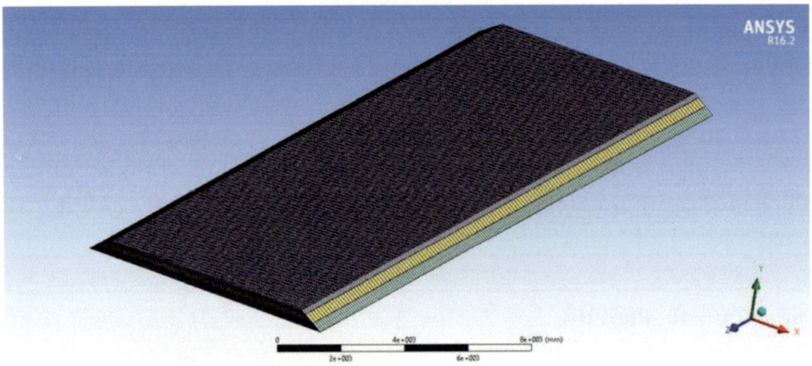

Figure 9. Finite element model (FEM) of road [34].

7. Results and discussion

7.1. Experimental results

Field and laboratory test works were conducted on the super-structure of the road for a time period of 1 year under traffic and environmental conditions. Drilling core samples were taken from the super-structure of the road in certain periods. The experimental part of the study was carried out in two stages. In the first stage, the physical properties of the aggregates and the neat and SBS modified bitumen which are using the asphalt mixture were determined. In the second stage, various tests performed HMA samples. The mechanical properties of the samples collected from the different asphalt pavement were examined in the 1st, 4th, 8th and 12th months for a one-year period.

7.1.1. Properties of bituminous binders and aggregates

The binder was used in B50/70 class and constituted the building block of the HMA, which was used in the road platform. The characteristics of the pure (neat) and HMA-modified bitumen are given in **Table 5** that the neat and SBS modified bitumen are a little susceptible to temperature in relation to the penetration index and they have a value above the limits of the specification as softening.

The gradation characteristics of the mixture are given in **Table 6**. The physical properties of the aggregate are shown in **Table 7**. It is understood that the aggregates used in the coating layer of HMA are within the limits of the related specifications in terms of their physical properties.

Property	Test method	Neat bitumen	SBS modified bitumen
Specific gravity (gr/cm^3)	EN 15326	1.022	1.017
Penetration (25°C, 0.1 mm)	EN 1426	B-61	B-68
Softening point (°C)	EN 1427	51.7	54.0
Ductility (25°C, cm)	EN 13589	>100	>100
Fraas breaking point (°C)	EN 12593	−17	−17.5

Table 5. The properties of the neat bitumen [34].

Sieve Size (mm)	12.5	9.5	4.75	2.00	0.425	0.18	0.075
Passed (%)	100	90.4	56.6	36.6	18.2	13.0	10.3

Table 6. Aggregate gradation.

Tested property	Standard	Coarse	Fine	Filler	Specification limit
Abrasion Loss % (Los Angeles)	ASTM C 131	20.5	–	–	Max 35
Frost action % (with Na_2So_4)	ASTM C 88	1.20	–	–	Max 10
Peel strength (%)	ASTM D903	60–70			Min 50
Flatness index (%)	BS 812	16.1			Max 30
Water absorption (%)	ASTM C127	0.38	0.88	–	
Specific bulk density (gr/cm^3)	ASTM C127	2.733	–	–	
Specific bulk density (gr/cm^3)	ASTM C128	–	2.678	–	
Specific bulk density (gr/cm^3)	ASTM D854	–	–	2.764	

Table 7. Aggregate characteristics [35].

7.1.2. HMA design properties

In order to determine the HMA design properties, the pure and SBS-modified asphalt concrete coating sample road platform was prepared by taking the aggregate gradation as the basis. The pure HMA optimum bitumen rate was determined as 4.95%, and the SBS-added HMA optimum bitumen rate was determined as 5.24%. The properties of the design criteria are given in **Table 8**, and it is observed that they meet the Conditions List criteria.

From the pavement spread and compacted according to the predetermined design the core samples were taken at three different time periods on the vehicle wheel passing line and banquette of the road. The core samples were sized to those of Marshall samples and they were tested by Stiffness Modulus, Indirect Tensile, Fatigue and Marshall tests.

The average volumetric characteristics of the core samples are given in **Table 9**.

In **Figure 10**, It is indicated that the alteration in the stability of the samples over time. According to the result of the experiment, there have been remarkable increases in the stability with the hardening in the asphalt coating over time. SBS modified specimens were found to have 24% greater stability at 1st month and 72% greater at 12th months, respectively, when compared to neat specimens. The Stability increased 34% in neat mixtures and 76% more in SBS modified mixtures, respectively for 12th month compared to 1st month.

Property	Pure HMA value	SBS modified HMA value
Optimum bitumen rate (%)	4.95	5.24
Practical specific gravity (Gmb, gr/cm^3)	2.415	2.411
Marshall stability (kgf)	1222	1170
Flow (mm)	3.03	3.58
Aggregate void ratio (Vma, %)	15.10	13.9
Asphalt void ratio (Vf, %)	73.93	75.4
Air void ratio (Vh, %)	3.94	3.10

Gmb, bulk specific gravity; Vh, air voids; Vf, voids filled with asphalt; Vma, the void volume between the aggregates.

Table 8. Design properties of pure and SBS-modified HMA [34, 35].

Specimen	Gmb	Vh (%)	Vma (%)	Vf (%)
N	2.267	8.929	19.362	53.899
NK	2.280	8.43	18.92	55.480
S	2.271	8.369	19.457	57.032
SK	2.280	8.004	19.136	58.188

Gmb, bulk specific gravity; Vh, air voids; Vf, voids filled with asphalt; Vma, the void volume between the aggregates.

Table 9. The volumetric characteristics of the core samples in the 1st month [35].

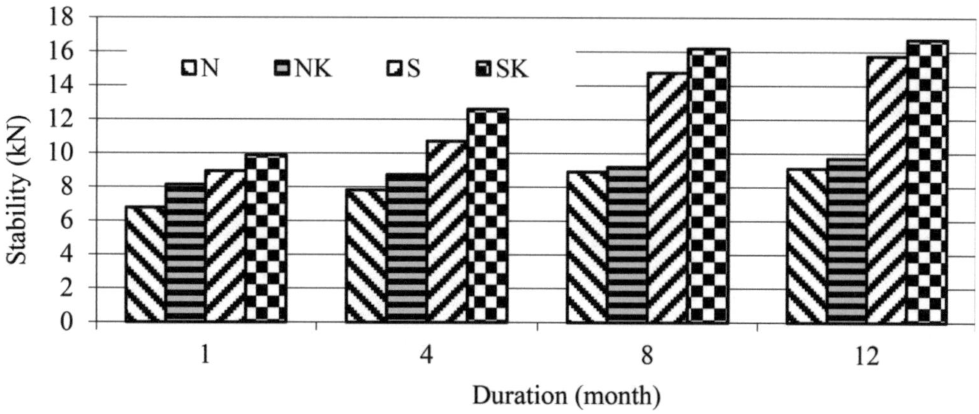

Figure 10. The change of stability in different period [35].

The change of the stiffness modules is illustrated in **Figure 11**.

In **Figure 11**, the stiffness modulus values show significant increase over time. In addition, the stiffness module of the samples taken from the edges and the vehicle wheel pass line of neat and SBS modified asphalt pavement were similarly affected by the time factor. SBS modified sample values increased by 8–27% more than neat sample values. The edge asphalt samples which are taken from banquette value were higher than the vehicle wheel passing. This difference is seen 2.3% to 11.8% in neat specimens and 1.8–10% in modified specimens. The reason for this increase is due to the fact that the samples from the edge are exposed to more water due to the transverse slope of the road [35].

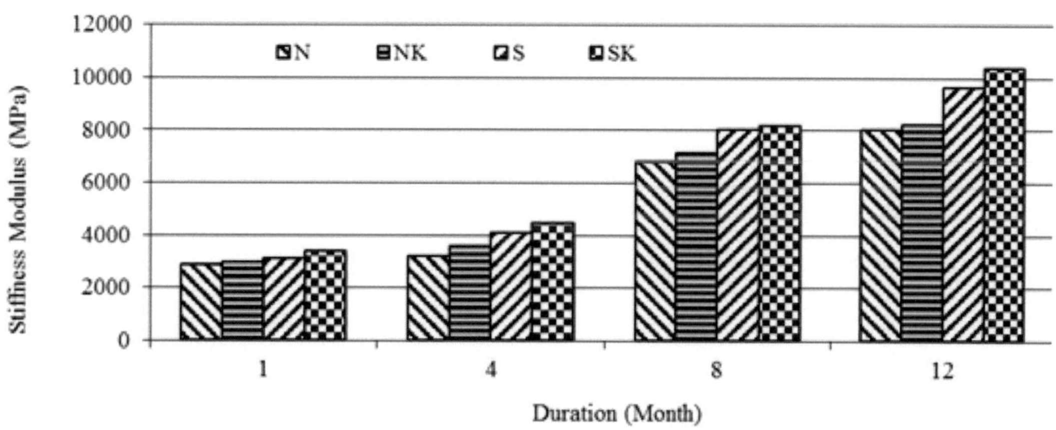

Figure 11. The change of the stiffness modules in different period [35].

The graph in **Figure 12** shows the average test results of the Indirect Tensile Strength Test.

In **Figure 12**, ITS values were found to be lowest at 1st month and highest at 12st months. In terms of monthly periods, ITS values of S and SK type core samples were more than N and NK type of core samples. It was realized the lowest values in the N type and the highest values in the SK type. The increase of samples from 1st to the 12th month were seen 42.68% NK, 49.89% N, 20.4% SK and 26.93% S type of samples, respectively.

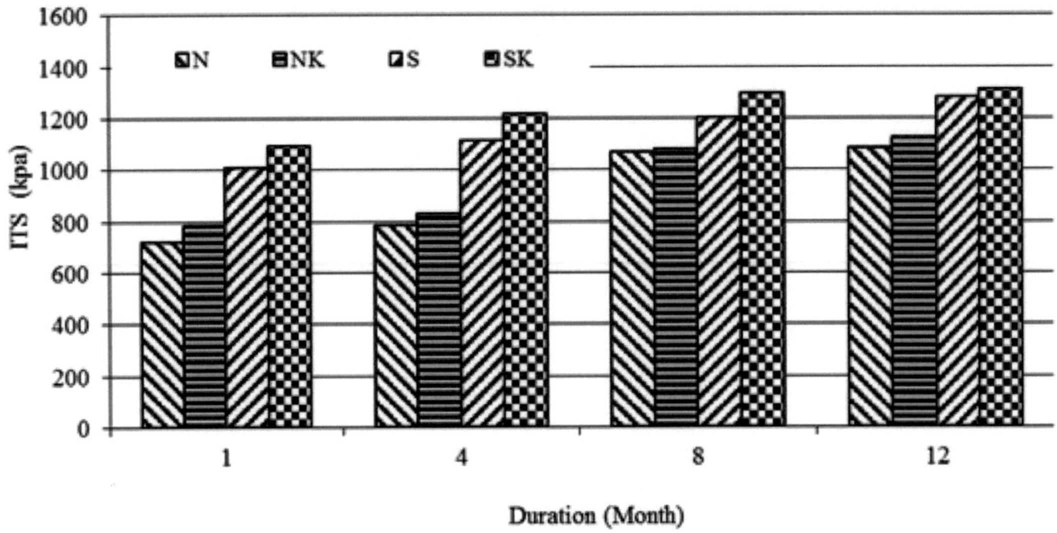

Figure 12. The change of the indirect tensile strengths in different period [35].

The change of fatigue strength in different period is given in **Figure 13**.

In **Figure 13**, it was observed that the loading repetition required for the samples to reach a level of deformation of about 2 mm was significantly increased using the SBS additive material. It was seen that the maximum loading repetition number (Nmax) values were increased after time period. The fatigue resistance of the samples with neat was lower than the SBS ones. The lowest deformation was gained from SK type samples. In the same manner, the highest Nmax values were achieved from SK type samples. It was clearly understood that

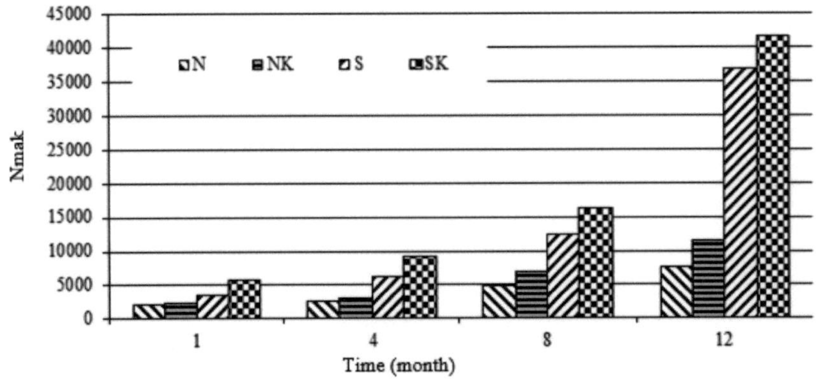

Figure 13. The change of fatigue strength in different period [35].

the flexibility of bituminous hot mixtures decreased after aging and for this reason a more brittle fracture may occupy [35].

7.2. Numeric analyses result

FEM models have been solved to determine the Von Mises stresses besides vertical deformations in super structures of the road, as well.

The FEM models of the Von Mises stresses for one of sample material is given in **Figure 14**.

Numerical analysis was done by applying the finite element method on the super-structure of the road. Von Mises Stress properties of the road super-structure are evaluated in **Figure 15**.

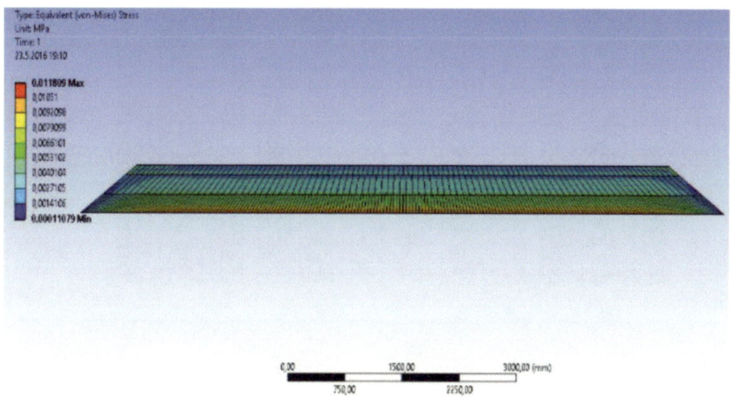

Figure 14. Von Mises FEM model [34].

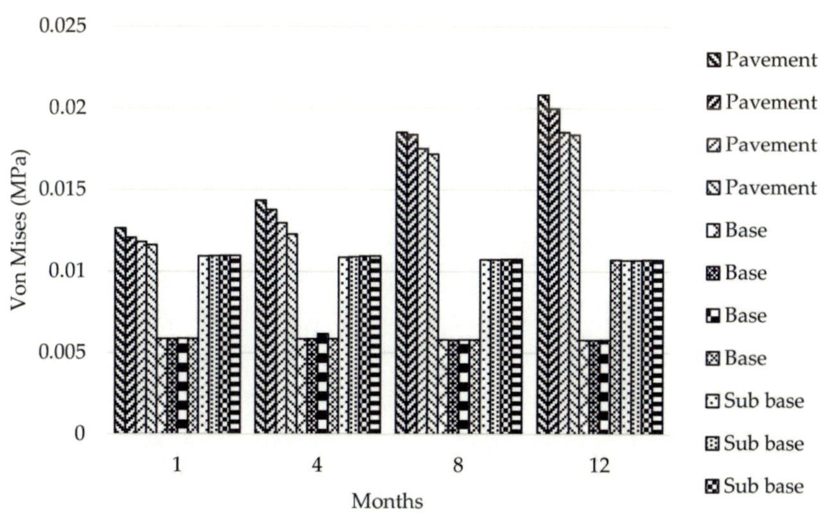

Figure 15. Von Mises tension-time changes of the super-structure of the road [34].

According to **Figure 15**; the highest Von Mises tensions were occurred the coating layer and the lowest values were observed in the base layer. It is fixed that The Von Mises tensions increased in the coating layer within 1 year period. It has been determined that the highest Von Mises tension were in SK type road coating whereas, the lowest Von Mises tensions were in N type coating. The Von Mises tensions of base layer indicated similarity properties and no significant changes were observed [34].

Vertical deformation (y)-time changes of the super-structure of the road is given in **Figure 16**.

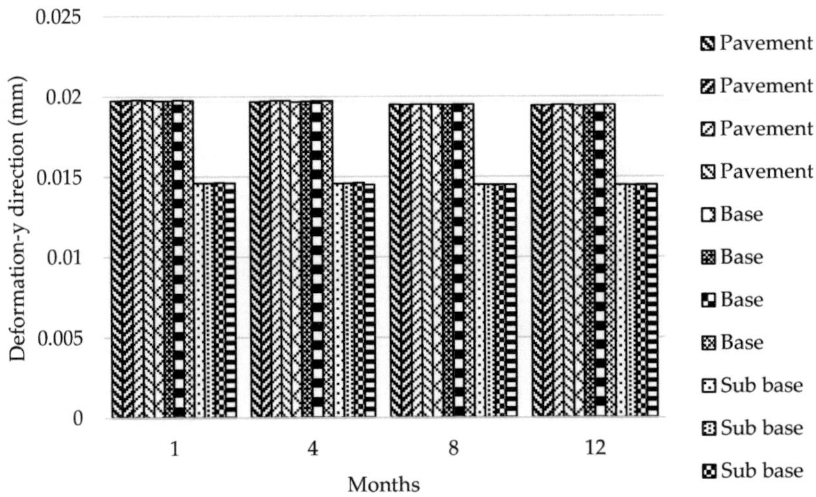

Figure 16. Vertical deformation (y)-time changes of the super-structure of the road [34].

When **Figure 16** is evaluated, it is seen that the highest vertical deformations are found at the base and lower base layers and the lowest deformations are found at the coating layer. Within a year, there have been small deviations in the amount of deformation. The amounts of deformation were confirmed as 0.014492 mm in SK-type, and 0.0145 mm in NK-type in the 12th month [34].

Von Misses stress and vertical deformation function and R^2 values are submitted in **Table 10**.

Layer types	Regression models	Function type			
		Von Mises stress function	R2	Vertical deformation function	R2
Coating	SK	$y = 0{,}0001x^2 + 0{,}0021x + 0{,}0101$	0,977	$y = 1E\text{-}06x^2 – 0{,}0001x + 0{,}0199$	0,9678
	S	$y = -3E\text{-}05x^2 + 0{,}003x + 0{,}0088$	0,9583	$y = -0{,}0001x + 0{,}0199$	0,9329
	NK	$y = 0{,}0099e^{0{,}1654x}$	0,9291	$y = 8E\text{-}06x^2 – 0{,}0002x + 0{,}02$	0,8994
	N	$y = 0{,}0001x^2 + 0{,}0019x + 0{,}0092$	0,9095	$y = -0{,}0001x + 0{,}0199$	0,8744
Base	SK	$y = 6E\text{-}07x^2 - 3E\text{-}05x + 0{,}0059$	0,9683	$y = -0{,}0001x + 0{,}0199$	0,9686
	S	$y = -4E\text{-}07x^2 - 3E\text{-}05x + 0{,}0059$	0,929	$y = -3E\text{-}06x^2 - 1E\text{-}04x + 0{,}0199$	0,9301
	NK	$y = -7E\text{-}05x^2 + 0{,}0003x + 0{,}0057$	0,4428	$y = 6E\text{-}06x^2 – 0{,}0001x + 0{,}0199$	0,9059
	N	$y = 0{,}0012x^2 – 0{,}0047x + 0{,}0096$	0,9282	$y = -3E\text{-}06x^2 - 9E\text{-}05x + 0{,}0199$	0,8785

http://dx.doi.org/10.5772/intechopen.75698

Layer types	Regression models	Function type			
		Von Mises stress function	R2	Vertical deformation function	R2
sub base	SK	$y = 5E\text{-}06x^2 – 0{,}0001x + 0{,}0111$	0,9683	$y = 2E\text{-}06x^2 – 5E\text{-}05x + 0{,}0147$	0,9645
	S	$y = 5E\text{-}06x^2 – 0{,}0001x + 0{,}0111$	0,946	$y = –1E\text{-}06x^2 – 4E\text{-}05x + 0{,}0147$	0,923
	NK	$y = 3E\text{-}06x^2 – 0{,}0001x + 0{,}0111$	0,8963	$y = 7E\text{-}07x^2 – 5E\text{-}05x + 0{,}0147$	0,8528
	N	$y = –2E\text{-}07x^2 – 9E\text{-}05x + 0{,}0111$	0,8926	$y = 2E\text{-}05x^2 – 0{,}0002x + 0{,}0147$	0,9271

Table 10. Function and regression analyses of the relation between the super-structures of the road vertical deformation time [34].

When **Table 10** is examined, It is understood that R^2 values are close to 1 for all road layers. It has been inferred that numerical analyzes have supported empirical studies because of the strong relationship [34].

8. Conclusion

It came out that the performance characteristics of the SBS-added the HMA are improved due to the use of the SBS polymer material as compared to the unmodified ones. This is because when the SBS polymer is used, the adhesion between the aggregates increases. It has been seen that the flexible structure of asphalt is stiffened due to environmental conditions due to oxidation, which is caused by temperature changes and precipitation during transportation, storage, mixing and production processes. In addition to, as the air void ratio increases, the hardening time decreases due to air contact. The hardening of the asphalt road coating was found to be higher in the coating without additive material and it was appeared that the polymer increased the deformation resistance at high temperatures and increased the tendency to fracture at low temperatures.

Analytical calculations from different material types and different strata in the flexible road coating are complex. The traffic loads and the distribution of pressure on the superstructures of the roads are varied, and different materials are used on each layer, and therefore the mechanical properties and the load distribution ability vary. Climate and various environmental conditions affect the road. Because of these reasons, empirical studies must be promoted with numerical analyzes to establish performance characteristics of the road.

Acknowledgements

The author would like to thank Transportation Laboratory, Department of Civil Engineering, Firat University; Geotechnical &Transportation Laboratory, Department of Civil Engineering, Dicle University, Istanbul University and Bimtas Research Laboratories, Istanbul.

Author details

Ahmet Sertac Karakas

Address all correspondence to: skarakas@istanbul.edu.tr

Istanbul University, Division of Construction and Technical Maintenance, Istanbul, Turkey

References

[1] Shingo K, Shigeru T, Zhang XM, Dong DW, Inagaki N. Compatibilizer role of styrene-butadiene-styrene tri-block copolymer in asphalt. Polymer Journal. 2001;**33**(3):209-213

[2] Airey GD. Styrene butadiene styrene polymer medication of road bitumens. Journal of Materials Science. 2004;**39**(3):951-999

[3] Becker Y, Mendez MP, Rodriguez Y. Polymer modified asphalt. Vision Technologica. 2001;**9**(1):39-50

[4] Behbahani H, Hassan Z, Shams N. The use of polymer modification of bitumen for Durant hot asphalt mixtures. Journal of Applied Sciences Research. 2008;**4**(1):96-102

[5] Awwad MT, Shbeeb L. The use of polyethylene in hot asphalt mixtures. American Journal of Applied Sciences. 2007;**4**(6):390-396

[6] Dong F, Zhao W, Zhang Y, Wei J, Fan W, Yu Y, Wang Z. Influence of SBS and asphalt on SBS dispersion and the performance of modified asphalt. Construction and Building Materials. 2014;**62**(15): 1-7

[7] Prowell BD, Zhang J, Brown ER. Aggregate Properties and the Performance of Superpave-Designed Hot Mix Asphalt. NCHRP Report 539, Transportation Research Board, 2005

[8] Cpatt OV. Investigation of polymer modified asphalt by shear and tensile compliances. Material Characterization for Inputs into AASHTO 2002 Guide Session of the 2004 Annual Conference, Transportation Association, Canada, Québec City; ; 2004. pp. 145–213

[9] Lucena MCC, Soares SA, Soares JB. Characterization and thermal behavior of polymer-modified asphalt. Materials Research. 2004;**7**(4):529-534

[10] Newman K. Polymer-modified asphalt mixtures for heavy-duty pavements: fatigue characteristics as measured by flexural beam testing. 2004 FAA Worldwide Airport Technology Transfer Conference, Atlantic City, New Jersey, USA, 2004

[11] Mohammad LN, Negulescu LL, Wu Z, Daranga C, Daly WH, Abadie C. Investigation of the use of recycled polymer modified asphalt binder in asphalt concrete pavements. Journal of the Association of Asphalt Paving Technologists. 2003;**72**:551-594

[12] Zubeck H, Raad L, Saboundjian S, Minassian G, Ryer J. Performance of polymer-modified-asphalt-aggregate mixtures in Alaska. Journal of Cold Regions Engineering. 2002;**16**:170-190

[13] Masson JF, Pelletier L, Collins P. Rapid FTIR method for quantification of styrene-butadiene type copolymers in bitumen. Journal of Applied Polymer Science. 2001;**79**: 1034-1041

[14] Lu X, Isacsson U. Influence of styrene-butadiene-styrene polymer modification on bitumen viscosity. Fuel. 1997;**76**(14–15):1353-1359

[15] Cavaliere MG, Diani E, Dia MD. Dynamic mechanical characterization of binder and asphalt concrete. In: Proceedings of the Euroasphalt and Eurobitumen Congress, No.5551; 1996

[16] Behbahani H, Ziari H, Noubakhat S. The use of polymer modification of bitumen for Durant hot asphalt mixtures. Journal of Applied Sciences Research. 2008;**4**(1):96-102

[17] Lu X, Isacsson U. Laboratory study on the low temperature physical hardening of conventional and polymer modified bitumens. Construction and Building Materials. 2000;**14**:79-88

[18] Navarro FJ, Partal P, Martinez-Boza F, Valencia C, Gallegos C. Rheological characteristics of ground tire rubber-modified bitumens. Chemical Engineering Journal. 2002;**89**:53-61

[19] Airey GD. Rheological properties of styrene butadienee styrene polymer modified road bitumens. Fuel. 2002;**82**(14):1709-1719

[20] Khattak MJ, Baladi GY. Engineering properties of polymer- modified asphalt mixtures. Transportation Research Record. 1998;**1638**:12-22

[21] Al-Qadi IL, Abuawad IM, Dhasmana H, Coenen AR. Effects of Various Asphalt Binder Additives/Modifiers on Moisture Susceptible Asphaltic Mixtures. Civil Engineering Studies. Illinois Center for Transportation Series No. 14–004. ISSN: 0197–9191. UILU-ENG-2014–2004

[22] Zhu J, Kringos N. Towards the development of a viscoelastic model for phase separation in polymer-modified bitumen, Road Mater. Pavement Design. 2015;**16**:39-49

[23] de Carcer ÍA, Masegosa RM, Viñas MT, Sanchez-Cabezudo M, Salom C, Prolongo MG, Contreras V, Barceló F, Páez A. Storage stability of SBS/sulfur modified bitumens at high temperature: Influence of bitumen composition and structure. Construction and Building Materials. 2014;**52**:245-252

[24] Larsen DO, Alessandrini JL, Bosch A, Cortizo MS. Micro-structural and rheological characteristics of SBS-asphalt blends during their manufacturing. Construction and Building Materials. 2009;**23**:2769-2774

[25] Fang H, Haddock JE, White TD, Hand AJ. On the characterization of flexible pavement rutting using creep model-based finite element analysis. Finite Elements in Analysis and Design. 2004;**41**(1):49-73

[26] Abed AH, Al-Azzawi AA. Evaluation of rutting depth in flexible pavements by using finite element analysis and local empirical model. American Journal of Engineering and Applied Sciences. 2012;**5**(2):163-169

[27] Whiteoak D, Read J. The Shell Bitumen Handbook. London: Thomas Telford Ltd; 2003

[28] Wu J. The Influence of Mineral Aggregates and Binder 1068 Volumetrics on Bitumen Ageing. England: The University of Nottingham; 2009

[29] Ilica M, Tayfur S, Ozen H. Asphalt and Application. Istanbul, Turkey: ISFALT; 2001

[30] Iskender E. Evaluation of permanent deformation problem in asphalt concrete [Master Science Thesis]. Trabzon, Turkey: Blacksea Technical University; 2004

[31] Dogan M. Effect of Polymer Additives on the Physical Properties of Bitumen Based Composites, a Thesis Submitted to the Graduated School of Natural and Applied Science. Ankara: Middle East Technical University; 2006

[32] Yildirim Z. Use of waste plastics in the production of polymer modified bitumen materials [Master Science Thesis]. Institute of Science and Technology, Aegean University; 2006

[33] Karakas AS, Sayin B, Kuloglu N. The changes in the mechanical properties of neat and SBS-modified HMA pavements due to traffic loads and environmental effects over a one-year period. Construction and Building Materials. 2014;**71**:406-415

[34] Karakas AS, Ortes F. Comparative assessment of the mechanical properties of asphalt layers under the traffic and environmental conditions. Construction and Building Materials. 2017;**131**:278-290

[35] Karakas AS, Kuloglu N, Kok BV, Yilmaz M. The evaluation of the field performance of the neat and SBS modified hot mixture asphalt. Construction and Building Materials. 2015; **98**:678-684

[36] Brinkgreve RBJ et al. Plaxis Finite Element Code for Soil and Rock Analyses. Netherlands: Delft University of Technologhy & Plaxis b.v; 2002

[37] Tunc A. Flexible Coating Materials Handbook. Ankara: Asil publication distribution; 2004

[38] Tunc A. Road Materials and Applications. Istanbul: Atlas publisher; 2001

Chapter 3

Novel Applications with Asphaltene Electronic Structure

Eva M. Deemer and Russell R. Chianelli

Additional information is available at the end of the chapter

http://dx.doi.org/10.5772/intechopen.78379

Abstract

Asphaltenes are the molecular components of Asphalts and have polyaromatic hydrocarbon (PAH) structures similar to nanographenes. Thus, organic-based dye sensitized solar cells can be produced from asphaltenes. In addition, graphene based structures doped with transition metals can be synthesized from asphaltenes. These materials are simple to synthesize and inexpensive relative to other methods for doping graphene. Thus, what is considered a "Waste Material - Tar" can now be utilized in several important applications. These novel materials from asphaltene are also termed Discotic Liquid Crystals (DLC) and now have great potential in many areas, because of novel and valuable properties, easy synthesis and low cost.

Keywords: synthetic asphaltene, organic photovoltaics (OPVs), polyaromatic hydrocarbons (PAHs), discotic liquid crystals (DLCs), dye sensitized solar cells (DSSC), light harvesting molecules, nanographene, doped graphene

1. Introduction

Carbon-based systems including polymers, nanotubes, graphene and quantum dots are active materials in electronic, photonic or magnetic devices. Their properties depend on chemical structure and size and thus they can be chemically or physically tuned. Extended π-conjugated scaffolds that can be decorated with functional groups are the main feature related in these structures. Their functionalization can be optimized in order to facilitate their self-assembly into highly ordered supramolecular architectures, that are driven by non-covalent $\pi - \pi$ stacking interaction, hydrogen bonding, etc. Small molecules which can be characterized by their π-conjugated networks (pyrenes, pentacenes, coronenes, polythiophenes, etc.) possess

IntechOpen

© 2018 The Author(s). Licensee IntechOpen. This chapter is distributed under the terms of the Creative Commons Attribution License (http://creativecommons.org/licenses/by/3.0), which permits unrestricted use, distribution, and reproduction in any medium, provided the original work is properly cited.

Device	Area Cross sectional (nm^2)	Current density (electrons/nm^2-sec)
1 mm copper wire	3×10^{12}	2×10^6
Polyphenylene wire	0.05	4×10^{12}
Carbon nanotube	3	2×10^{11}

Table 1. Current density of molecular wires expressed in nanoscale units compared to current density of macroscopic copper [14].

electronic bandgaps that are well defined and shapes such as fibers, nanocrystals or uniform monolayers which enable molecular self-assembly into highly ordered low-dimensional architectures. Conjugated molecules have been explored for use in organic electronics and vary widely, however conjugated discotic materials and their self-assembly have been of interest to researchers due to their unique liquid crystalline properties [1, 2].

Experiments done in the mid and late 1990s demonstrated individual molecules can possess physical phenomena which previously thought to have been limited to semiconductors and revolutionized the material science. In 1996, it was still questionable whether or not individual molecules could actually conduct electricity but experiments performed from 1995 to 1997 [3–8] determined that individual molecules can conduct and also switch small electrical currents [9–13]. These revolutionary works demonstrated the use and assembly of molecular electronic devices using aromatic organic molecules, nanotubes, biomolecules or nanowires (**Table 1**) but more importantly, they gave the promise of low cost, carbon-based electronics at the smallest of scales which could be printed onto plastic.

2. Asphaltene and synthetic asphaltene: the theoretical platform

Asphaltenes are a relatively exotic class of hydrocarbon component in crude oils [15, 51]. Naturally occurring crude oil contains an enormous range of compounds that spans hydrocarbon compositions from dry natural gas to tar. The density and viscosity increases dramatically and color changes from clear to deep brown as asphaltene content increases from 0 to nearly 20% across that range. Researchers agree that the asphaltene structure is comprised of some carbon and hydrogen atoms which are bound in ring-like, aromatic groups, and also contains the heteroatoms. Alkane chains and cyclic alkanes include the rest of the carbon and hydrogen atoms, linked to ring groups. Within this arrangement, asphaltenes exhibit a range of molecular weight and composition. These characteristics of asphaltenes are acceptable to nearly all researchers and specialists, but leaves sufficient room for discussion regarding the structure or size of individual asphaltene molecules.

Extensively studied by Chianelli et al., asphaltene structure has been shown to have a structure similar to that of the synthetic compound tetraoctyl carboxylate perylene (TOPC) [16]. X-ray diffraction (XRD) studies performed on TOPC (shown in **Figure 1**) differentiate two peaks

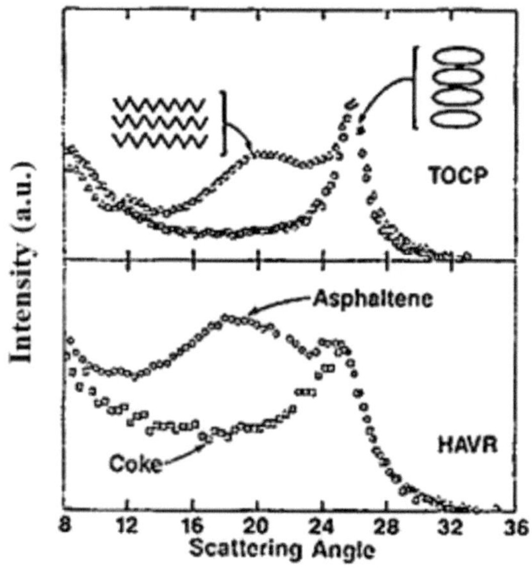

Figure 1. X-ray diffraction of asphaltene –high aromatic vacuum residue (HAVR) compared to synthetic molecule tetraoctyl carboxylate perylene (TOPC) [16].

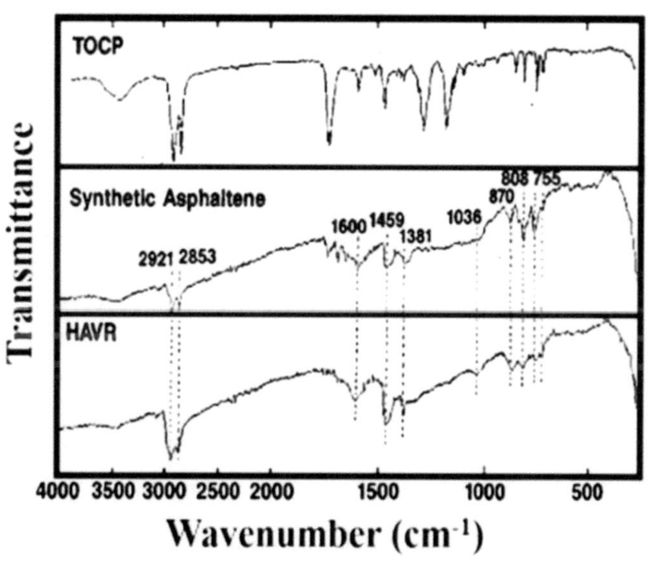

Figure 2. Infrared spectra of the TOCP precursor, synthetic asphaltene and HAVR (heavy Arab vacuum reside) asphaltene showing the significant similarity between the synthetic and the HAVR asphaltene [16].

that can determine distinction between alkane chains and stacked aromatic cores. Using these synthetic structures as a model, the structure of asphaltene before and after thermal processing was explored using XRD and the results show that the production of coke is actually

Chemical properties of synthetic asphaltene and HVAR				
	H/C	% C Aromatic	% C Aliphatic	MW
Synthetic asphaltene	1.06	58	42	2750
HVAR asphaltene	1.10	33–45	55–65	1.000–10.00

Table 2. Comparison of some chemical properties of synthetic asphaltene and HVAR [16].

the self-assembly of stacked aromatic cores. This work helped researchers understand the fundamental physical characteristics and the process of coke formation through self-assembly using of synthetic asphaltene.

When the temperatures increased above 350°C during thermal processing, the aliphatic ester of TOCP becomes the weakest region for thermal cleavage. When temperature reaches above 420°C in the absence of oxygen, TOCP completely thermally converts into a sea of free radicals containing aromatic cores and long-chain alkyl fragments. After Chianelli et al. successfully prepared synthetic asphaltene, the Infrared (IR) spectrum of TOCP, synthetic asphaltene, and HVAR (Heavy Arab Vacuum Residuum) were compared (**Figure 2**) and it is clear that absorption bands of synthetic asphaltene match well with band position and intensity of the IR spectrum of HVAR [17]. The comparison in chemical properties between synthetic asphaltene and HVAR illustrated in **Table 2**.

3. Organic photovoltaics using asphaltenes

3.1. Introduction to organic photovoltaics

The photovoltaic effect represents the generation of a potential difference at the junction of two different materials in response to various forms of radiation. The whole field of converting solar energy into electricity is the field of "photovoltaics" [18]. The photovoltaic process combines four stages which comprise the following: (1) light absorption, (2) charge generation, (3) charge transport, and (4) charge collection [19]. When the material has a semiconducting property that responds to incident waves, the first stage light absorption occurs. The absorption characteristic of the material is based on the intrinsic extinction coefficient and the band gap of the semiconducting material. Once light has absorbed by a material, the second stage occurs. As the incident photon hits electrons at the ground state, inorganic semiconductors generate free carriers (i.e. holes and electrons) and charge generation ensues. However, in organic semiconductors, excited electrons form an exciton, which is a bounded electron and hole pair. To make an efficient organic photovoltaic cell, the dissociation of excitons is an important matter as the binding energy of the exciton is relatively large (0.3 eV) [20].

Dye sensitized solar cells (DSSC) are low cost alternatives for the conventional solar cells, and commercially promising because of the promise of low-cost materials and roll-to-roll manufacturing. The DSSC can be classified as a photo electrochemical (PEC) solar cell because of its utilization of photons, charges, and electrolyte for its basic operation [21]. O'Regan and Grätzel built and demonstrated the first DSSC by 1991, [22] and attracted widespread academic and industrial interest. Compared to conventional solar cells, DSSCs had the promise of easy fabrication at a low-cost in environment friendly and have high-energy conversion efficiency [23]. One more advantage of DSSCs is that changes in temperature do not affect the performance, while temperature changes can affect the performance of conventional silicon solar cells [24].

Ooyama and Harima [30] have done an extensive literature review on the main requirements for efficient organic molecule (dye) used so far for DSSCs. Their conclusions are summarized in the following four statements:

1. The organic molecule (dye) should have a minimum of one anchoring group like COOH, $-SO_3H$, $-PO_3H_2$, –OH for adsorption onto the TiO_2 surface to provide a tightly held dye and suitable electron interaction between them.
2. To achieve efficient electron injection from the excited dye to the conduction band of the TiO_2, the dye has to have an energy level of the lowest unoccupied molecular orbital (LUMO) which is higher (more negative) than that of the conduction band of the TiO_2 electrode. However, the energy level of the highest occupied molecular orbital (HOMO) of the dye must be lower (more positive) than the I_3^-/I^- redox potential in the electrolyte to achieve efficient regeneration of the oxidized state by electron transfer from I_3^-/I^- redox couple.
3. The organic molecule should have high molar absorption coefficients over the wide region of sunlight extending into near infra-red, to provide a large photocurrent to give the high light harvesting efficiency.
4. To achieve a stable DSSC, the organic molecule should have chemical stability in its photo excited state and the redox reactions throughout the reaction cycle.

Dye molecules (sensitizers) are the key component of a DSSC that increase efficiency through their abilities to absorb visible light photons Despite the fact that the highest efficiency of a dye sensitized solar cells is achieved from a combined effect of numerous physiochemical nanoscale properties, the main focus in these studies is the dye sensitization of large band-gap semiconductor electrodes. In DSSCs, this is accomplished by coating the internal surfaces of porous TiO_2 electrodes with dye molecules specifically tuned to absorb the incoming photons [23]. The dye is the light absorber and the photoreceptor sensitizing the semiconductor and so some conditions must be fulfilled. Without a doubt, a dye that absorbs nearly all the sunlight radiation incident on earth, like a black-body absorber, is highly desirable. Organic dyes consisting of natural pigments and synthetic organic dyes have a donor-acceptor structure called push-pull design, which improves the absorption in red and infrared region by increasing short circuit current density. Natural pigments, like chlorophyll, carotene, and anthocyanin, are available in plant leaves, flowers, and fruits and fulfill these requirements. Experimental results showing that natural-dye sensitized solar cells achieved the efficiency of 7.1% and high stability [24].

Co-sensitization, the use of two dyes combination that complements each other in their spectral features is another strategy to get a broad optical absorption expanding throughout the visible and near IR region. This strategy enhances photo-absorption through optical effects which allow the two sensitizers to be additive. In particular, there was no negative interference between co-adsorbed chromophores, making way for testing a multitude of other dye combinations [25].

To increase improvement of the performance of DSSC, different kinds of additives have been added into dye solutions. For example, Kay and Grätzel found that the addition of cholic acid (CA) with porphyrin derivatives improved both the photocurrent and photovoltage of the solar cell [26]. Additionally, other co-adsorbents such as hexadecylmalonic acid (HDMA) [27], deoxycholic acid (DCA) [28], and chenodeoxycholic acid (CDCA) [29] were also used in the dye solution when fabricating sensitized TiO2 films into devices, among which CDCA is the most popular co-adsorbent.

Dye aggregation on the TiO_2 surface should be avoided. This aggregation leads to reduction in electron-injection yield from the dyes to the CB of the TiO_2 due to intermolecular energy transfer causing low conversion efficiency of the DSSC [30]. The main parts of the dye sensitized solar cell are illustrated schematically in **Figure 3**. The transparent and counter conducting electrodes, wide bandgap semiconducting nanostructured layer, the sensitizer, and the electrolyte are the main four elements of the cell. The transparent conducting electrode and counter-electrode are coated with a conductive and transparent thin film such as fluorine doped tin dioxide (SnO_2) [31].

The photocurrent density calculated at short circuit (J_{SC}) the open-circuit photovoltage (V_{oc}) the fill factor (FF), and the power of the incident light (P_{in}) are used to determine the total conversion efficiency (%η) of the dye-sensitized solar cell. These values can be taken from the photocurrent density-voltage characteristics (IV curves) under full sunlight (P_{in} = 100 $mWcm^{-2}$). The relationship between voltage (V) and current (I) is determined by varying the resistance of the outer circuit and considered J_{SC} when resistance of the outer circuit is zero (thus voltage is zero) and Voc when resistance is maximum (thus photocurrent is zero). The power output of the device equals the product of J and V, and the fill factor states the efficiency of the device compared to that of an ideal cell. P_{max} relates to the maximum value that can reach the output power and is usually reported as the output power of the commercial device. The performance of DSSC can be projected using the following Equations [32]: (101)

$$\eta\,(\%) = \frac{Jsc * Voc * FF}{Pin} \tag{1}$$

where:

$$FF = \frac{Jm\ Vm}{Jsc\ Voc} \tag{2}$$

3.2. Results: The resurrection of an ancient leaf

Dr. Rajab E. Abujnah built the first ever asphaltene DSSC (dye-sensitized solar cell) as a graduate student in the lab of Dr. Chianelli, which demonstrated a photo-conversion efficiency (PCE) of 1.8%. Four natural organic asphaltene fractions were extracted and tested as dyes using various parameters with TiO_2 based dye-sensitized solar cell DSSC. The photovoltaic performance of the cells was determined by parameters such as open-circuit voltage (V_{oc}), short-circuit current (I_{sc}), fill factor (FF), and series resistance. The overall energy conversion efficiency was also measured to correlate the different effects of asphaltene fraction as well as concentration and influence on the improvement of the solar cell parameters.

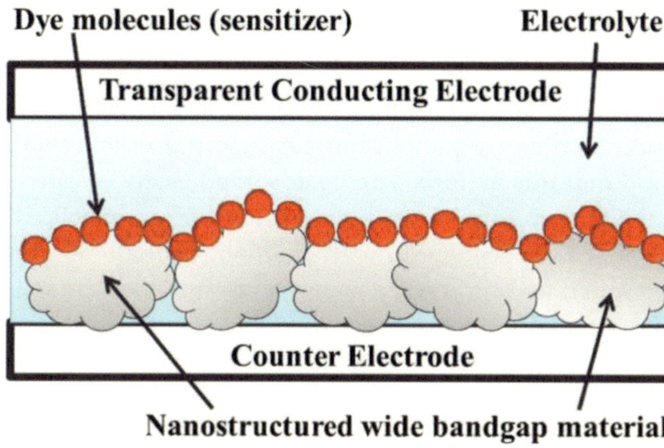

Figure 3. Schematic of the structure of the dye sensitized solar cell.

The sensitization of TiO_2 electrode using asphaltene obtained with a 90/10 ratio of toluene to pentane ratio resulted in an energy conversion efficiency of 1.8%. Un-fractionated asphaltene was shown to produce comparable results as Fraction 3 as sensitizer in DSSC. A strategy to minimize series resistance, improve photo current and open circuit voltage was also studied. Purifying asphaltene using the RTV mask, employing the TiO_2 compact layer, in addition to UV-Ozone cleaning treatment of the transparent electrode have been shown to improve the performance of asphaltene DSSCs [33].

The asphaltene dye covered TiO_2 electrode and Pt-counter electrode were assembled into a sandwich type and assembled with using a binder clamps as shown in **Figure 4**. The clamps used to attach the two electrodes and to provide space for the electrolyte between two electrodes. Sealant also can be used to assemble the sandwich of electrodes. It is can be used to prevent leakage of electrolyte. For good results, sealant was put between two electrodes and put under the hot press machine. Temperature and pressure were adjusted and then the hot press was heated up to join both electrodes by using the sealant. A drop of the electrolyte solution was

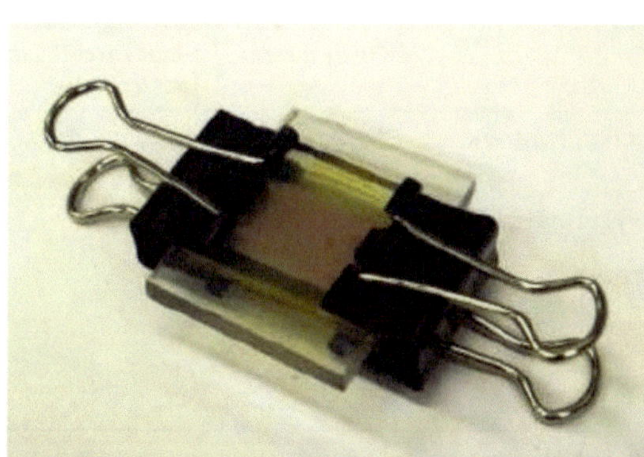

Figure 4. Completed synthetic asphaltene DSSC.

injected into the hole in the back of the counter electrode and to prevent leakage of electrolyte, the hole was covered by tape. The electrolyte was introduced into the cell by using vacuum.

Synthetic asphaltene has shown experimental evidence and preliminary results to suggest it can fulfill the requirements of organic semiconductors due its ability to absorb visible light. It was also found that this material can be adsorbed easily on the surface of the nanocrystalline particles TiO_2, therefore synthetic asphaltene is an excellent candidate for use in DSSC. Dr. Hassan Sharif experimented with synthetic asphaltene samples subsequently to Dr. Abujnah and applied as-synthesized asphaltenes into various solar cells with different parameters in the TiO_2 layer of the DSSC. DSSCs were fabricated using the synthetic asphaltene as dye.

To further research and optimize the cells, Porphyrin was added to the synthetic asphaltene to improve the efficiency of the cell. Porphyrin is already used as organic dye in DSSC. The new compound is prepared in the lab by dissolving the same ratio of synthetic asphaltene with free metal Porphyrin in Toluene for 24 hrs under heating; the mixture was kept under the hood to evaporate the Toluene. The final product was collected and used in this experiment. The table below shows the effects of mixing Porphyrin with synthetic asphaltene on the conversion efficiency of synthetic asphaltene dye. From the table it is clear that the new dye is increasing both current density J_{SC} and V_{OC}, and then increasing the performance efficiency.

Asphaltene has shown to be able to produce voltage, and according to the results outlined in **Table 3**, could be considered a good candidate in photovoltaic in part, due to its low cost [34]. New material was prepared by dissolving synthetic asphaltene and crude oil asphaltene. This asphaltene was extracted from crude oil asphaltene and purified by using Toluene. Real asphaltene was dissolved with synthetic asphaltene in Toluene in the ration of 1:1 and kept for 24 hours for dissolving under heating. The solvent kept under the hood until the Toluene completely evaporated, and the solid product was collected and used for the experiment.

During his work with synthetic asphaltene as new dye in the fabrication of DSSC, Dr. Sharif noticed some precipitates when synthetic asphaltene dissolved in Toluene. These precipitates prevent the good absorbance of synthetic asphaltene dye into the surface of TiO2. So, synthetic asphaltene was filtrated to remove any impurities. The table below compares the

Experiment	Open current Voltage	Short Circuit Current (mA/cm²)	Fill Factor %	PCE (%)
Synthetic asphaltene mixed with porphyrin and dissolved in toluene	0.429	6.57	54	1.85
Real Asph. mixed with synthetic asphaltene	0.435	7.28	46	1.40
Unpurified synthetic asphaltene	0.445	4.19	56	1.0
Purified synthetic asphaltene	0.455	7.05	61	1.83
Purified synthetic asphaltene with chenodeoxycholic acid (CDCA)	0.506	7.00	0.61	2.08

Table 3. The performance of synthetic asphaltenes in DSSCs using different parameters.

results before and after purifying the synthetic asphaltene. There is a high increase in the current density and also fill factor, but the small increase in the voltage. The efficiency was improved from 1.0 to 1.83% which illustrates the importance of processing parameters on devices performance and optimization.

The effects of chenodeoxycholic acid (CDCA) in a dye solution as a co-adsorbent on the photovoltaic performance of dye-sensitized solar cells (DSSCs) based organic dye containing purified synthetic asphaltene dye was also studied. It was observed that the co-adsorption of CDCA can eliminate and reduce the formation of dye aggregates and improve electron injection yield and thus, the short-circuit current (*J*sc). This has also led to a rise in photovoltage, which is attributed to the decrease of charge recombination. The DSSC, based on synthetic asphaltene dye with CDCA, overall conversion efficiency was further improved to 2.0% (J_{sc} = 7.00 mA/cm^2, V_{oc} = 0.506 V, FF = 0.61) upon addition of 1:1 ratio of CDCA to the dye solution.

3.3. Summary and discussion

In general, we can conclude from these studies that asphaltene was successfully prepared in the lab and the characterization results were similar as in the literatures. Even primary results show that synthetic asphaltene has good ability to absorb sunlight, but the results of the first use of synthetic asphaltene in DSSC were not satisfied. More work was carried during the study to increase the performance efficiency from 0.4 to 2.0%. We conclude also synthetic asphaltene can give good results by adding some additives, but it also can improve the conversion efficiency just by purifying process which increase efficiency from 1.0 up to 1.83%. Instead of additives, there is another parameters can increase the efficiency of synthetic asphaltene as dye in DSSC for example, the number of layers of the paste and using UV-O treatment. Therefore, a total overall efficiency of 3.5–5.0% could be easily attained with good engineering design and material process.

The first result obtained with using synthetic asphaltene as dye was 0.4%. The maximum PCE obtained using synthetic asphaltene as dye after adjusting some variables and adding some additives reached 2.0%. The benefit to this study and in using synthetic material will help researchers to find new and inexpensive dyes as well as to understand better the characterization of crude oil asphaltene [35].

4. Doped graphene from asphaltene

4.1. Introduction to doped graphene applications

The electronic and magnetic properties of graphene can be modified through combined transition-metal (TM) and nitrogen decoration of vacancies. Additional modes of functionalization that are currently being explored for a wide range of applications in nanoelectronics [36–38], spintronics [39] and electrocatalysts [40]. Researchers previously discovered that TMs bind to graphene strongly in a porphyrin ring coordinated by four nitrogens [41]. The stability in the presence of the defects associated with TMs can be attributed to the reduced electrostatic repulsion between nitrogen lone-pair electrons due to the hybridization between N and TM. Results from DTF studies have predicted these types of structures to be particularly promising candidates for graphene-based ferromagnets, which could find applications in nanoelectronics and nanomagnetism.

Graphene-type molecules, typically large polycyclic aromatic hydrocarbons (PAHs), have gained enormous interest because of their unique self-organization behavior and promising electronic properties for applications in organic electronics [42]. Asphaltene is a part of crude oils that contain a large number of structures, specifically high molecular weight bonded aromatic hydrocarbons components with hetero-atoms. Several metals (e.g., Ni, V, Fe, Al, Na, Ca, and Mg) shown to accumulate in the asphaltenes fraction of crude oil, typically in concentrations less than 1% w/w [43–45]. Vanadium and nickel, the most abundant of the trace metals, present mainly as chelated porphyrin compounds, and they linked to catalyst poisoning during upgrading of heavy oils [46, 47]. The concentrations of other trace metals not bound in porphyrin structures (e.g., Fe, Al, Na, Ca, and Mg) indicated to change in deposits as a function of well depth [48], and among sub fractions of asphaltene.

Graphene produced from chemical vapor deposition is viable for use in technologies such as touch screens, sensors, transistors, novel optical/electronic/photonic devices such as IR cameras, radiation shielding and camera lenses. Because they are expected to induce novel magnetic and superconducting behavior, TM adatoms and graphene are a topic of great interest are expected to induce novel magnetic and superconducting behavior. Because of this interest, there have been extensive theoretical studies but the experimental exploration of TM/graphene systems is very limited. Graphene with Boron and Nitrogen (BCN) is a sought after material due to the fact that graphene has no band gap [49]. Graphene in its single crystal form is a conductor, not a semi-conductor, so while it has amazing properties sought after by so many researchers, it has yet to make a significant impact on electronic industries for this reason. BCN is currently the subject of doped graphene research because it can be introduced by vapor. In contrast, TMs cannot be introduced by vapor. Even if they could be introduced at high temperatures, it has been determined that they would not be stable in an in-plan configuration due to the high differences in energy between TMs and carbon. Boron and Nitrogen both neighbor carbon on the periodic table and do not have a large differences in electron structure. As a carbon source, asphaltene is generated from waste crude oils and remains unsuitable for hydrocracking because of the presence of TMs. This study attempts to utilize that waste as a valuable source of metalloporphyrins for placing TMs doping graphene materials.

Asphaltene materials were extracted from crude using different n-alkanes and dissolved in toluene for deposition. Asphaltene/toluene (1 mg/ml) solutions were then deposited using drop coating onto previously prepared copper foil and experiments proceeded according to **Table 4**. Experiment 1 deposited asphaltene onto copper using spin coating. A 4 inch tube furnace was used to anneal the material under a reducing environment (H_2). The system was equipped with a turbo pump with a direct line-of-sight to the sample and the substrate and allowed for base pressures below 10^{-8} Torr. A turbo pump was used and all the flanges on the high vacuum side of the system were constructed with conflat flanges; capable of achieving ultra-high vacuum (UHV). Reaching low background pressures allowed the removal of residual gasses like water

http://dx.doi.org/10.5772/intechopen.78379

Experiment	Temperature [°C]	Time [min]	Material	Gas
1	1050	5	C7-spin coated	H (10 ccm)
2	1050	5	C7-drop coated	H (10 ccm)
3	1050	5	Synthetic	H (10 ccm)

Table 4. Experimental conditions and parameters and (above) figure of furnace used in experiment.

and oxygen from the system before growth. The use of a turbo pump also allowed for the system to reach pressures low enough (below 10^{-5} Torr) to operate a residual gas analyzer (RGA) instrument. In a typical growth, the sample was loaded into the furnace and allowed to pump down using the turbo pump for 2 hours prior to heating and gas flow (**Figure 5**).

Raman was taken using Witec Alpha 300 micro-Raman confocal microscope after graphene had been transferred to SiO_2/Si wafers. Scanning Electron Microscope (SEM) Images were taken using a FEI Quanta 650 SEM equipped with Bruker EDX system for chemical analysis. Optical Microscope images of asphaltenes were taken on copper foils before and after growth. Transfers were performed by spin coating polymer, dissolving copper foil and transferring graphene to SiO_2/Si wafers. Zeiss Axiovert 100A Light Microscope was used to take images of asphaltenes before and after growth.

4.2. Single-layer asphaltene

Conditions for graphene growth were applied for all experiments. The resulting averaged Raman from all experiments show heavy carbonization and some graphitization of the solid carbon source. We were able to find metals in more than one sample of the post growth heptane extracted asphaltene. When drop coating was used for application before the growth process, the Raman from **Figure 6** indicates mostly carbon with some graphitization. Only

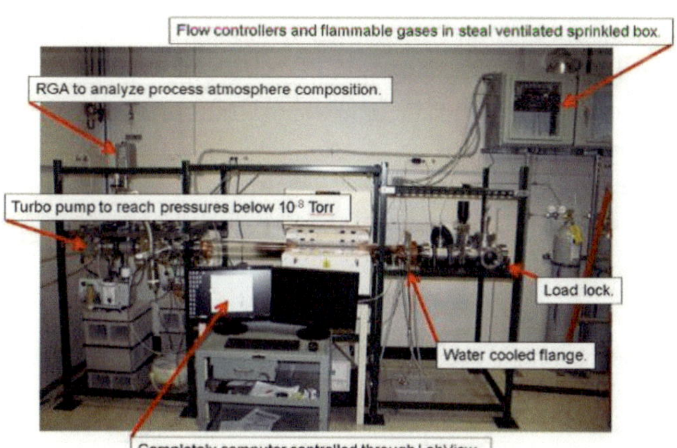

Figure 5. CVD set-up at the J.J. Pickle Research Center Austin, TX [50].

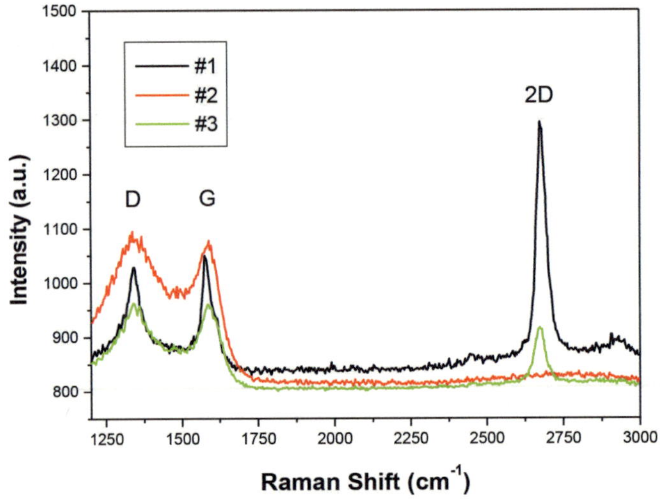

Figure 6. Raman summary of experiment 1, 2 & 3.

experiment 1 exhibited the ratio of (> 2:1) 2D to G and D peaks that would indicate the presence of quality graphene rather than sp^2 hybridized carbon.

Metals identified in post growth samples from experiment 1 and 2 included Al, Fe, Zr and adatoms included sulfur. Electron Diffraction X-ray (EDX) analysis was taken of the copper foil before any etching in **Figure 7** and foil which had been sonicated in acetone before applying asphaltenes shows no metals or adatoms present before growth indicating that metals and adatoms indeed came from asphaltene samples. After growth, EDX spectra taken of sample from Experiment 1 show the presence of Aluminum and Iron.

To further study asphaltene adatoms, EDX elemental analysis was taken of synthetic asphaltene containing no metals applied to copper foil. Needless to say from data in **Figure 8** no TM or adatoms could be found present after growth.

Images taken using optical microscope **Figure 9** show copper coated with heptane extracted asphaltenes before and after Experiment 1 growth. There are clustered discotic structures before growth and curiously there are areas after growth that show clearly more than a few layers graphitized **Figure 10**. There are areas that can be seen where copper can be seen as clear orange from underneath sheets of graphitized carbon in **Figure 9(c)** and **(d)**.

4.3. Summary

Because they are expected to induce novel magnetic and superconducting behavior, TM adatoms and graphene are a topic of great interest. There have been extensive theoretical studies because of these interesting properties but the experimental exploration of TM/graphene systems is very limited. Graphene with Boron and Nitrogen (BCN) is a sought after material due to the fact that graphene has no band gap. Graphene in its single crystal

http://dx.doi.org/10.5772/intechopen.78379

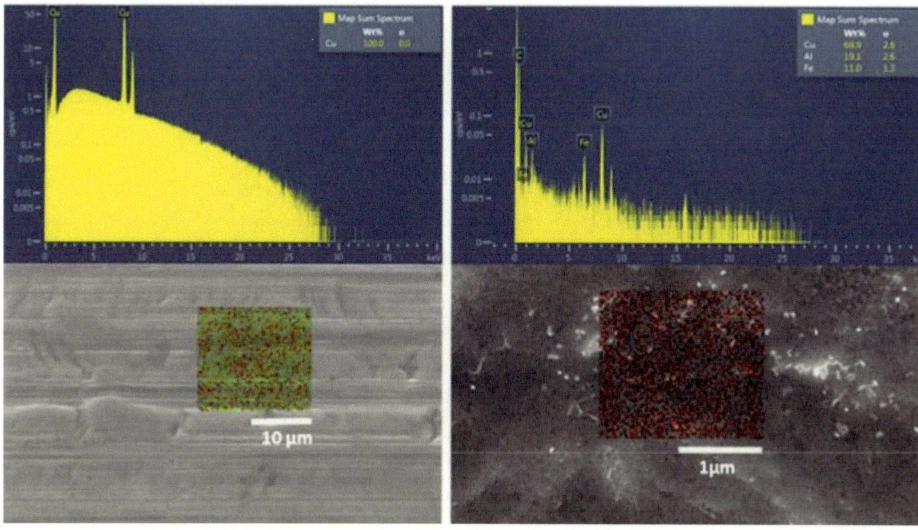

Figure 7. EDX analysis (top left) of copper foil sonicated in acetone before application of solid carbon source with corresponding SEM image (bottom left) EDX elemental analysis respectively (top right) from SEM section (bottom right) from asphaltene extracted using heptane in Experiment 1.

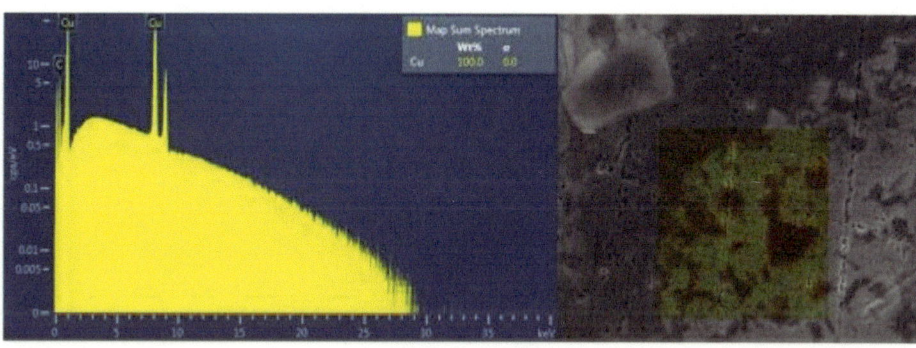

Figure 8. EDX (left) copper with drop coated synthetic asphaltene (right) after growth.

form is a conductor, not a semi-conductor, so while it has amazing properties sought after by so many researchers, it has yet to make a significant impact on electronic industries for this reason. BCN is currently being researched because it can be introduced by vapor. In contrast, TMs cannot be introduced by vapor. Even if they could be introduced at high temperatures, it has been determined that they would not be stable in an in-plan configuration due to the high differences in energy between TMs and carbon. Boron and Nitrogen both neighbor carbon on the periodic table and do not have a large difference in electron structure. Our carbon source generated from waste crude oils is unsuitable for hydrocracking because of the presence of TMs. We were able to utilize that waste as a valuable source of metalloporphyrins for placing TMs directly inside the graphene lattice.

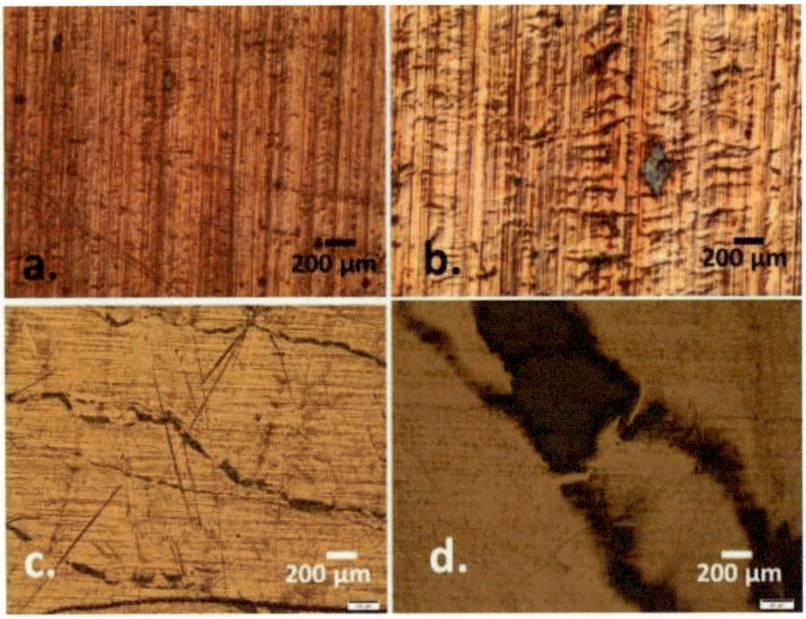

Figure 9. Optical microscope images of heptane asphaltene on copper foil from experiments before (a,b) and after growth (c,d).

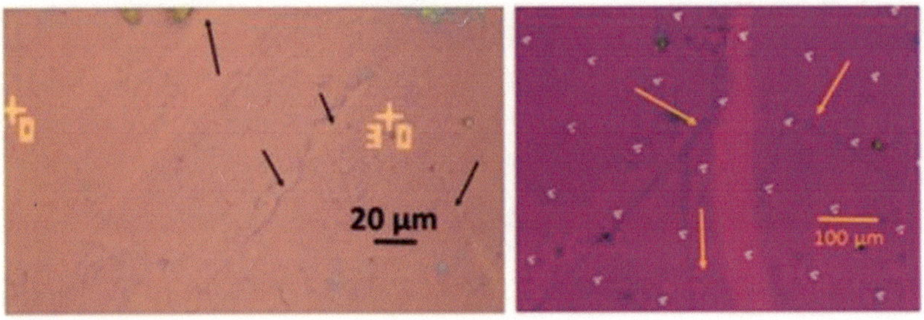

Figure 10. Optical microscope image of experiment 1 transferred to SiO2/Si wafer with rips and wrinkles (left) compared to OM image of a graphene sheet (purple) on SiO_2/Si wafer (pink). Arrows show wrinkles and tears in the material (right). (graphene provided courtesy Ruoff group at UT).

5. Conclusions

All electronic processes in nature occur using carbon-based molecular structures; from signal transduction in the brain all the way to photosynthesis in a leaf. Carbon Molecular Electronics (CMEs) as a field is the result of developments in an interdisciplinary field with increasing numbers of synthetic and quantum chemists, physicists and engineers. New devices have been designed and investigated for the exciting realization of CMEs for photovoltaics, transistors, switches and other non-linear electronic components.

The results demonstrated in these works illustrate the use of asphaltene as a light-harvesting active molecular dye effectively in dye sensitized solar cells for application in the field of organic photovoltaics. Higher-power conversion efficiencies can be achieved by further optimizing the device structure. In addition, they also show asphaltene as a promising candidate for doping multi-layer and potentially single layer graphene for future novel nanoscale magnetic or optoelectronic uses.

Acknowledgements

We would like to acknowledgeDr. Harry Chou's assistance with graphene growth and characterization in Section 4. Doped Graphene from asphaltene at The University of Texas (Austin, TX). Thanks To Dr. Carl Magnuson for assistance and instruction on tube furnace construction and apparatus.

Author details

Eva M. Deemer[1] and Russell R. Chianelli[1,2]*

*Address all correspondence to: chianell@utep.edu

1 Materials Research and Technology Institute, The University of Texas at El Paso, TX, USA

2 Department of Chemistry, The University of Texas at El Paso, El Paso, TX, USA

References

[1] Sergeyev S, Pisula W, Geerts YH. Discotic liquid crystals: A new generation of organic semiconductors. Chemical Society Reviews. 2007;**36**:1902. DOI: 10.1039/b417320c

[2] Laschat S, Baro A, Steinke N, Giesselmann F, Hägele C, Scalia G, Judele R, Kapatsina E, Sauer S, Schreivogel A, Tosoni M. Discotic liquid crystals: From tailor-made synthesis to plastic electronics. Angewandte Chemie International Edition. 2007;**46**:4832-4887. DOI: 10.1002/anie.200604203

[3] Bumm LA, Arnold JJ, Cygan MT, Dunbar TD, Burgin TP, Jones L, Allara DL, Tour JM, Weiss PS. Are single molecular wires conducting? Science (80). 1996;**271**:1705-1707. DOI: 10.1126/science.271.5256.1705

[4] Reed MA, Zhou C, Muller CJ, Burgin TP, Tour JM. Conductance of a molecular junction. Science. 1997;**278**:252-254. DOI: 10.1126/science.278.5336.252

[5] Joachim C, Gimzewski JK. Analysis of low-voltage I(V) characteristics of a single C60 molecule. Europhysics Letters. 1995;**30**:409-414. DOI: 10.1209/0295-5075/30/7/006

[6] Dorogi M, Gomez J, Osifchin R, Andres RP, Reifenberger R. Room-temperature Coulomb blockade from a self-assembled molecular nanostructure. Physical Review B. 1995;**52**:9071-9077. DOI: 10.1103/PhysRevB.52.9071

[7] Dai H, Wong EW, Lieber CM. Probing electrical transport in nanomaterials: Conductivity of individual carbon nanotubes. Science (80). 1996;**272**:523-526. DOI: 10.1126/science.272.5261.523

[8] Dekker C, Tans SJ, Verschueren ARM. Room-temperature transistor based on a single carbon nanotube. Nature. 1998;**393**:49-52. DOI: 10.1038/29954

[9] Zhou C, Deshpande MR, Reed MA, Jones Ii L, Tour JM. Nanoscale Metal/Self-Assembled Monolayer/Metal Heterostructures. Applied Physics Letters. 1997;**71**:611 DOI: org/10.1063/1.120195

[10] Reed MA. Molecular-Scale Electronics. In: Proceedings of the IEEE. April 1999;**87**(4): 0018-9219

[11] Joachim C, Gimzewski JK. An electromechanical amplifier using a single molecule. Chemical Physics Letters. 1997;**265**:353-357. DOI: 10.1016/S0009-2614(97)00014-6

[12] Metzger RM, Chen B, Höpfner U, Lakshmikantham MV, Vuillaume D, Kawai T, Wu X, Tachibana H, Hughes TV, Sakurai H, Baldwin JW, Hosch C, Cava MP, Brehmer L, Ashwell GJ. Unimolecular electrical rectification in hexadecylquinolinium tricyanoquinodimethanide. Journal of the American Chemical Society. 1997;**119**(43):10455-10466. DOI: 10.1021/JA971811E

[13] Chen R, Rawlett T. Large on-off ratios and negative differential resistance in a molecular electronic device. Science. 1999;**286**:1550-1552. http://www.ncbi.nlm.nih.gov/pubmed/10567259 [Accessed: March 24, 2017]

[14] Ellenbogen JC, Love JC. Architectures for molecular electronic computers. I. Logic structures and an adder designed from molecular electronic diodes. In: Proc. IEEE. 2000;**88**:386-426. DOI: 10.1109/5.838115

[15] Mullins OC, Sheu EY, Hammami A, Marshall AG, editors. Asphaltenes, Heavy Oils and Petroleomics. New York City: Springer; 2007

[16] Chianelli RR, Siadati M, Mehta A, Pople J, Ortega LC, Chiang LY. Self-assembly of asphaltene aggregates: Synchrotron, simulation and chemical modeling techniques applied to problems in the structure and reactivity of asphaltenes. In: Asphaltenes, Heavy Oils, and Petroleomics. New York, New York, NY: Springer; 2007. pp. 375-400. DOI: 10.1007/0-387-68903-6_15

[17] Zeman M. Introduction to photovoltaic solar energy. Solar Cells. Coll. 2003:1-139. http://paginas.fisica.uson.mx/horacio.munguia/aula_virtual/Cursos/Teoria de Control/Solar Cells Miro Zeman.pdf [Accessed: December 10, 2017]

[18] Knier G. How do Photovoltaics Work? | Science Mission Directorate, Scientific Beta. 2016. https://science.nasa.gov/science-news/science-at-nasa/2002/solarcells [Accessed: December 10, 2017]

[19] Knupfer M. Exciton binding energies in organic semiconductors. Applied Physics A: Materials Science & Processing. 2003;**77**:623-626. DOI: 10.1007/s00339-003-2182-9

[20] Yu K, Chen J. Enhancing solar cell efficiencies through 1-D nanostructures. Nanoscale Research Letters. 2009;**4**:1-10. DOI: 10.1007/s11671-008-9200-y

[21] O'Regan B, Grätzel M. A low-cost, high-efficiency solar cell based on dye-sensitized colloidal TiO_2 films. Nature. 1991;**353**:737-740. DOI: 10.1038/353737a0

[22] Grätzel M. Conversion of sunlight to electric power by nanocrystalline dye-sensitized solar cells. Journal of Photochemistry and Photobiology A: Chemistry. 2004;**164**:3-14. DOI: 10.1016/j.jphotochem.2004.02.023

[23] Grätzel M. Dye-sensitized solar cells. Journal of Photochemistry and Photobiology C:Photochemistry Reviews. 2003;**4**:145-153. DOI: 10.1016/S1389-5567(03)00026-1

[24] Campbell WM, Jolley KW, Wagner P, Wagner K, Walsh PJ, Gordon KC, Schmidt-Mende L, Nazeeruddin MK, Wang Q, Gra M, Officer DL. Highly efficient porphyrin sensitizers for dye-sensitized solar cells. Journal of Physical Chemistry C. 2007;**111**(32):11760-11762. DOI: 10.1021/jp0750598

[25] Sayama K, Tsukagoshi S, Mori T, Hara K, Ohga Y, Shinpou A, Abe Y, Suga S, Arakawa H. Efficient sensitization of nanocrystalline TiO_2 films with cyanine and merocyanine organic dyes. Solar Energy Materials & Solar Cells. 2003;**80**:47-71. DOI: 10.1016/S0927-0248(03)00113-2

[26] Kay A, Graetzel M. Artificial photosynthesis. 1. Photosensitization of titania solar cells with chlorophyll derivatives and related natural porphyrins. The Journal of Physical Chemistry. 1993;**97**:6272-6277. DOI: 10.1021/j100125a029

[27] Wang P, Zakeeruddin SM, Comte P, Charvet R, Humphry-Baker R, Grätzel M. Enhance the performance of dye-sensitized solar cells by co-grafting amphiphilic sensitizer and hexadecylmalonic acid on TiO_2 nanocrystals. Journal of Physical Chemistry B. 2003;**107**(51):14336-14341. DOI: 10.1021/JP0365965

[28] Wang Z-S, Cui Y, Dan-oh Y, Kasada C, Shinpo A, Hara K. Thiophene-functionalized coumarin dye for efficient dye-sensitized solar cells: Electron lifetime improved by coadsorption of deoxycholic acid. Journal of Physical Chemistry C. 2007;**111**(19):7224-7230. DOI: 10.1021/JP067872T

[29] Qu S, Wu W, Hua J, Kong C, Long Y, Tian H. New diketopyrrolopyrrole (DPP) dyes for efficient dye-sensitized solar cells. Journal of Physical Chemistry C. 2010;**114**:1343-1349. DOI: 10.1021/jp909786k

[30] Ooyama Y, Harima Y. Molecular designs and syntheses of organic dyes for dye-sensitized solar cells. European Journal of Organic Chemistry. 2009;**18**:2903-2934. DOI: 10.1002/ejoc.200900236

[31] Jasim KE. Dye sensitised solar cells-working principles, challenges and opportunities. Dye Sensitized Solar Cells - Working Principles, Challenges and Opportunities. 2007:171-204. http://cdn.intechopen.com/pdfs/23333.pdf [Accessed: December 10, 2017]

[32] Colodrero S, Calvo ME, Míguez H. Solar Energy, 2010. http://cdn.intechweb.org/pdfs/8565.pdf [Accessed: December 10, 2017]

[33] Abujnah RE. Asphaltene as Light Harvesting Material in Dye Sensitized Solar Cell. Texas, El Paso: ETD Collect. Univ; 2011. http://digitalcommons.utep.edu/dissertations/AAI10118135 [Accessed: December 10, 2017]

[34] RE A, H S, B T, K C, V G, RR C. Asphaltene as light harvesting material in dye-sensitized solar cell: Resurrection of ancient leaves. Journal of Environmental & Analytical Toxicology. 2016;**6**. DOI: 10.4172/2161-0525.1000345

[35] Sharif HM. Synthetic Asphaltene as a Novel Dye in Dye Sensitized Solar Cells DCCs. Texas, El Paso: ETD Collect. Univ; 2013. http://digitalcommons.utep.edu/dissertations/AAI3597255 [Accessed: December 10, 2017]

[36] Krasheninnikov AV, Nieminen RM. Attractive interaction between transition-metal atom impurities and vacancies in graphene: A first-principles study. Theoretical Chemistry Accounts. 2011;**129**:625-630. DOI: 10.1007/s00214-011-0910-3

[37] Wei D, Liu Y, Wang Y, Zhang H, Huang L, Yu G. Synthesis of N-doped graphene by chemical vapor deposition and its electrical properties. Nano Letters. 2009;**9**:1752-1758. DOI: 10.1021/nl803279t

[38] Wu M, Cao C, Jiang JZ. Light non-metallic atom (B, N, O and F)-doped graphene: A first-principles study. Nanotechnology. 2010;**21**:505202. DOI: 10.1088/0957-4484/21/50/505202

[39] Krasheninnikov AV, Lehtinen PO, Foster AS, Pyykkö P, Nieminen RM. Embedding transition-metal atoms in graphene: Structure, bonding, and magnetism. Physical Review Letters. 2009;**102**:126807. DOI: 10.1103/PhysRevLett.102.126807

[40] Tripkovic V, Vanin M, Karamad M, Björketun ME, Jacobsen KW, Thygesen KS, Rossmeisl J. Electrochemical CO_2 and CO reduction on metal-functionalized porphyrin-like graphene. Journal of Physical Chemistry C. 2013;**117**:9187-9195. DOI: 10.1021/jp306172k

[41] Kattel S, Atanassov P, Kiefer B. Stability, electronic and magnetic properties of in-plane defects in graphene: A first-principles study. Journal of Physical Chemistry C. 2012;**116**:8161-8166. DOI: 10.1021/jp2121609

[42] Pisula W, Feng X, Müllen K. Charge-carrier transporting graphene-type molecules. Chemistry of Materials. 2011;**23**:554-567. DOI: 10.1021/cm102252w

[43] Spiecker PM, Gawrys KL, Kilpatrick PK. Aggregation and solubility behavior of asphaltenes and their subfractions. Journal of Colloid and Interface Science. 2003;**267**:178-193. http://www.ncbi.nlm.nih.gov/pubmed/14554184 [Accessed: March 24, 2017]

[44] Pena ME, Manjarréz A, Campero A. Distribution of vanadyl porphyrins in a Mexican offshore heavy crude oil. Fuel Processing Technology. 1996;**46**:171-182. DOI: 10.1016/0378-3820(95)00053-4

[45] Thomas J, Kaminski HSF, Wolf N, Wattana P, Mairal A. Classification of Asphaltenes via fractionation and the effect of heteroatom content on dissolution kinetics. Energy Fuels. 1999;**14**(1):25-30. DOI: 10.1021/EF990111N

[46] Yang X, Hamza H, Czarnecki J. Investigation of subfractions of athabasca asphaltenes and their role in emulsion stability. Energy Fuels. 2004;**18**(3):770-777. DOI: 10.1021/EF0301654

[47] Ali MF, Perzanowski H, Bukhari A, Al-Haji AA. Nickel and vanadyl porphyrins in Saudi Arabian crude oils. Energy & Fuels. 1993;**7**:179-184. DOI: 10.1021/ef00038a003

[48] Chouparova E, Lanzirotti A, Feng H, Jones KW, Marinkovic N, Whitson C, Philp P. Characterization of petroleum deposits formed in a producing well by synchrotron radiation-based microanalyses. 2004. DOI: 10.1021/EF030108A

[49] Ozturk B, de-Luna-Bugallo A, Panaitescu E, Chiaramonti AN, Liu F, Vargas A. Atomically thin layers of B–N–C–O with tunable composition. Science Advances. 2015;**1**(6):1790-1798. DOI: 10.1126/sciadv.1500094

[50] Magnuson CW, Chemical vapor deposition graphene on polycrystalline copper foil, [thesis]. 2014. http://citeseerx.ist.psu.edu/viewdoc/download?doi=10.1.1.835.2607&rep=rep1&type=pdf [Accessed: December 11, 2017]

[51] Mullins OC, Sheu EY. Structures and dynamics of asphaltenes. New York City: Plenum, 1998. In: Chilingarian GV, Yen TF, editors. Bitumens, Asphalts, and Tar Sands. New York City: Elsevier Scientific Publishing Co; 1978

Fatigue Destruction of Asphalt Concrete Pavement: Self-Organization and Mechanical Interpretation

Bagdat Teltayev

Additional information is available at the end of the chapter

http://dx.doi.org/10.5772/intechopen.75536

Abstract

This chapter analyzes the fatigue failure of the asphalt concrete pavement of the highway on a macroscopic level. Based on the actual data, obtained on the sections of the real highways, the principle was formulated for the staged fatigue failure of the asphalt concrete pavement. During the analysis of the existing classifications of the fatigue cracks, it was determined that in Kazakhstan, the stage behavior of the fatigue cracks was not considered, and in the USA, the relation between stages of failure was not established. Similarly to the known phenomena of self-organization in thermodynamics of irreversible processes and dynamics of nonlinear systems (synergetics)—Benar's effects and division of biological cell, it was proposed to consider the parts of the asphalt concrete pavement as the specific dissipative structures, occurring in critical conditions, and a new regularity of fatigue failure was formulated. The formulated regularity of the staged fatigue failure of the asphalt concrete pavement was explained based on the new proposed scheme of bifurcation with the use of the results of experimental determination of the occasional, cyclic, long-term and residual strength.

Keywords: asphalt concrete pavement, staged fatigue failure, consequential change of deformation types, asphalt concrete strength, self-organization, dissipative structures, bifurcation

1. Introduction

Fatigue is one of the main types of failure for the asphalt concrete pavement of a highway. As it determines, in association with other types of failure (rutting, low-temperature cracking), the service life of pavement in accordance with the requirements of standard document [1], all pavement structures of the highways of capital and lightweight types on the design stage

IntechOpen

© 2018 The Author(s). Licensee IntechOpen. This chapter is distributed under the terms of the Creative Commons Attribution License (http://creativecommons.org/licenses/by/3.0), which permits unrestricted use, distribution, and reproduction in any medium, provided the original work is properly cited.

should be calculated for strength under the criterion of tensile bending strength of the cast-in-situ (asphalt concrete) layers.

It is considered that the fatigue cracking on the asphalt concrete pavement of highways occurs at frequently repeated load impact of the vehicles' wheels [2].

Hveem F.N. of the far abroad country has been one of the first researchers, who mentioned the phenomenon of fatigue in the asphalt concrete pavement of a highway [3]. The first investigations of the fatigue properties for the asphalt concretes in laboratory conditions were performed by Sall R.N.J., Pell P.S., Taylor I.F. in the UK [4, 5] and Monismith C.L. in the USA [6].

The study of the fatigue of the asphalt concretes in the former Soviet Union was started by the works of Sall A.O. (Leningrad), Radovskiy B.S. (Kiev), Zolotaryov V.A. (Kharkov), Rudenskiy A.V. and Kalashnikova T.N. (Moscow) [7–11].

It was found out that the phenomenon of fatigue failure for the asphalt concrete pavement of a highway was a complex one. In spite of the fact that the specialists of many countries of the world have been studying the phenomenon to the present day, the issue of the fatigue life for the asphalt concretes and asphalt concrete pavement remains actual.

2. Field observation

Figure 1 represents the photos showing fatigue failure of the asphalt concrete pavement of Karagandy-Shakhtinsk highway. These photos show clearly that the fatigue failure occurs by stage: first, parallel quasi-straight line longitudinal cracks occur on the patch line, between which quasi-straight line asphalt concrete strips have been formed, and then these quasi-straight line asphalt concrete strips are divided into cells of small sizes due to the occurrence of quasi-transversal cracks. Thus, in the considered case, the formation of a grid of alligator

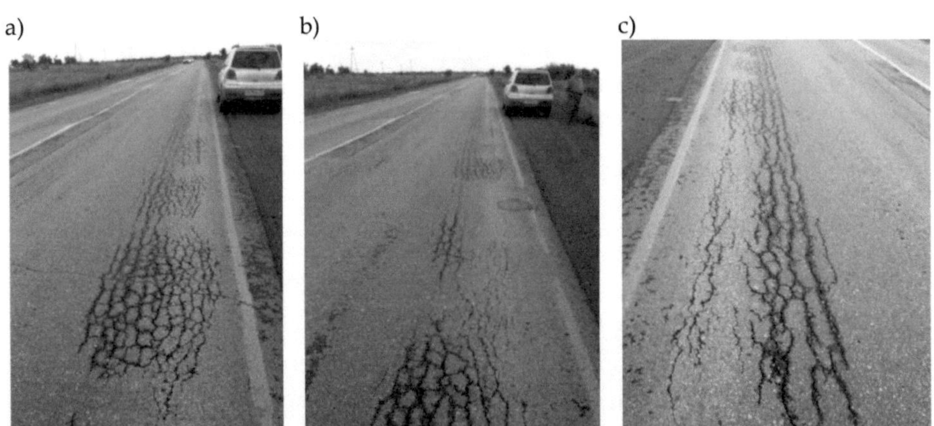

Figure 1. Fatigue failure of asphalt concrete pavement (860 km) of Karagandy-Shakhtinsk highway (Kazakhstan, Karagandy oblast, July, 2016).

cracks on the asphalt concrete pavement represents by itself a two-staged process, each of which has been realized during less or more prolonged period.

The fragments of another sequence for crack occurrence in the staged process of fatigue failure of alligator type on the asphalt concrete pavement are shown in **Figures 2** and **3**. As it is seen from **Figure 2**, there are only specific (separated from each other) quasi-parallel transverse cracks on the asphalt concrete pavement within the patch line. At first sight, they are similar to the low-temperature cracks. But that is not the case. The considered highway is located in Sharjah city (United Arab Emirates), where there is actually no winter. But, **Figure 3** shows the patterns of fatigue cracks of alligator type, which are formed at the further occurring of longitudinal cracks, connecting the existing transverse ones. During the period of observation, it has been determined that a large number of multiaxle heavy vehicles run along the highway, and it suggests that this type of staged fatigue failure has been connected with dense traffic of multiaxle heavy vehicles. Based on the abovementioned types of the staged failure for asphalt concrete pavements of the highways, we can formulate the following principle: "The process of fatigue failure for the asphalt concrete pavements of the highways occurs stage by stage and according to various types of stages. Type 1: stage I—quasi-linear parallel longitudinal cracks occur on the patch lines, between which quasi-straight line asphalt concrete strips are formed; stage II—formation of patterns of fatigue cracks of the alligator type by division of quasi-straight line strips of the asphalt concrete into the cells with small sizes by quasi-transverse cracks. Type 2: stage I—occurrence of insulated quasi-parallel transverse cracks; stage II—formation of alligator type cracks with relatively large dimensions of cells due to the occurrence of longitudinal cracks, connecting the existing transverse ones; stage III—decrease of dimensions for the cells of crack patterns due to the sequential occurrence of transverse and longitudinal cracks within each cell".

We suppose that the staged progressing of the fatigue failure is mechanically and thermodynamically "of benefit" to the system—to the asphalt concrete pavement, and which type of the stage is progressed—"the system will choose itself" depending on specific conditions: volume and type of traffic, regime and speed of traffic, weather and climatic conditions, design features, properties of the materials, including asphalt concretes, and so on.

a) b)

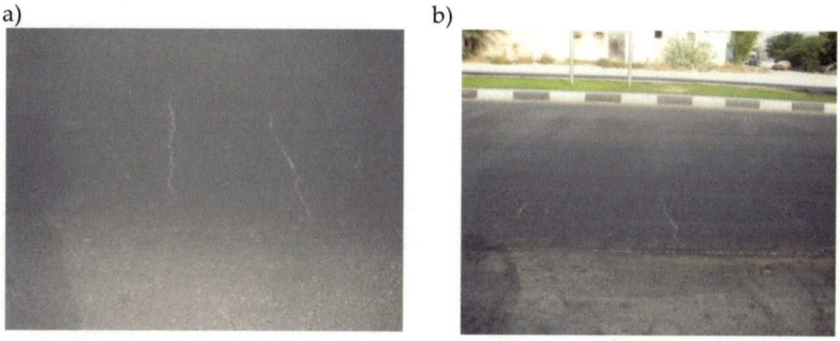

Figure 2. Fatigue failure of asphalt concrete pavement: transverse fatigue cracks (Sharjah city, United Arab Emirates, August, 2010).

a) b)

Figure 3. Fatigue failure of asphalt concrete pavement: alligator cracks (Sharjah city, United Arab Emirates, August, 2010).

3. Fatigue cracks classification

In Kazakhstan, the fatigue and other types of cracks on an asphalt concrete pavement for diagnostics and evaluation of road condition are considered in the standard [12], in which all defects on the pavements are divided into two groups: the defects, certifying inadequate strength and the defects, which do not certify inadequate strength in explicit form. Analysis of these defects shows that:

- in spite of the fact that they have different causes for their occurrence and progress character, cracks of various types (fatigue, thermal, reflected, sagging) are not identified separately;
- the staged development nature is not reflected for fatigue cracks;
- the maximum allowable measures are not contained for the cracks, including the fatigue ones.

The largest and wide scale program on investigation of performance for road structures (highway pavements) was started within the so-called Strategic Highway Research Program (SHRP) in the USA in 1987. The road agencies of the American States and 15 other countries have been collecting the data for 20 years for state of repair of pavements, climate, volume and density of traffic on more than 1000 experimental sections of the highways.

A special guide has been developed to perform data collection under the unique method, which was published three more times in the following years [13]. This Guide gives the following definition for the fatigue cracks in asphalt concrete pavement: "They occur in the areas subjected to repeated traffic loadings (wheel paths). They can be a series of interconnected cracks in early stages of development. They develop into many-sided, sharp-angled pieces, usually less than 0.3 m on the longest side, characteristically with a chicken wire/alligator pattern, in later stages. The fatigue cracks are divided into three levels. Low level: an area of cracks with no or only a few connecting cracks; cracks are not spalled or sealed; pumping is not evident. Moderate level: an area of interconnected cracks forming a complete pattern; cracks may be slightly spalled; cracks may be sealed; pumping is not evident. High level: an area of moderately or severely spalled interconnected cracks forming a complete pattern; pieces may move when subjected to traffic; cracks may be sealed; pumping may be evident."

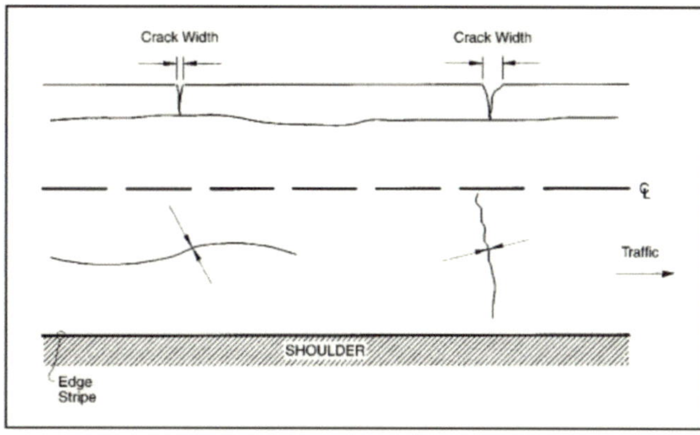

Figure 4. All levels of fatigue failure of asphalt concrete pavement for the highway according to the Guide [13].

a)

b)
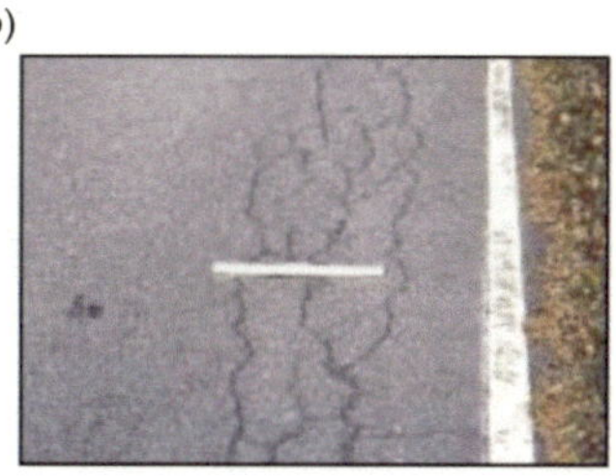
c)

Figure 5. Fatigue failure of asphalt concrete pavement for the highway according to the Guide [13]: (a) low level; (b) middle level and (c) high level.

Figures 4 and **5** show the photos from the Guide [13], which demonstrate visually the levels of the fatigue failure for the asphalt concrete pavement under the adopted classification.

As it is seen, contrary to the Kazakhstan Guide, the American Guide identifies the fatigue cracks separately from other types of cracks and three levels have been established for their development. Another American standard document [14] subdivides the fatigue cracks into two types: surface-down fatigue cracking and bottom-up fatigue cracking; admissible limit values have been shown for these types of cracks for surface-down—1000 ft./mile = 190 m/km and for bottom-up—25–50% of the lane area.

4. Self-organization

The works [15–17] based on provisions of thermodynamics of irreversible processes and non-linear dynamics (synergetics) show that the asphalt concrete pavement with low-temperature cracks is a specific dissipative structure, which is the form of adaptation for a thermodynamic system to the external conditions and each time, when air temperature reaches the critical temperature of pavement, the crack occurs. This is regularity, determined by collective behavior (self-organization) of structural elements of the asphalt concrete pavement in critical conditions.

In thermodynamics [18, 19], the systems, which exchange their energy and mass with the environment, are considered as open ones, and they are structurally complex. Due to the complexity of open systems, the various forms of structure occur in them in critical conditions. Energy dissipation plays the constructive role in the formation of these structures. To emphasize that I. Prigozhin introduced the term "dissipative structures" [20–23], and H. Haken introduced the term "synergetics" to stress the role of collective behavior for substructural elements in formation of dissipative structures [24, 25].

Prigozhin I. showed that the entropy variation ds for open thermodynamics system can be considered as the sum of two summands [19–21]:

$$ds = ds_e + ds_i, \tag{1}$$

where ds_e is entropy variation, connected with its inflow or outflow; ds_i is an amount of entropy, produced inside a system.

For short, ds_i is simply called "entropy production."

Component ds_e can have as positive sign, as well as negative one depending on the fact if the system receives or give energy as the result of interaction with environment. According to the second law of thermodynamics, entropy production ds_i is positive or equal to zero:

$$ds_i \geq 0, \tag{2}$$

Equal-zero entropy production, that is, $ds_i = 0$ will occur only under condition of balance.

4.1. Benar's effect

It is known that Benar's effect [26–28] is one of the famous examples for formation of dissipative structures in an open thermodynamic system. It occurs at critical difference of temperatures ΔT_{cr} of bottom and upper surfaces of the thin layer of the viscous liquid (for example, in silicon oil) in a dish, heated from below. When reaching T_{cr}, the behavior of the liquid varies dramatically—convection occurs, and the liquid is divided into hexagonal cells (**Figure 6**). The new structure is created by joint cooperative molecular motion of the liquid. As it is seen from **Figure 7**, the sharp break occurs for a dependence of heat transport rate dQ/dt on temperature difference ΔT at ΔT_{cr} and formation of a new structure occurs. The outflow (export) of entropy is precisely compensated by entropy production inside the liquid up to ΔT_{cr}, and when reaching ΔT_{cr}, the heat transport rate increases due to the convective mechanism of the heat exchange.

4.2. Cell separation

The work of M.V. Volkenstein [26] showed one more example of the formation for the dissipative structure in the open thermodynamics system. This is a cell separation of the living organism.

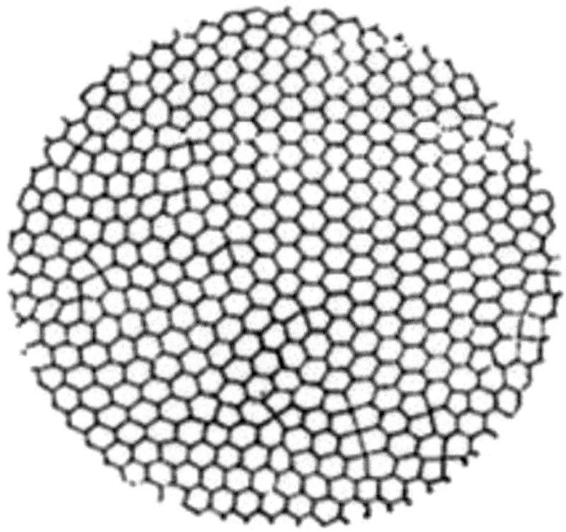

Figure 6. Benar's effect.

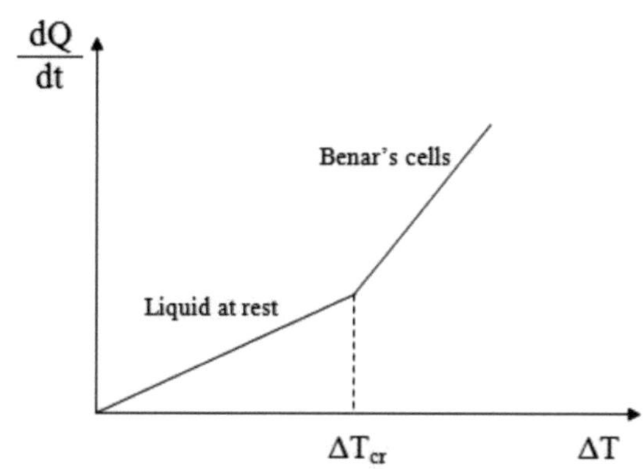

Figure 7. Dependence of heat transport rate on temperature difference.

For the simplicity, the cell is considered as a sphere with radius R. Entropy production inside the cell ds_i is proportional to its volume $V = \frac{4}{3}\pi R^3$, and entropy outflow from the cell ds_e is proportional to the area of its surface $S_{пов.} = 4\pi R^2$. Then, according to the expression (Eq. (1)), we have:

$$ds = A \cdot \frac{4}{3}\pi R^3 - B \cdot 4\pi R^2, \tag{3}$$

where A and B are the parameters of proportionality, which have appropriate dimensions.

The cell grows with the growth of the organism, and radius of the sphere R increases. The cell under the mechanism of self-organization tries to remove the excess of the accumulated entropy. As the entropy production ds_i increases proportionally to the cube of the radius R, that is, R^3, and the entropy outflow increases proportionally to the square of the radius R, that is, R^2, then the gradual accumulation of entropy occurs under the expression (Eq. (3)).

Stationary state is achieved at $R = \frac{3B}{A}$, that is, $ds = 0$. And at $R > \frac{3B}{A}$, that is, $ds > 0$, therefore, at $R_{cr} > \frac{3B}{A}$ (R_{cr}: critical size of the cell) the cell should be separated, otherwise, it will die. The volumes of the mother cell and two daughter cells are similar, and the total area of the surfaces of new cells is bigger.

The abovementioned examples for self-organization in thermodynamics systems—Benar's cells and cell separation can be used further for the explanation of the fatigue failure phenomenon for the asphalt concrete pavement.

Fatigue failure of the asphalt concrete pavement, of course, has been directly connected with the asphalt concrete strength.

5. Asphalt concrete strength

5.1. Bitumen

Bitumen of grade BND 100/130, produced by Pavlodar Petrochemical Plant (PPCP), was used for the preparation of fine-grained dense asphalt concrete in laboratory conditions in this work. Bitumen complies with the requirements of Kazakhstan standard ST RK 1373-2013 [29]. Standard indicators for bitumen are represented in **Table 1**. Content of bitumen in the asphalt concrete was 4.8% by the mass of the dry filler.

5.2. Asphalt concrete

Hot dense fine-grained asphalt concrete of Type B was adopted for test, which satisfies the requirements of the Kazakhstan standard ST RK 1225-2013 [30], and it was prepared with the use of aggregate of fractions 5–10 mm (20%), 10–15 mm (13%), 15–20 mm (10%) from Novo-Alekseevsk rock pit (Almaty oblast), sand fraction 0–5 mm (50%) from the plant "Asphaltconcrete-1" (Almaty city) and mineral powder (7%) from Kordai rock pit (Zhambyl oblast). The main standard indicators for asphalt concrete are represented in **Table 2**. The grading curve of mineral part of hot mix asphalt concrete is shown in **Figure 8**.

5.3. Test methods

In this study, asphalt sample tests have been performed according to the following methods:

1. Determination of asphalt concrete strength at direct tension at various temperatures has been performed in thermal chamber of TRAVIS, manufactured by Infratest GmbH (Germany). Sample tests have been performed at deformation with constant rate 1 mm/min in accordance with European standard pr EN 12697-46 [31]. Samples had dimensions of 5 × 5 × 16 cm.

2. Cyclic (fatigue) asphalt concrete strength at various temperatures has been determined by sample testing with dimensions 5 × 5 × 38 cm in thermal chamber of four-point bending device under European standard EN 12697-24 [32]. Loading frequency was f = 10 Hz. The stress, equal to $\sigma = 1400$ kPa, was kept as constant prior to the sample failure.

3. Asphalt concrete samples in the form of beam with dimensions 4 × 4 × 16 cm have been tested at various temperatures on mechanical press with the use of special device under transverse bending scheme according to standard ST RK 1218-2003 [33]. The deformation rate was 3 mm/min.

4. Asphalt concrete sample strength of various shapes (cylindrical and rectangular), various dimensions and at various temperatures at direct compression has been determined by their testing on the mechanical press under standard ST RK 1218-2003 [33]. The deformation rate was 3 mm/min.

Indicators	Unit	Requirement of ST RK 1373-2013	Value of indicators
Depth of needle penetration			
25°C	0.1 mm	101–130	110
0°C		30	37
Penetration index	—	−1.0... + 1.0	−0.82
Ductility			
25°C	cm	≥90	135
0°C		≥4.0	6.6
Softening point	°C	≥43	44.0
Fraas point	°C	≤−22	−30.2
Dynamic viscosity, 60°C	Pa·s	≥120	121.0
Kinematic viscosity, 135°C	mm^2/s	≥180	329.0

Table 1. Main standard indicators for bitumen.

Indicators	Unit	Requirements of ST RK 1225-2013	Value of indicators
Average density	g/cm^3	—	2.38
Water saturation	%	1.5–4.0	3.4
Air voids of mineral filler	%	≤19	15.1
Air voids of asphalt concrete	%	2.5–5.0	3.8
Compression strength			
0°C	MPa	≤13	7.4
20°C		≥2.5	3.5
50°C		≥1.3	1.38
Water resistance	—	≥0.83	0.80
Shear resistance	MPa	≥0.38	0.39
Crack resistance	MPa	4.0–6.5	4.5

Table 2. Main standard indicators for asphalt concrete.

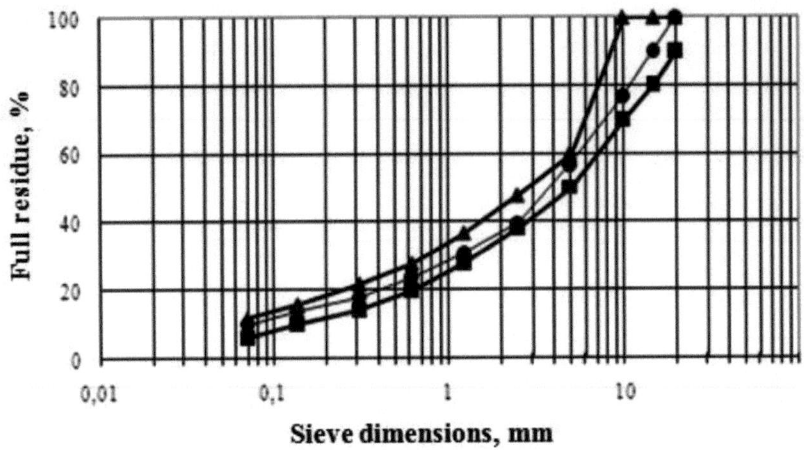

Figure 8. Asphalt mixture grading curve.

5.4. Sample preparation

Asphalt concrete samples of cylindrical shape, designed for direct compression, were prepared under Kazakhstan standard ST RK 1218-2003 [33] by compaction of the asphalt concrete mix in special mold. The samples of rectangular shape and in the shape of beam of various dimensions were prepared in the following way. First, the asphalt concrete samples in the shape of square slab with dimensions 5 × 30.5 × 30.5 cm were prepared by roller compactor (model CRT-RC2S, company Cooper, UK) in accordance with European standard EN 12697-33 [34]. Then, the samples with the shape of rectangular prism of various dimensions were obtained from square slabs.

5.5. Single loading, cyclic and long-term strength of asphalt concrete

Figure 9 represents the graphs, showing the dependence of the asphalt concrete strength at various types of loading—tension, compression and bending. As it is seen, in the considered temperature interval (0–50°C), the asphalt concrete has the least tensile strength, and the largest—at compression. Bending strength occupies the intermediate location between tension and compression. Meanwhile, compression and bending strength of the asphalt concrete with the temperature increase decreases nearly with the similar rate within the whole temperature interval considered, and at tension, the rate of decrease is higher than two times compared with compression and bending strength. It is also seen that the difference between temperature curves of bending strength and compression strength is kept as constant in the whole temperature interval considered and it is equal to, at average, 1.0 MPa. The maximum difference between temperature curves of tensile strength and bending (compression) strength occurs at low temperatures (0–10°C), which is equal to 2.5 MPa (3.5 MPa) and decreases with the temperature increase; these differences at temperature 50°C are equal to 0.8 and 1.6 MPa, respectively.

It is generally accepted that the fatigue cracks occur due to the repeated impact of the tensile stress in the bottom surface of the asphalt concrete pavement [2, 35–39], the more the value of

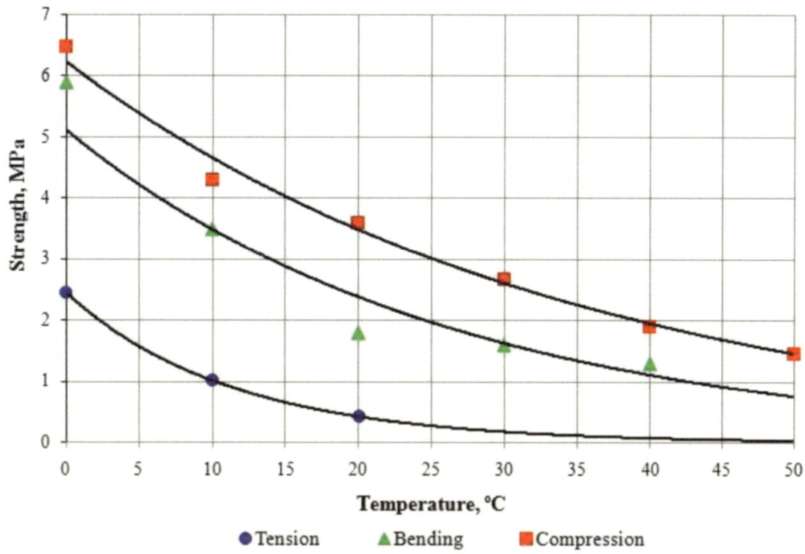

Figure 9. Strength of fine-grained asphalt concrete (BND 100/130, PPCP) at various types of stressed condition.

which at any other equal conditions, the more is the ratio of elasticity modulus of the asphalt concrete layers to the elasticity modulus of the under layers of the pavement base and subgrade soil [10, 40]. It is considered that the abovementioned ratio of elasticity moduli in pavement structure is the biggest one in the spring season, when the upper part of subgrade has been defrosted and loosened, and the asphalt concrete pavement has a big stiffness due to the relatively low air temperature, which is within 0 and +10°C [2, 41–43].

As it is seen from **Figure 10**, namely within the temperature interval 0 and +10°C, the difference in the asphalt concrete tensile strength and bending strength (compression) is the biggest one!

Figure 10 shows the graphs of cyclic strength for fine-grained dense asphalt concretes at bending and tension at the temperature of 20°C. The upper graph has been constructed at testing of asphalt concrete samples under scheme of bending on the four-point bending device in Kazakhstan Highway Research Institute, and the bottom one has been constructed according to the experimental data, obtained in the University of North Carolina State (USA) at direct tension [44]. It is clearly seen that the cyclic strength of the asphalt concrete at bending is considerably higher than at tension. Similar regularity can be seen on the curves of the long-term strength of the asphalt concrete, as shown in **Figure 11** [45, 46].

Thus, single loading, cyclic and long-term strength of the asphalt concretes at tension is considerably less, than at bending and compression.

To answer the question: "Does the asphalt concrete strength depend on the tested sample dimensions at compression?", the test of samples for the fine-grained dense asphalt concrete of type B (BND 100/130) has been performed for direct compression. The thickness of all the samples was similar and equal to 5 cm, and the length and the width of the samples had the values

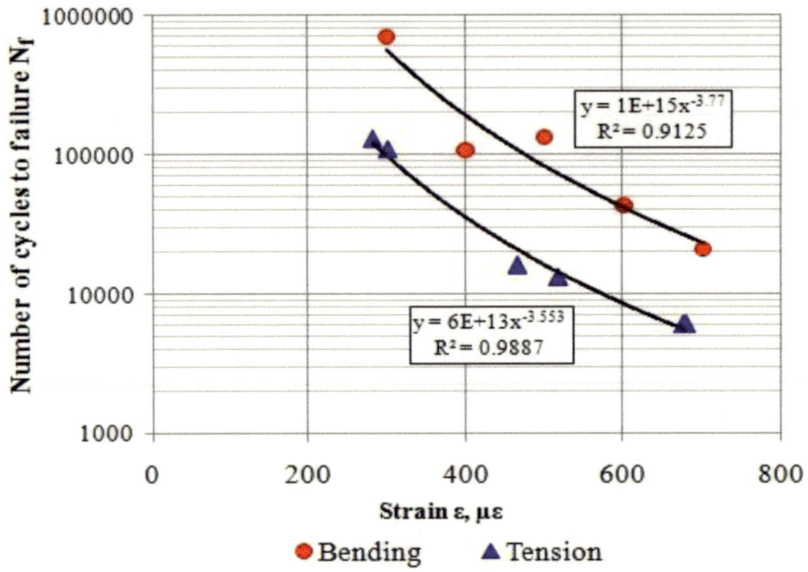

Figure 10. Cyclic strength of the asphalt concrete at bending and direct tension at the temperature of 20°C.

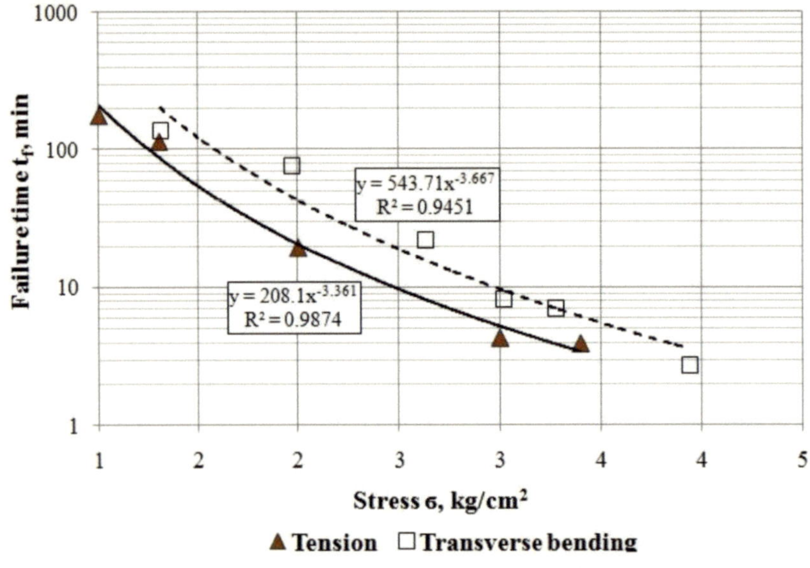

Figure 11. Long-term strength of the fine-grained asphalt concrete of type B (BND 60/90) at transverse bending and direct tension.

equal to 2, 5, 7, 10, 12 and 15 cm. The test has been performed at the temperatures of 0, 10 and 20°C. Three parallel tests have been performed at each temperature and dimensions of samples.

As it is seen from **Figure 12**, the asphalt concrete strength at compression depends considerably on sample dimensions. The biggest strength occurs at the temperatures of 10 and 20°C

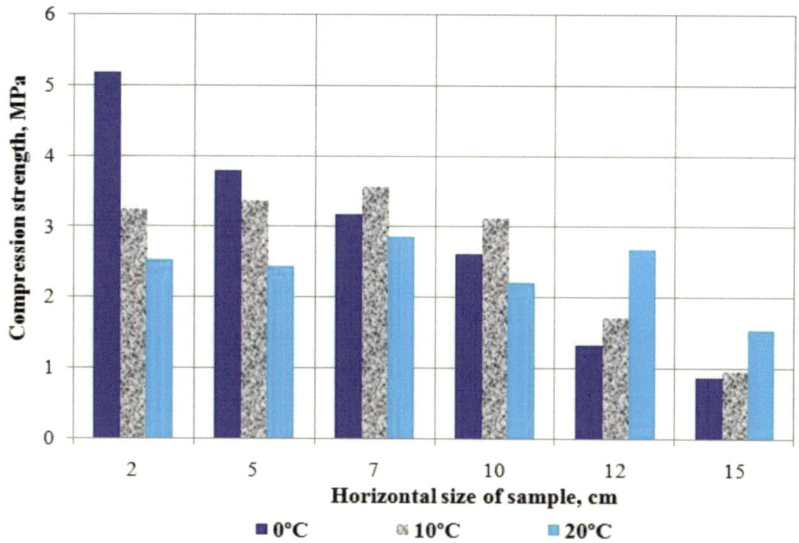

Figure 12. Compression strength of the fine-grained asphalt concrete samples of type B (BND 100/130, PPCP) of various dimensions at different temperatures.

with the length of sample side equal to 7 cm, and the strength increases almost linearly at the temperature of 0°C with the decrease of sample dimensions.

These results serve as the reliable explanation for gradual decrease of the horizontal dimensions of the asphalt concrete pavement cells with progressing of its fatigue failure.

5.6. Residual strength of asphalt concrete

The fatigue crack on the asphalt concrete pavement occurs when it almost completely lost its tensile strength (tension at bending). Let us raise the question: can such asphalt concrete have residual compression strength? To clarify the issue, we have carried out a special experiment. The same fine-grained dense asphalt concrete of type B (bitumen grade BND 100/130) has been adopted. First, the asphalt concrete samples with dimensions 5 × 5 × 38 cm have been tested on the four-point bending device for bending fatigue to failure (stiffness reduction to 10% of the initial one) at the temperatures of 10, 20 and 30°C. Then, the samples with dimensions

T, °C	Number of cycles to failure N_f			
	Parallel 1	Parallel 2	Parallel 3	Average
+10	5187	4965	10,180	6777
+20	568	512	534	538
+30	565	219	222	335

Table 3. Test results of the fine-grained dense asphalt concrete of type B (BND 100/130, PPCP, f = 10 Hz, σ = 1400 kPa) at fatigue on the four-point bending device.

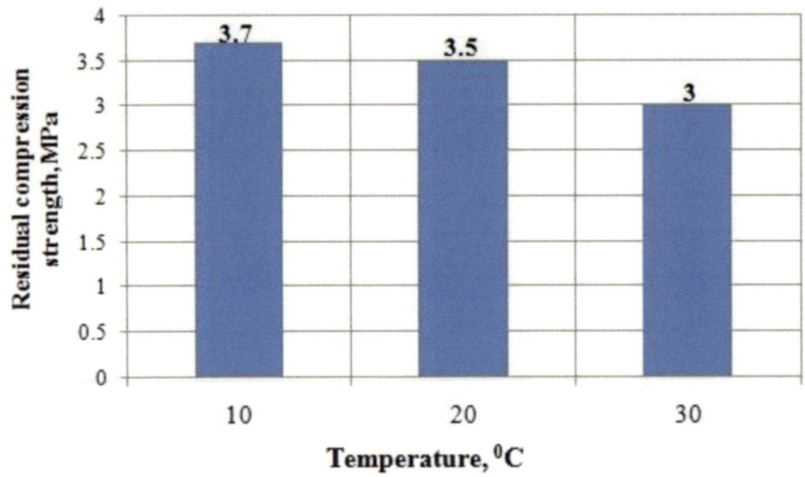

Figure 13. Residual strength of the asphalt concrete at compression after cyclic bending at various temperatures.

5 × 5 × 5 cm have been prepared from these samples, and they have been tested for direct compression at the same temperatures. Three asphalt concrete samples have been tested at each temperature. The results of the initial tests of the asphalt concrete samples at cyclic bending have been represented in **Table 3**, and their further test at direct compression—in **Figure 13**. As it is seen, the asphalt concrete has a high residual compression strength after the cyclic bending to failure, comparable with the strength of a new asphalt concrete (**Figures 9** and **12**).

6. Principle of consequential change of deformation types

As it is known, the self-organization phenomenon occurs in complicated open thermodynamics systems and new structures occur in them in critical conditions such as Benar's cells, separated cells of living organisms, laser ray, and so on. Occurrence of specific dissipative structures in critical low-temperature conditions on the asphalt concrete pavement has been shown in the works [15–17, 47]. The provided above staging of the fatigue failure for the asphalt concrete pavement, possibility in principle for occurrence of dissipative structures in it at critical conditions as a result of self-organization of its structural elements, significant dependence of single loading, cyclic and long-term strength of the asphalt concrete on deformation type (stressed condition) and existence of residual strength at another deformation type after failure allow formulating of a new regularity for the fatigue failure:

Fatigue failure of an asphalt concrete pavement under repeated load impact is realized according to the consequently changing stages in every of which pavement parts function as specific dissipative structures with characteristic deformation type, which interchange in the sequence of: tension-bending-compression.

http://dx.doi.org/10.5772/intechopen.75536

7. Bifurcation

The principle of consequential change of deformation types at fatigue failure for an asphalt concrete pavement formulated earlier can be explained on the basis of provisions for thermodynamics of irreversible processes and nonlinear dynamics (synergetics).

In short description of the examples for dissipative structure occurrence—Benar's cells and cell separation, it has been mentioned earlier that the action of systems in critical conditions in both cases is a benefit for them: liquid flow along the hexagonal cells allows including additional convective mechanism of heat exchange with environment; separation of the cell into two saves it from "death."

We also consider that the realization of the fatigue failure according to the consequent stages, changing deformation type from "tension" into "bending" and from "bending" into "compression" at the continuous mechanical impact is a benefit to the asphalt concrete pavement, as:

1. The asphalt concrete strength at bending is bigger than at tension, and at compression, it is bigger than at bending.
2. Residual strength of asphalt concrete at compression is relevant after its failure at deformation under the scheme of bending (tension).

Such staged failure with consequential change of deformation type prolongs the existence time ("life cycle") of separate parts of the asphalt concrete pavement.

The formulated principle can be visually demonstrated by the bifurcation scheme proposed (**Figure 14**).

In thermodynamics and synergetics, it is accepted to consider that the system away from the equilibrium condition acquires new properties. The system in the strong nonequilibrium condition becomes more active and all substructural elements of the system work jointly, consistently, fluctuations are synergized and new structures occur at the critical moment [18–26, 48]. In addition, the system has a choice in critical conditions—what scenario of evolution to follow further.

In accordance with the proposed bifurcation scheme, the asphalt concrete pavement works as continuous medium under scheme of volumetric stressed-deformed condition since the moment of starting of operation to the moment of losing of the tension resistance (0–1). At the time moment of the complete losing of the tension resistance (point 1), the thermodynamics system (substructural elements of the asphalt concrete pavement) has a choice—which thermodynamics branch (branch A and branch B)—to function further. If the system in point of bifurcation chooses the thermodynamics branch A, the parallel cracks occur on the patch lines in point 1, and the asphalt concrete strips work as a long beam between points 1 and 2, and they are deformed under the scheme of bending. The transverse cracks occur in point 2, long asphalt concrete strips are divided into more short parts, each of the obtained parts for the period of

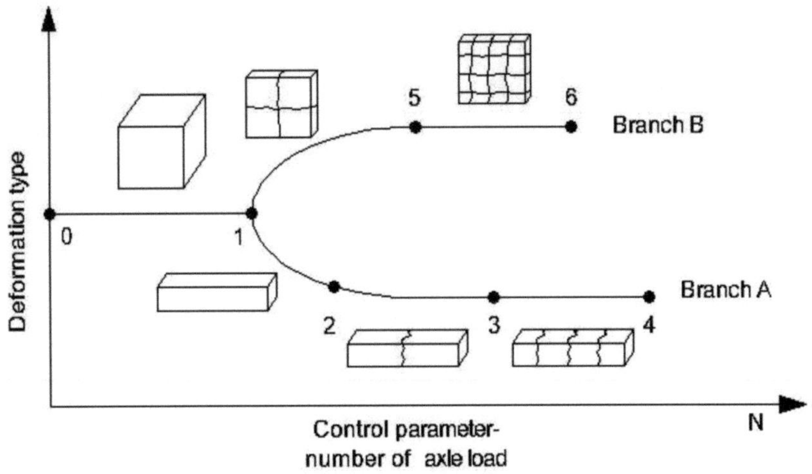

Figure 14. Bifurcation scheme at fatigue failure of an asphalt concrete pavement.

2–3, work as a short beam, and it is also deformed under the scheme of bending. In point 3, the number of the occurred cracks increases and for the period of 3–4, separate failure fragments of pavement work under scheme of direct compression. Complete failure of the asphalt concrete pavement occurs at the time moment 4. The road section of the highway "Karagandy-Shakhtinsk," described earlier can serve as an example of the practical realization for the fatigue failure of the asphalt concrete pavement according to the thermodynamics branch A.

If the system in point of bifurcation 1 chooses the thermodynamics branch B, then, first, the transverse cracks occur on pavement, then longitudinal fatigue cracks occur between them, and for the period of time 1–5, separate pavement blocks function as big and short slabs, and they are deformed under scheme of bending. Additional transverse and longitudinal cracks occur in point 5, grids of cracks become more intensive until each of the pavement fragments is not deformed under scheme of direct compression (time period 5–6). Complete failure of the asphalt concrete pavement occurs in time moment 6. The road section of the highway located in Sharjah city (UAE) can serve as an example of the practical realization for the fatigue failure of the asphalt concrete pavement under the thermodynamics branch B.

8. Conclusion

The results of this study allow drawing the following conclusions for the fatigue failure of the asphalt concrete pavement of a highway:

1. In Kazakhstan, the cracks of various types on the asphalt concrete pavements (fatigue, thermal, reflected and sagging) are not identified separately. The staged character of the fatigue failure was not considered. Maximum allowable characteristics, including the fatigue ones, were not determined as well. In the USA, the fatigue cracks are identified separately

from other types of cracks, three levels for their development have been determined, but the relation is not considered between these levels.

2. Fatigue failure of the asphalt concrete pavement is realized stage by stage. Change of stages for failure occurs under mechanism of self-organization for the substructural elements of the pavement material—asphalt concrete in critical conditions. Similar to the known phenomena of self-organization—Benar's effect and biological cell separation, it is proposed to consider the parts of the asphalt concrete pavement as specific dissipative structures. They work as specific dissipative structures on each stage of the fatigue failure.

3. Comparison of the results for the performed and known tests of the asphalt concretes for determination of single loading, cyclic, long-term and residual strength for tension, bending and compression has shown that the strength at bending is always more than at tension; and it is more at compression than at bending.

4. The determined staged character of the fatigue failure for the asphalt concrete pavement, possibility in principle for occurrence of dissipative structures in it, dependence of asphalt concrete strength on deformation type (stressed condition), moreover, its increase in the sequence of "tension-bending-compression," and also the existence of residual strength for the asphalt concrete at compression after failure at tension have served as a basis for formulating of a new regularity for the fatigue failure: *fatigue failure of an asphalt concrete pavement under repeated load impact is realized according to the consequently changing stages in every of which pavement parts function as specific dissipative structures with characteristic deformation type, which interchange in the sequence of tension-bending-compression.*

Author details

Bagdat Teltayev

Address all correspondence to: bagdatbt@yahoo.com

Kazakhstan Highway Research Institute, Almaty, Kazakhstan

References

[1] CN RK 3.03-19-2006. Design of flexible pavements. Astana; 2007. 87 p

[2] Ivanov NN, editor. Design and Calculation of Flexible Pavements. Moscow: Transport; 1973. 328 p

[3] Hveem FN. Pavement defections and fatigue failures. Highway Research Board. Bulletin 114

[4] Saal RNJ, Pell PS. Fatigue of bituminous road mixes. Kolloid-Zeitschrift. 1960;**171**:61-71

[5] Taylor IF, Pell PS. Could fatigue be a problem in flexible pavements? Roads and road Construction. 1969;**47**(560):236-243

[6] Monismith CL, Secor KE, Blackmer W. Asphalt mixture behavior in repeated flexure. Proceedings of Association of Asphalt Paving Technologists. 1961;**30**:188-222

[7] Sall AO. Peculiarities of asphalt concrete pavement bending at short-term load. Transactions of SoyuzdorNII. 1965;**3**:19-26

[8] Sall AJ, Zolotaryov VA, Radovskiy BS, Iliyev EB. Design parameters of asphalt concretes as applied to VSN 46-72. Automobile Roads. 1977;**5**:28-30

[9] Radovskiy BS, Rudenskiy AV. Regarding impact of characteristics of the materials structure on their fatigue and long-term strength. Transactions of SoyuzdorNII. 1975;**79**:43-48

[10] Radovskiy BS, Suprun AS, Kozakov II. Pavement Design for Trucks Traffic. Budivelnyk: Kiev; 1989. 168 p

[11] Rudenskiy AV, Kalashnikova TN. Analysis of asphalt concrete fatigue. Transactions of GiprodorNII. Road Construction Materials. 1973;**7**:3-15

[12] PR RK 218-27-2014. Instruction on diagnostics and evaluation of transport maintenance condition of the automobile roads. Astana; 2014

[13] Distress identification manual for the long-term pavement performance program. Report No. FHWA-RD-03-031. Federal highway administration. Georgetown Pike McLean; 2003

[14] Guide for mechanistic-empirical design of new and rehabilitated structures. Final report. Part 3. Design analysis. Chapter 3. Design of new and reconstructed flexible pavements. ARA, Inc., ERES Consulting division; 2004

[15] Teltayev BB. Regularities of self-organization of the low temperature cracking of asphalt concrete pavement. Reports of National Academy of Science of Republic of Kazakhstan. 2015;**4**:40-65

[16] Teltayev BB. Regularities of the increase for the number of cracks on the road asphalt concrete pavement. Report of National Academy of Science of Republic of Kazakhstan. 2015;**5**:35-57

[17] Teltayev BB. Road asphalt concrete pavement as dissipative structure. Report of National Academy of Science of Republic of Kazakhstan. 2016;**2**:11-37

[18] Bazarov IP. Thermodynamics. Saint-Petersburg: Publisher "Lan"; 2010. p. 384

[19] Prigozhin I. Introduction to Thermodynamics of Irreversible Processes. Moscow: Foreign Literature; 1960. 128 p

[20] Glansdorff P, Prigozhin I. Thermodynamic Theory of Structure, Stability and Fluctuations. Moscow: Mir; 1973. 281 p

[21] Prigozhin I. Time, structure and fluctuations (Nobel lecture in chemistry of 1977 year). Successes of Physical Sciences 1980;**131**(2):185-207

[22] Prigozhin I. From being to becoming. Moscow: Science; 1985

[23] Prigozhin I, Stengers I. Order out of Chaos: Man's New Dialogue with Nature. Moscow: Progress; 1986. 432 p

[24] Haken H. Synergetics. Moscow: Mir; 1980. 405 p

[25] Haken H. Synergetics: The Hierarchy of Instabilities in Self-organizing Systems and Devices. Moscow: Mir; 1985

[26] Wolkenstein MV. Entropy and Information. Moscow: Nauka; 1986. 192 p

[27] Klimontovich YuL. Introduction to physics of open systems. Soros Education Journal. 1996;**8**:109-116

[28] Osipov AI. Thermodynamics yesterday, today and tomorrow. Part 2. Nonequilibrium thermodynamics. Soros Education Journal. 1999;**5**:91-97

[29] ST RK 1373-2013. Bitumens and bitumen binders. Oil road viscous bitumens. Technical specifications. Astana; 2005

[30] ST RK 1225-2003. Hot mix asphalt for roads and airfields. Technical specifications. Astana; 2003

[31] PR EN 12697-46. Bituminous mixtures. Test methods for hot mix asphalt. Part 46: Low temperature cracking and properties by uniaxial tension tests. 2004

[32] EN 12697-24. Bituminous mixtures. Test methods for hot mix asphalt. Part 24: Resistance to fatigue. 2004

[33] ST RK 1218-2003. Materials based on the organic binders for road and airfield construction. Test methods. Astana; 2003

[34] EN 12697-33. Bituminous mixtures. Test methods for hot mix asphalt. Part 33: Specimen prepared by roller compactor. 2003

[35] Yoder EJ. Principles of Pavement Design. New York: John Wiley & Sons, Inc.; 1959

[36] Yoder EJ, Witczak MW. Principles of Pavement Design. New Jersey: John Wiley & Sons, Inc.; 1975. 736 p

[37] Birulya AK. Design and Calculation of Flexible Pavements. Transport: Moscow; 1964. 168 p

[38] Batrakov OT, editor. Strengthening of Flexible Pavements. Moscow: Transport; 1985. 144 p

[39] Privarnikov AK. Polydimensional deformation of multilayer base course. In: Stability and Strength of Structural Elements. Dnepropetrovsk. 1973. pp. 27-45

[40] Radovskiy BS. Experimental research of the stress strain behavior of road pavements as the layerwise visco-elastic base course at flexible loading. Applied Mechanics. 1980; **16**(4):131-135

[41] Krasikov OA. Evaluation of Strength and Calculation of Strengthening for Flexible Pavements. Almaty: KazgosINTI; 2006. 308 p

[42] Krasikov OA. Regularities for deforming of flexible pavements. Causes for occurring of deformations and failures. Transport and Roads of Kazakhstan. 2007;**1**:17-21

[43] Vasiliyev AP, Sidenko VM. Maintenance of Automobile Roads and Traffic Management. Transport: Moscow; 1990. 304 p

[44] Zhang J, Sabouri M, Guddati MN, Kim R. Development of a failure criterion for asphalt mixture under fatigue leading. Journal of the Association of Asphalt Paving Technologists. 2013;**82**:1-22

[45] Aitaliyev SM, Iskakbayev AI, Teltayev BB. Experimental research of regularities for creep and long-term strength of the asphalt concrete at cross bending. Bulletin of National Academy of Science of Republic of Kazakhstan. 1998;**1**:32-36

[46] Aitaliyev SM, Iskakbayev AI, Isayev D, Teltayev BB. Regularities of long-term strength for asphalt concrete samples at pure tension. Bulletin of National Academy of Science of Republic of Kazakhstan. 1998;**6**:23-25

[47] Teltayev BB. Regularity of self-organization for asphalt concrete pavement during low temperature cracking. Certificate No. 495 for scientific discovery. Moscow; 2016

[48] Prigozhin I. End of certainty. Time, Chaos and New Courses of Nature. Izhevsk: RDC "Regular and chaotic dynamics"; 2000. 208 p

Chapter 5

The Enhancement of Asphalt Concrete Surface Rigidity Based on Application of Shungite-Bitumen Binder

Podolsky Vladislav Petrovich,
Lukashuk Alexandr Gennadievich,
Tyukov Evgeny Borisovich and
Chernousov Dmitry Ivanovich

Additional information is available at the end of the chapter

http://dx.doi.org/10.5772/intechopen.76877

Abstract

Physical interaction of organic binder depends on bituminous film ability to absorb the mineral particles on the surface. The thick film formation begins with the bitumen amalgamation with mineral powder grains and finishes during the processes of application, seal arrangement and cooling down of asphalt concrete. Active surface centers are on every mineral material and they supply their reactive ability and interaction with bitumen. For proving experimentally and to study the physical and mechanical properties in real operation conditions that were arranged, the blanket was arranged from stone-mastic asphalt with shungite bituminous binder of 500 m long. The monitoring conducted during 3 years indicates that asphalt concrete with shungite binder is subjected to calibration to a lesser degree in comparison with asphalt concrete based on standard powder. For operation reliability of nonrigid road base, it is necessary to estimate and prognosticate asphalt concrete fatigue properties in a blanket. The characterized property appears in such a way that the effect of loads which is significantly smaller than the destructive ones leads to the gradual degradation of fatigue and destruction of the blanket. The results proved that asphalt concrete with shungite bituminous binders has higher operation characteristics.

Keywords: shungite, fullerenes, mineral powder, scan probe microscope, fatigue destruction

IntechOpen

© 2018 The Author(s). Licensee IntechOpen. This chapter is distributed under the terms of the Creative Commons Attribution License (http://creativecommons.org/licenses/by/3.0), which permits unrestricted use, distribution, and reproduction in any medium, provided the original work is properly cited.

1. Introduction

The increase of interrepair period of life duration of road blanket from asphalt concrete is one of the important problems in the road industry. In the conditions of existing trend for the growth of intensity and traffic volume, it is necessary to find new materials and technologies to provide road integrity for the whole period of life cycle. The existing shortage stimulates the finding of the new materials and rock for dolomite and dolomitized limestone replacement during the process of asphalt concrete mixes replacement without their poorer operation quality. One of these rocks is shungite as a unique carbonaceous composite. That is why the research of fine dispersed powder, the substantiation of its application while fabricating asphalt concrete mixes, innovation technologies development of construction of blankets with better operation characteristics is an actual problem.

The high-cost of dolomite mineral powder and stable tendency to its rise stimulate the finding of new materials and technologies which can be used without asphalt concrete operation abilities deterioration.

Shungite is a black rock (flaky stone) from the field near Shunga. Strong strata of this mineral go out onto the shores of the lake Onega near the Tolvuya settlement. There are known five fields: Nizkoozyornoe, Shungskoe, Vikshozerskoe, Myagkozernoe and Krasnaya Selga, the raw materials of which were used for shungite production as a light filler for bitumen.

Shungite is a unique rock consisting of amorphous carbon. It is very active in redox reactions, silicon dioxide, ferric oxide, cobalt, vanadium, titanium and other elements.

Antimicrobial properties of shungite were known even in the seventeenth–eighteenth century that is why Russian tsar Peter I ordered the soldiers to drink water from the vessels with the piece of flaky stone in them.

American professor Semyon Tzipursky from Arizona University determined the uniqueness of carbon in Karelian shungite where it is evenly distributed as conglomerations of 200–300 angstrom. Therefore, they are able to migrate into water in appreciable volumes especially from powder shungite. According to the radiocarbon analysis, shungite age is about 2 billion years.

Carbon (C) from the fourth group of the periodic table is the only element for which valency and coordination number coincide. Thanks to this peculiarity, carbon is able to form the compounds practically with any atoms number in the chain in which there can be any numbers of multiple bonds at any combinations.

The work by Robert F. Curl and Richard Errett Smalley is of great importance. There, a hollow carbon molecule is described as cluster with 60 atoms in it [1]. Theoretically, the structure of 60 atoms (C_{60},) is calculated. The most stable structure of spherical shell is the combination of pentagons and hexagons. Under thermal graphite decomposition, there was synthesized a new substance, the molecules of which have spherical form. So there appeared fullerenes. Different compounds of fullerene molecules with hydroxyl groups were synthesized. Diagram of fullerene molecules is given in **Figure 1**. Fullerene is named by Buckminster Fuller who

http://dx.doi.org/10.5772/intechopen.76877

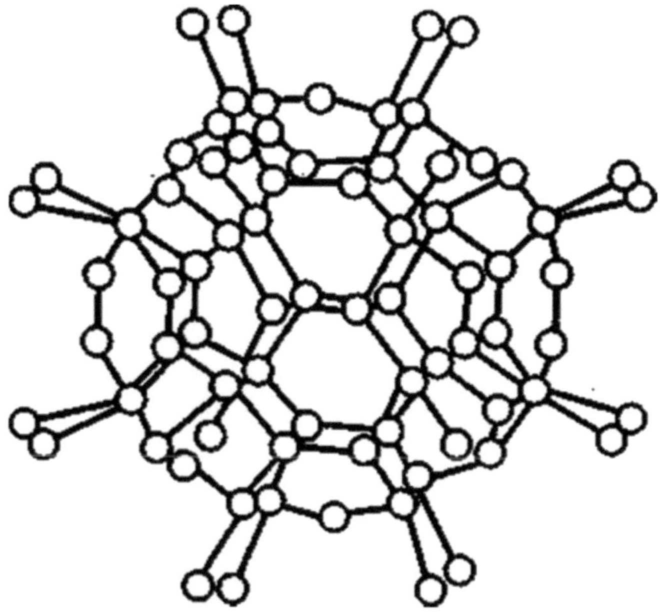

Figure 1. Schematic structure of fullerene is molecule C_{60} with affixed to it radicals (OH).

applied such forms while constructing dome-shaped structures [19]. Schematic structure of fullerene is molecule C_{60} with affixed to it radicals (OH). It is demonstrated in **Figure 1**.

There is a real possibility to obtain different organic compounds based on fullerenes in different spheres of man activity. Particularly, the researches of shungite affecting asphalt concrete surface operating performances are conducted in Voronezh State Technical University (Russia) [2–8].

The attention of the scientists from different countries to the new road construction materials can lead to the appearance of new carbon containing technologies based on fullerenes and graffenes and spreading our conceptions which will influence the civilization development.

2. Researches of materials surface on scanning tester microscope

The effect of shungite mineral powder particles surface nature on asphalt concrete binding agent performances is done by scanning tester microscope NanoEducator. It is for visualization, diagnosis and modification of substance with micro- and nanodimensional level of spatial resolution.

The complex of scanning tester microscope NanoEducator consists of measuring head of scanning tester microscope, electronic block, video camera, connecting cables, computer, eatching probes and also a set of test samples, expendable materials and toolwares. Software Mac OS X is used.

Main characteristics of C3M NanoEducator are presented in **Table 1**.

Characteristics	Quantitative characteristics
Scanning mode	ACM, CTM, lithography
Scanning area	70 × 70 × 10 mkm
Spatial resolution X-Y Z	~ 50 nm ~ 2 nm (depends on radius of probe curvature)
Minimal scanner step	10 nm
Scanning carrent	100–200 nm
Radius of probe curvature	10–100 nm
Time of scanning	30–40 min (depends on scanning area)
Time of set-up	Not more than 10 min
Tested sample size	12 × 12 mm
Sample height	Not more than 5 mm
Environment temperature температура	25 ± 5°C
relative humidity	Not more than 60%
Atmospheric pressure	760 ± 30 mm. of mercury column.
Electric network voltage	220 B
Frequency	50 Hz
	grounding

Table 1. Main characteristics of CZM (C3M) NanoEducator.

Fine powder from limestone, shungite and bitumen binding up to 70 μm was analyzed on the first stage [8].

Samples for scanning were prepared in the following way:

- dispersed powder-like material spreading on an adhesive tape. Before testing there were applied test metering lattices of the height from 21 to 500 nm ((± 1.2–2.5 nm) and the period of 3.0 ± 0.01 μm.

Then, the sample is fixed to the magnetic padding and placed on the magnetic table in the measuring head. After that, the area is chosen for investigation and the surface is scanned.

The state of mineral particles surface plays a great role as the chips, defects and microcracks promote the stronger internal frictions and consequently the better interaction of organic binding and mineral powder.

Mineral powder plays a role of asphalt concrete structure forming component which provides the transmission of voluminous bitumen into film state. Together with bitumen it forms asphalt concrete binding agent substance.

As a result of the analysis, it was determined that limestone powder (**Figure 2**) is characterized with compound surface relief by alternation of peaks and depressions with heights difference from −371 to +429 nm on the scanning area of 8.5 × 8.5 μm. Some evident peaks (light areas) and depression zones (dark areas) are observed on the surface under research. The surface unevenness is 0.43 μm. Fractal dimensionality of surface D is 1.73. The structure of shungite surface (**Figure 2** – limestone powder in 3D picture) has less compound relief on the area of 0.46 micromicrone and height difference from −342 to +457 nanomicrone. Shungite surface in 3D is shown in **Figure 3**.

Such structural nature indicates the more dense powder nature and its high value of specific surface and particles aggregation.

The structure of shungite surface has less compound relief on the area of 0.46 micromicrone and height difference from −342 to +457 nanomicrone. Surface fractal dimensionality D = 1.43.

Such nature of the structure indicated that the powder has more dense structure, its specific surface and particles aggregation is high.Bitumen has rather even uniform surface without any defects in structure (**Figure 4**). The scan sample size is 9.5 × 9.5μm, heights difference is from −821 to 820 nm. Fractal dimensionality of surface is D = 1.97. Bitumen surface in 3D is demonstrated in **Figure 4**.

Composite systems are studied on the second stage. The results of scanning show the interaction of shungite and organic binding agent (**Figure 5**). The results of the research by Chernousov D.I. testify the fact, that shungite with its high adsorptive activity relative to organic binding agent promotes its structurization. Bitumen particles penetrate into shungite porous space and fill it.

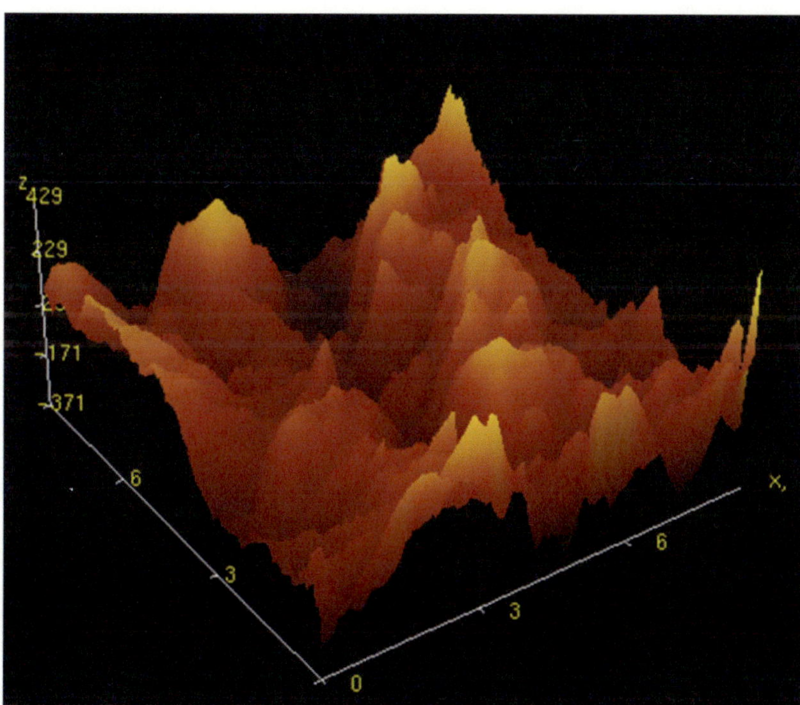

Figure 2. 3D picture of limestone powder. Surface fractal dimensionality D = 1.43.

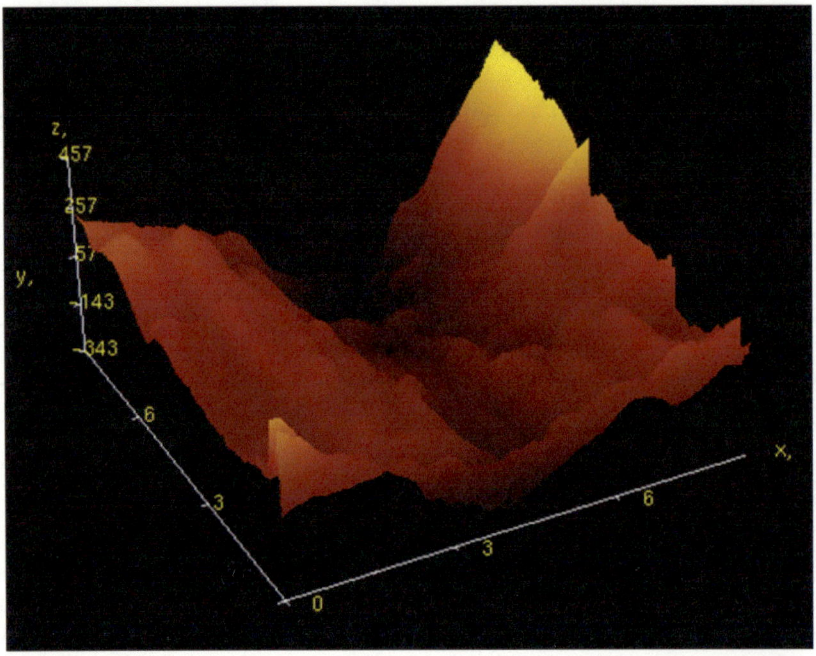

Figure 3. 3D picture of shungit surface.

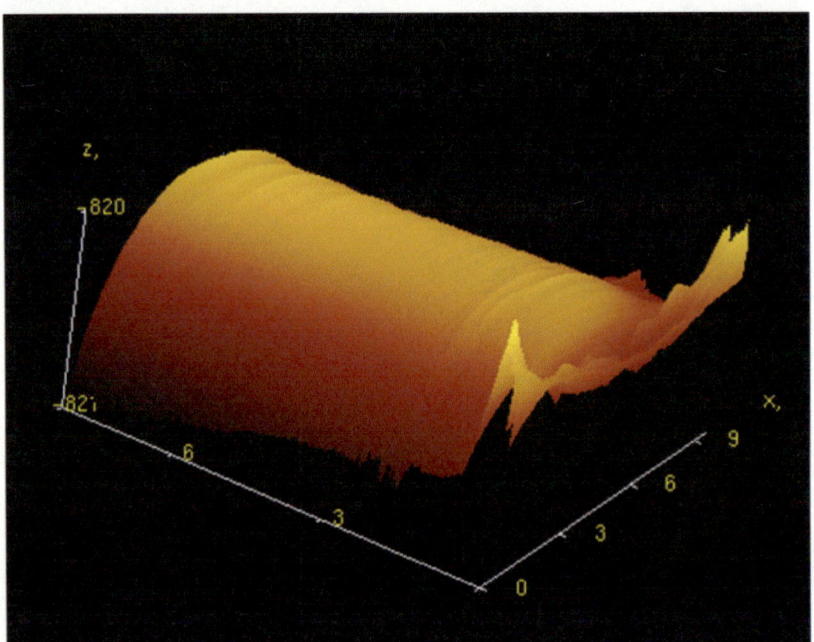

Figure 4. 3D picture of bitumen surface.

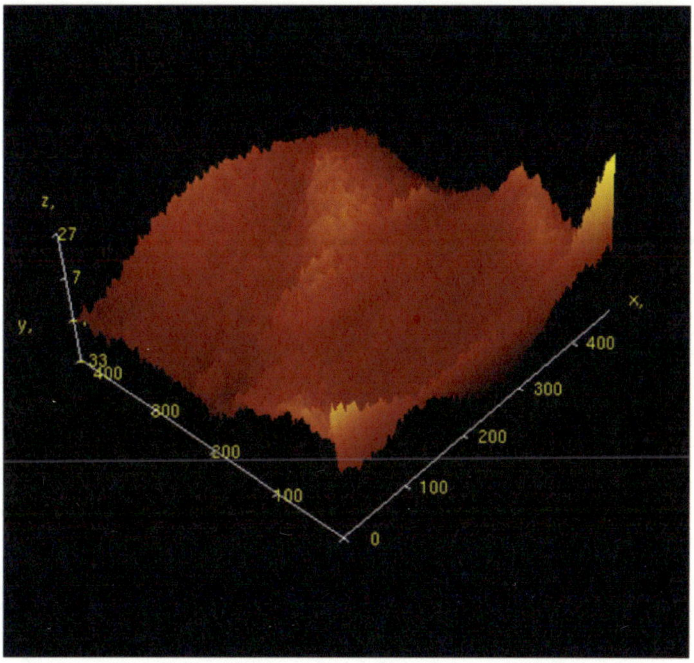

Figure 5. 3D picture of the composite "bitumen-shungite surface".

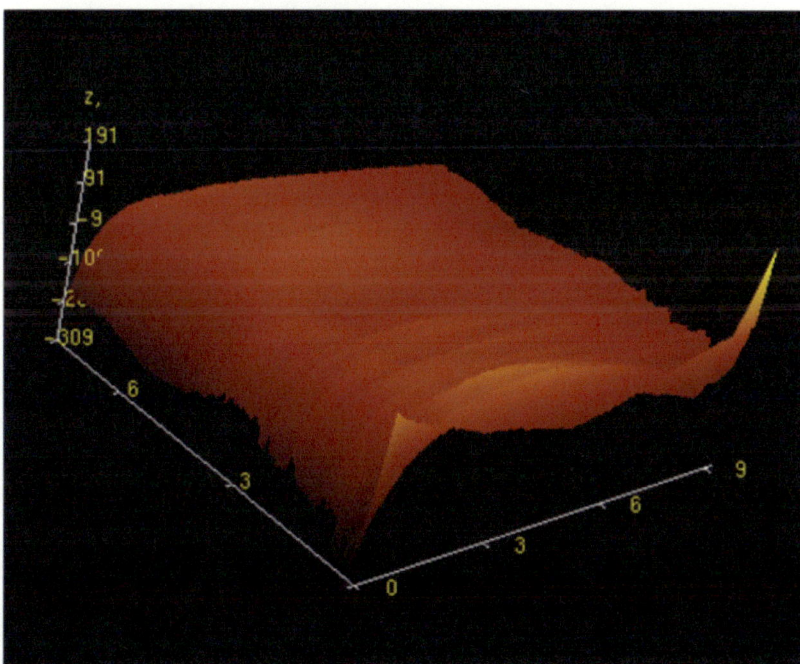

Figure 6. 3D picture of the composite "bitumen-lime stone" surface.

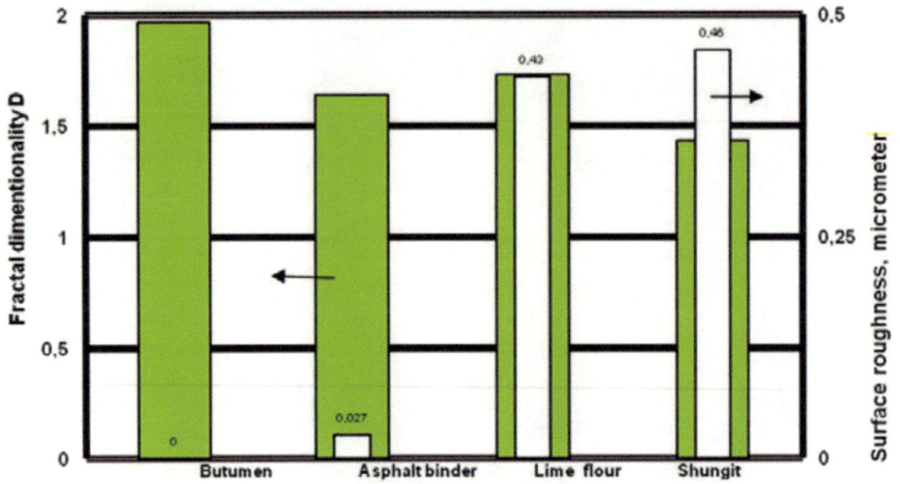

Figure 7. Correlation of the surface roughness and fractal dimensionality characteristics.

Thus the maximal degree of roughness of asphalt concrete binding agent with shungite is 0.027μm (the scan size is 5.0 × 5.0 μm). Fractal dimensionality of the surface is D = 1.64. Composite "bitumen-shungite" surface in 3D is presented in **Figure 5**.

The less surface roughness the nearer the value of surface fractal dimensionality to the criteria "2" that is the material is more dense and homogeneous. Composite "bitumen-lime stone" surface in 3D is given in **Figure 6.**

Correlation of the indicators of the surface roughness and fractal dimensionality characteristics is shown in **Figure 7**.

3. Comparison of reactive ability of shungite and lime stone powders during amalgamation with bitumen

Mineral powder together with bitumen form oriented bitumen layers on mineral grains. These oriented bitumen layers are developed in different degrees. These layers are followed by free bitumen. The qualitative structure of bitumen changes occur in oriented layer as a result of the adsorption processes and bitumen or its separate components diffusion deep into mineral material.

Adsorption is the concentration of substances on a surface or inside a solid body. Not less than two components take part in the process of adsorption.

The nature of such interaction is conditioned not only by the group bitumen structure but also by the properties of mineral powder, its chemical mineralogical composition, grains form, structural peculiarities and surface nature.

Active surface centers are on the surfaces of any mineral material. They condition their capacity of reaction and take part in the interaction with organic binding agent.

The presence of the active centers which can absorb practically all organic compounds existing in bitumen provides strong adhesion contacts between binding agents and mineral materials surface.

Chemisorptions are substances adsorption from the surrounding environment by liquid or solid body. It is accompanied by the formation of chemical linkages. Absorption is considered as a substance chemical adsorption by solid body surface that is as chemical adsorption. At chemisorptions there is a linkage formed between atoms (molecules) of adsorbent and adsorbate. Therefore, it can be considered as chemical reaction which is limited by surface layer.

Interaction of binding agents and mineral materials is the most fully appeared during chemisorptions processes taking place on the interface. Herewith the binding components diffuse into pores spaces of mineral materials and there is a physical adsorption of binding surface film which adheres to the stone materials.

In the very work, the reactive properties of the mineral material were considered according to the bitumen amount adsorbed from benzene solutions [9].

The amount of bitumen adsorption from benzene solutions on the surface of mineral powders was defined by the following method:

Bitumen solutions were prepared on the pure chemical nonpolar benzene of four different concentrations: 1, 3, 6, 9 g/l. Weights of tested powders of 5 g each were put into with ground-in stoppers glass flasks of 200 ml capacity and then they were covered with 50 cm^3 of benzene solutions of designed concentrations and agitated on a special plant for an hour.

After agitating, the flasks with the contents were left alone for 24 h. Then the part of the solution was taken out from the flasks and the bitumen concentration was determined by photometer KFK-3 (KΦK-3).

The absorption was calculated by the formula:

$$A = \frac{(C_0 - C) \cdot V}{1000 \cdot m} \tag{1}$$

where

Co – initial concentration, kg/m^3;

C – equilibrium concentration, kg/m^3;

V – solution volume, m^3;

M – mineral powder mass, kg

The amount of bitumen chemically connected with the powder surface was defined according to the difference of adsorption and adsorbed bitumen values.

Desorption were conducted by the following method:

The powder left after measuring of adsorption value was covered by 50 sm^3 of pure benzene and agitated on a special plant for an hour. The flasks were left alone for 24 h. After that the samples were taken and the bitumen concentration was determined by photometer KFK-3 (KΦK-3).

According to the amount of the powder left on the mineral powder surface, it is possible to reason about the powder surface activity.

The conducted experiment allows to estimate the adsorption properties of the under test mineral materials. The results of the investigations are given in the **Figure 8**.

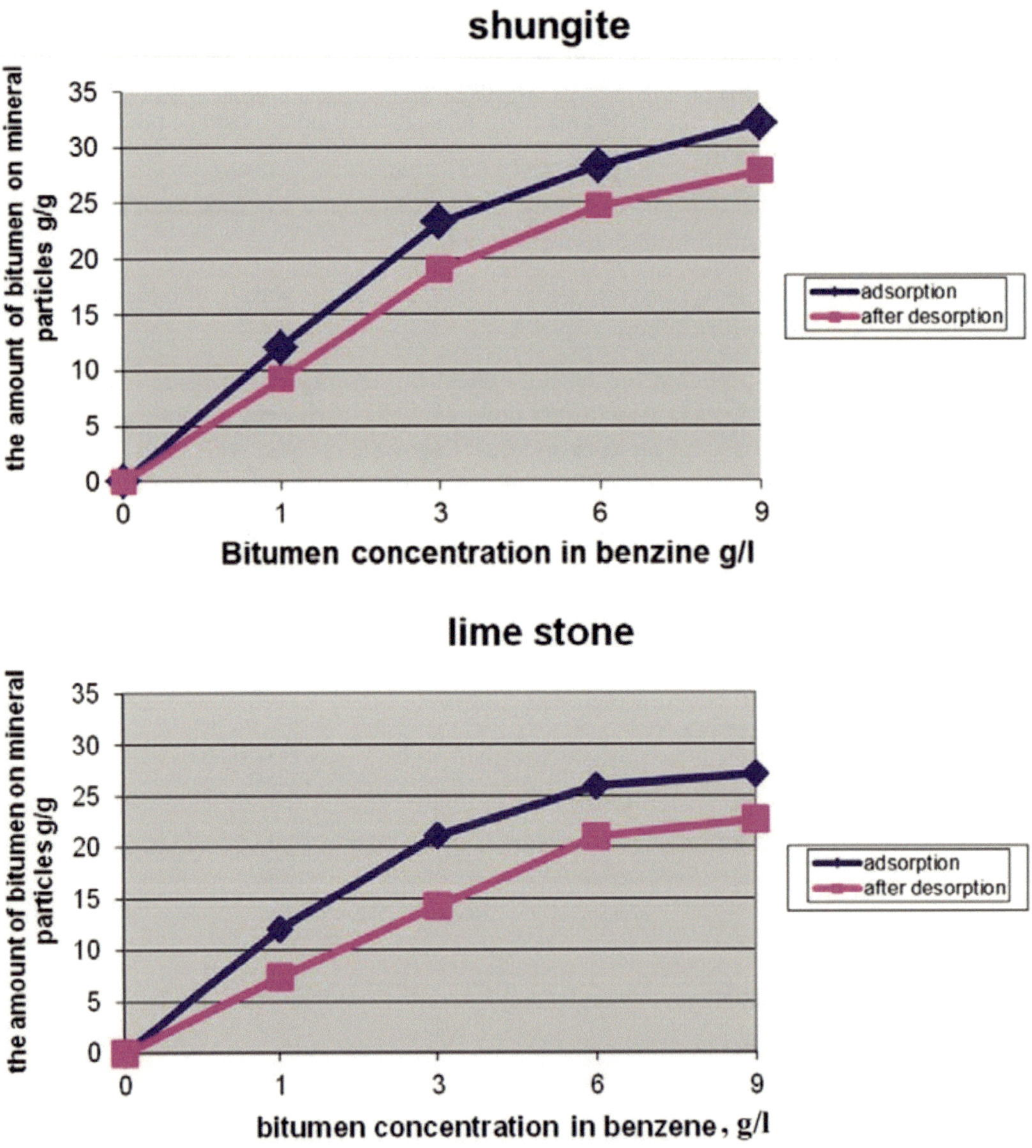

Figure 8. Adsorption–desorption of bitumen on shungite and lime stone surfaces on benzene solutions.

The bigger amount of bitumen is adsorbed on the shungite mineral powder surface than on the lime stone material surface. It was the same at the primary absorption and after the adsorption with benzene. For example, 19 g/g of bitumen are absorbed on shungite surface at bitumen concentration in benzene solution of 3 g/g but 14 g/g of bitumen are absorbed on a lime stone surface, respectively.

The influence of mineral powders on the bitumen structure changing after its desorption from the surface was researched by the spectroscopy method. This method allows investigating mechanism of powders interaction with bitumen, revealing the changes in its structure after adsorption on the mineral powder surface.

It was determined that after bitumen desorption from the shungite surface insignificant amount of organic compounds were left in binding agent. Therefore, they were adsorbed on the surface of shungite mineral powder while their great part on the surface of mineral powder from lime stone was desorbed with benzene. **Figure 8** demonstrate adsorption–desorption of bitumen on shungite and lime stone surfaces on benzene solutions.

The research results of bitumen adsorption and desorption on the surfaces of dispersed materials and bitumen infrared spectrums after the contact with mineral powders show that shungite mineral powder interacts with bitumen more actively than with the lime stone powder (**Figure 8**). The obtained results prove the formation of stable chemisorption bitumen bonds with shungite surface.

4. Infrared spectroscopy

During infrared radiation passing through the substance, the excitation of molecule oscillating movement or their separate fragments takes place. It is observed that the abatement of intensity of the light passed through the sample. The absorption is not in the whole spectrum of incident radiation but only at those wavelengths the energy of which relative to the energy of excitation of oscillations in the molecules under research. Therefore, the wavelengths (or frequency) at which there is maximal absorption of infrared radiation can indicate the presence of these or those functional groups or other fragments in the sample molecular.

5. Shungite mineral powder effect on the asphalt concrete structural-mechanical properties in time

The researchers studied theoretical and practical aspects of shungite mineral powder with account of its interaction with organic binding. The interaction of binding agent with mineral materials depends on chemisorptions processes on their interface. The strength of chemiadsorption bonds between bitumen and mineral powder depends not only on the chemical impact of the bitumen and mineral powder components but also on the mineral powder particles surface state, studied with the microscope (C3M) NanoEducator.

The higher shungite mineral powder porosity than limestone promotes the concentration of the considerable amount of resin in surface microspores. But some amount of oils penetrates the material because of the selective diffusion. That is why structured bitumen films on the shungite particles surface have strong cohesion with them.

Porous filler in shungite bitumen binding agent composition provides the lower temperature of the surface cracking. The tensile strength in such a composition is rather higher than the standard asphalt concrete tensile strength based on dense filler as the friction between the particles is higher due to the greater roughness and specific surface of shungite particles.

The presence of shungite bitumen binding agent in asphalt concrete mix allows increasing surface fracture strength. It is because the chemical activity of shungite particles after their pores filled with bitumen light fractions minimizes internal tension and improves thermal stability of the material.

Shungite particle porosity has better developed specific surface of different mix components as during dry mixing of mineral materials before their integration with bitumen the small mineral powder particles adhere to filler grains. At such processing, a mechanical "modification" of big grains surface which improve bitumen adhesion takes place with them.

Bitumen in thin layer on the mineral particles surface becomes structured, more dense and viscous changing its nature in time. That is why the nature of asphalt concrete mixes based on mineral powder from shungite was defined in different periods at samples conditioning in air: 2, 15, 30, 60 and 90 daily age. Changing of structural mechanical properties of asphalt concrete in time at t + 20°C is presented in **Figure 9**.

The analysis of received results allows determining the following:

in consequence of continuing processes of structure forming the compression strength limits of asphalt concrete based on shungite binding agent at t = 20° (**Figure 9**) and 50°C (**Figure 10**) increases in dependence on sample age conditioning in air. **Figure 10** shows the changing of asphalt concrete structural mechanical properties in time at t + 50°C.

The higher strength limit of asphalt concrete with mineral powder under compression at t = 20 and 50°C than standard asphalt concrete based on lime mineral powder indicates the higher

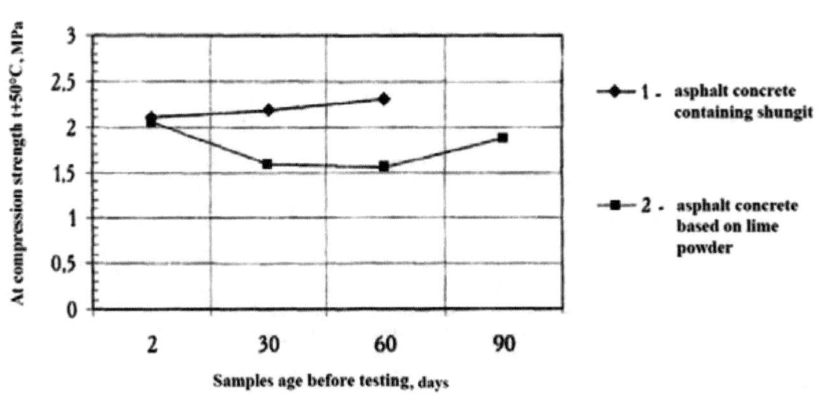

Figure 9. Changing of structural mechanical properties of asphalt concrete in time at t + 20°C.

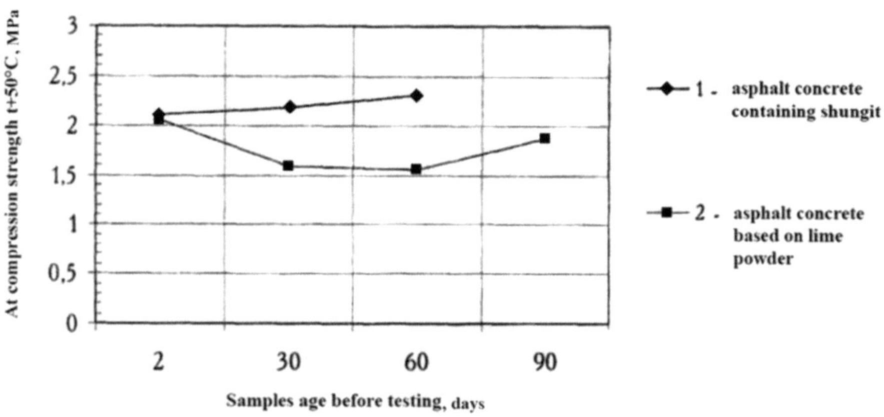

Figure 10. Changing of asphalt concrete structural mechanical properties in time at t + 50°C.

shear strength of asphalt concrete surfaces with shungite bitumen binding agent [4]. It has become usual that the asphalt concrete life service is considerably smaller than theoretically possible according to the abradability conditions. It shows the unconformity of structure – rheological performances of asphalt concrete and its operation conditions. Asphalt concrete during its life duration is affected by different factors: traffic loads, heating, cooling, humidification, and so on. But the normative documents recommend asphalt concrete designing only according to the strength in dependence on traffic load. The possibility of forming shear deformations on asphalt concrete surface in summer time or thermal cracks at negative temperatures is not considered. That is why the deformations excite premature surface destructions more often than the vehicle loads.

Asphalt concrete under different effects is elastic or elastic-viscous, or plastic substance. The theory of elasticity or the theory of plasticity describes only particular cases of asphalt concrete stress state and cannot create the full picture of asphalt concrete operation. Rheology describes asphalt concrete work in deflected mode more objectively.

In rheology, the possibility to describe the stress and deformation in time by differential equations appears. The sphere of these equations definition depends on conditions in which the given materials work. It is suggested that compositional viscous -elastic–plastic material has the total main properties: elasticity, viscosity and plasticity.

While differential equating the main material properties are shown as physically substantiated mechanical models, deformation laws of which are premeditated. In rheology, different simple and complicated models of physical bodies which should correspond to the differential equations situation are used.

While researching rheological characteristics of asphalt concrete based on mineral powders from shungite and limestone, the method by Boguslavsky [10] and Ya. N. Kovalyov was applied. For that target, the height h and the diameter d of asphalt concrete samples before and after the testing and the limit of compression strength were measured.

According to these data, the values of rheological parameters were calculated as follows:

Relaxation time:

$$\Theta = \frac{400 \cdot h \cdot \Delta h^2}{\Delta d^2} \quad (2)$$

Retardation time:

$$\tau = \frac{1000 \cdot \Delta d}{h} \quad (3)$$

Coefficient of elasticity:

$$K = \frac{R \cdot h}{\Delta h} \quad (4)$$

Viscose compliance coefficient:

$$\gamma = \frac{\tau}{K} \quad (5)$$

Viscose coefficient according to Mazwell' s equation:

$$\eta_m = \Theta \cdot K \quad (6)$$

Viscosity of undisturbed structure was defined by the formula.

$$\eta_m = \frac{400 \cdot R \cdot h_1^2 \cdot \Delta h}{\Delta d} \quad (7)$$

Water saturation duration, samples age and freezing cycle number influence rheological characteristics of asphalt concrete were investigated in the road laboratory.

The blanket from crushed stone-mastic mix with shungite mineral powder was constructed in order to satisfy the experiment and study physical-mechanical properties in real operation conditions. Before that, shungite mineral powder of gabbroic-diabase- crushed stone and crushed stone from gabbroic –diabase and bitumen were tested.

The additive based on cellulose filaments Viapor was used as stabilized additive preventing mix segregation and bitumen-shungite binding delimitation at high technologic temperatures. This additive increases the thickness of bitumen film on mineral particles. The mineral proportion of crushed stone-mastic asphalt concrete mix was matched during the research. The composition should provide as follows:

- shear resistance and rut forming resistance at high temperatures;
- water saturation of surface layers at small residual porosity.
- higher crack resistance at surface deformation and under mechanic effects from vehicles.
- high wear resistance.
- uneven surface texture and designed coefficient of adhesion;
- aging resistance.

For optimization of composition of crushed stone-mastic and shungite mineral powder, the specific peculiarities of them were considered.

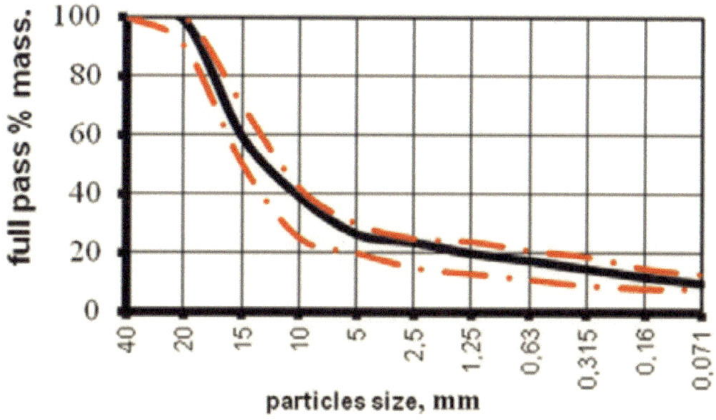

Figure 11. Grain proportion of mineral part of asphalt concrete.

Date of testing	Number of samples	Place of sampling	Layer thickness g(upper /lower) layer)	Surface samples			
			Fact.	Weight of dry sample, g	Weight of dry sample in 30 minutes of aging in water, g	Volume weight g/cm^3	Water saturation, % of volume
1	2	3	4	5	6	9	10
01.10.2011 (shungite mineral (powder)	1	1–1 (B)	5.0	572.20	576.93	2.53	2.36
		1–2 (B)	5.1	920.72	927.76	2.54	3.50
		1–3 (B)	5.0	972.90	976.12	2.58	2.11
	3	3–1 (B)	5.3	967.88	970.26	2.57	1.68
		3–2 (B)	5.8	973.60	976.79	2.57	1.26
		3–3 (B)	5.5	1169.82	1171.60	2.56	0.66
	4	4–1 (B)	5.0	902.55	905.24	2.58	1.01
		4–2 (B)	5.3	1041.18	1044.03	2.54	1.88
		4–3 (B)	5.1	879.19	884.64	2.54	3.10
01.10.2011 Lime stone mineral powder	5	5–1 (B)	5.1	868.72	871.10	2.59	1.66
		5–2 (B)	5.0	1152.31	1156.71	2.50	2.46
		5–3 (B)	5.0	871.29	872.98	2.56	1.79
	6	6–1 (B)	5.2	909.79	911.48	2.60	1.28
		6–2 (B)	5.7	1420.38	1422.85	2.56	0.81
		6–3 (B)	5.7	1363.83	1366.46	2.55	1.06

Table 2. Physical-mechanical indexes of asphalt-concrete from pilot and control sites.

Shungite	Shungite	Shungite	Shungite	Shungite	Lime stone	Lime stone	Lime stone	Lime stone	Lime stone
30.08.2011	19.10.2012	19.05.2012	03.10.2012	08.05.2013	30.08.2011	19.10.2012	19.05.2012	03.10.2012	08.05.2013
0.390	0.510				0.380	0.450			
0.420	0.560				0.420	0.400			
0.335	0.520				0.44	0.48			
0.382	0.53	0.451	0.389	0.365	0.413	0.443	0.401	0.325	0.309
0.333	0.5	0.437	0.361	0.338	0.379	0.398	0.359	0.315	0.246
0.43	0.56	0.466	0.418	0.393	0.448	0.489	0.443	0.335	0.372

Table 3. The dynamics of adhesion capacity on pilot and control sites.

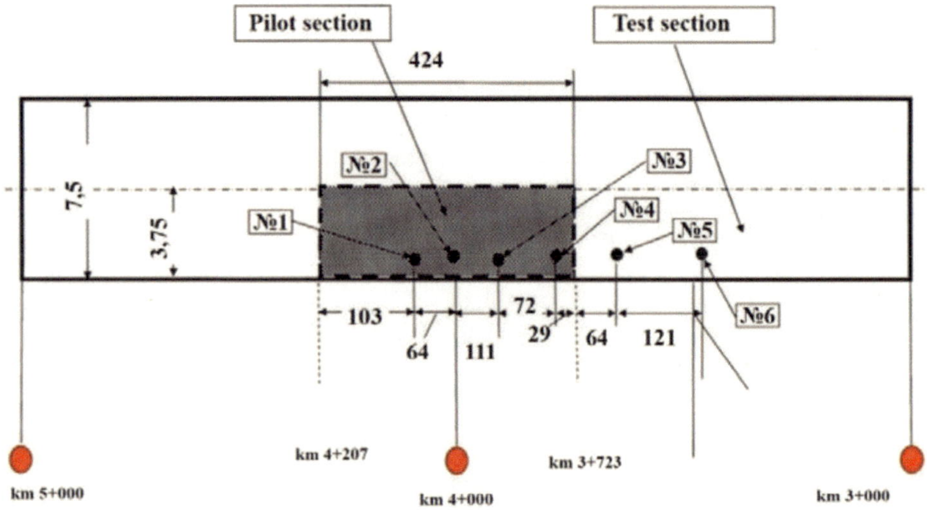

Figure 12. The diagram of sampling on pilot site.

The materials physical-mechanic properties which comply with demands of standard documents were used in asphalt concrete crushed stone-mastic composition.

Granularmetric proportion of crushed-stone asphalt-concrete is demonstrated in **Figure 11**.

After asphalt concrete paving on the pilot site, the blanket has been observed. The adhesion coefficient, evenness, rutting and physical-mechanical properties of asphalt concrete was defined. The results of physical-mechanical properties, adhesion indexes testing are in **Tables 2, 3**, and the diagram of sampling is in **Figure 12**. Physical-mechanical indexes of asphalt-concrete from pilot and control sites are in **Table 2**.

There is the diagram of sampling on pilot site demonstrated in **Figure 12**.

The coefficient of adhesion is defined at the same time on the pilot site of crushed stone surface with shungite mineral powder and on the control site with lime stone mineral powder. The

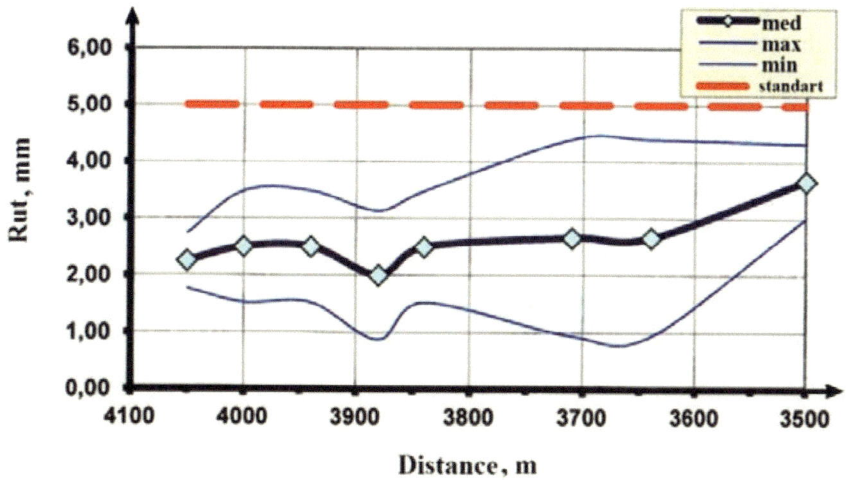

Figure 13. Ruts state on the site with shungite mineral power.

dynamics of adhesion capacity on pilot and control sites are given in **Table 3**, but the ruts state is demonstrated in **Figure 13**.

The monitoring for the state of crushed stone-mastic blanket shows that shungite containing asphalt concrete is in less degree rut prone in comparison with asphalt concrete on lime stone mineral powder.

6. The analysis of traffic load effect on a road blanket

The assessment and prediction of fatigue asphalt concrete properties are the actual problems in assurance of nonrigid road surface reliability.

The effect of fatigue is taken as a process of progressive material destruction by cracks. There are differed three forms of destruction: the beginning of cracks formation; the period of their stable growth and the stage of their intensive growth.

In comparison with the initial period of cracks formation which takes a significant time, the phase of stable cracks growth can be analyzed as the main fatigue process. Cyclic effect of moving vehicles (load) leads to the progressive cracks growing up to the critical depth. A crack increases spontaneously in the third stage which induces the fracture of the material. [11].

The analysis of fatigue performances based on the thermal fluctuation theory of fracture by S.N. Zhurkov shows that fatigue properties of asphalt concrete directly depends on the degree of the plasticity and there is a definite connection between strength and fatigue material properties. There are two different methods of fatigue tests [12]: the test with constant stress amplitude; the test with constant deformation amplitude.

The test of fatigue with constant deformation amplitude is more correct in comparison with the test with constant strain amplitude; as practically, it is not possible to determine the constant stress during the cracks growing and accordingly and the reduction of material area.

Asphalt concrete fatigue fracture in the blanket takes place as a result of cracks formation in the lower tension region and their constant diffusion onto the cover blanket surface. The number of repeated loads necessary for the cracks diffusion onto layer thickness exceeds the number of cycles before the appearance of the first fatigue crack in 20 times. Therefore, the asphalt concrete durability in a blanket exceeds the number of cycles "load-unload" without intervals in which asphalt concrete sample sustains up to fracture in the laboratory in 100 [12] or 200 [13] times.

Operation ability factors of asphalt concrete reliability can be divided into two groups: external and internal.

To the external factors we can refer the environment influence of three types: mechanical, physical-climatic and chemical; to the internal: the indexes of quality, determining the asphalt concrete ability to resist the mentioned types of external one.

The typical fatigue peculiarity is the following: the load factor which is less than the destructive one leads to the progressive durability and degradation of road surface.

Bearing capacity and operation durability of asphalt concrete blanket are more completely characterized by modulus of elasticity and tensile in bending strength.

Dynamic state of road blanket depends on traffic flow concentration and composition and also on the intervals between separate cars.

Traffic loads induce both deflections and oscillations in road blanket are conditioned by momentary elasticity and elastic deflection of road constructions.

While researching bitumen water resistance and adhesion to mineral powders in a laboratory it was proved that bitumen stresses exfoliation from the mineral material surface were comparatively small [14, 15]. In real operation conditions, the additional stresses appearing at traffic movement influence on the wet blanket exceed the thermodynamic ones. Practically, water presses into a blanket in front of the moving wheel and presses out from a blanket after a wheel. The results of such a pumping action are various and difficult for consideration.

At short-term contact of moving wheels with surface, the water pressure in pores of up to 0.05 cm radius is not enough for asphalt concrete fracture but for popcorn mixes it can reach destruction quantity.

If the prolong temperatures are >50°C the deformations of different types appear on an asphalt concrete blanket. The reason is the following: after heat accumulation in asphalt concrete layers, the degradation of bitumen viscosity takes place. According to the second law of thermodynamics (entropy), the grade of the temperature, concentration, barometric pressure always direct from the big value to small one. That is why bitumen starts to move upwards because its concentration is higher inside the asphalt concrete blanket than on the surface. Hereby there is its transition from structured state into the phase of free bitumen. The distance between the mineral material grains increases and the possibility of transmission as a result of wheel load application appears.

It was proved that the mechanism of accelerated fatigue fracture of water saturated asphalt concrete during surface under cyclic dynamic loads is conditioned in certain grade by the appearance of pulse hydrodynamic pressures in water saturated pores of between grains space. Experimental researches of fatigue properties of asphalt concrete while working in the water saturated state show that the service life before the fracture in water saturated state is shorter in 1.5–2.5 times in comparison with the asphalt concrete working in dry state [11, 16]. That is why the formation of pulse hydrodynamic pressures in pores of water saturated asphalt concrete under the moving cars dynamic loads is one of the main mechanisms of asphalt concrete fatigue in watering conditions [14].

7. Previous work reports of dynamic loads on fatigue properties of asphalt concrete

At the car speed of more than 30 km/h the surface load process at the reference point *d* during a small time interval can be represented as the static load *F* affecting on the definite blanket volume on the base under the harmonic vibrations. With account of the before mentioned assumption, the authors used the developed in the university (VGASU) method of the prediction of asphalt concrete operational properties based on the results of testing of the beams $4 \times 4 \times 16$ (cm) on the vibration exciter (**Figure 14**).

Vibration exciter provides the production of a frequency and vibrational amplitude from 10 up to 100 Hz and from 0.1 to 5 mm respectively gradually regulated during the operation time. The load onto the sample can change from 2 to 60 kg.

Figure 14. Vibration exciter at the moment of asphalt concrete beams testing.

Analytical dependences able to determine operation period duration under simulation load were taken as the base of method.

The vibrations similar to the harmonic form are modeled on a vibration exciter:

$$A = A_0 \sin(\omega t + \varphi_0) \tag{8}$$

where A – vibrational amplitude at the time moment t:

Ao – maximal amplitude;

φo – initial form.

$\omega = 2\pi f$ - cyclic frequency, where f – vibrational frequency of vibration exciter.

direction of the forces, effecting on asphalt concrete test beam in relaxation stage can be represented in **Figure 14** as:

where F1 – static load weight influencing the asphalt concrete beam.

F2 – elastic force in asphalt concrete beam;

F3 – inertial force during vibration exciter vibrations;

d – arbitrary point of load application.

The process of asphalt concrete fatigue fracture usually has three stages: internal fractures accumulation accompanied by a significant elasticity module reduction; microcracks appearing at small velocity of module elasticity reduction; propagation and progressive development of cracks with strong reduction of elasticity module [3].

The tension acting in the beam section in the bridge span center was calculated by the formula:

$$\sigma = \frac{3P_{\partial}l}{2bh^2} \tag{9}$$

where P_{∂} - dynamic load; l – span between piers; b, h – section size. The value of the relative deformation was defined by the formula:

$$\varepsilon_m = \frac{(n_1 - n_2)}{Km} \cdot 5 \cdot 10^{-6} \tag{10}$$

where n1 и n2 - strain transducer registration for the sample without loading and vibration, and after the loading and vibration; K*m*– coefficient of strain-sensitivity of a strain transducer.

The values of relative deformation were defined by a strain transducer registrations should be close to the design ones received by the formula:

$$\varepsilon_p = \frac{6fh}{l^2} \tag{11}$$

At adherence to the situation $\varepsilon_p = \varepsilon_m$ the module of elasticity is defined by the formula:

$$E_\partial = \sigma/\varepsilon_m \tag{12}$$

$$\sigma = \frac{3Pl\left(g + 4A_0\pi^2 f^2\right)}{2bh^2} \tag{13}$$

The final formula for dynamic module of elasticity determination is written down as:

$$E_\partial = \frac{3Pl\left(g + 4A_0\pi^2 f^2\right)K_m}{2bh^2(n_1 - n_2)\cdot 5\cdot 10^{-6}} \tag{14}$$

The assessment of operation reliability caused by operating –climatic effect (T) is implemented by the formula:

$$T = \frac{N_p}{N_Э} \tag{15}$$

where Np - number of asphalt concrete loading cycles up to its dynamic elasticity module abatement in two times; design number of wheels loading application during the year of road blanket service.

$$N_Э = N_{np}K_1K_2 \tag{16}$$

where N_{np} – equated movement intensity; K_1 – coefficient of a car track overlapping; K_2 – coefficient considering portion of design cars in a flow (0.3–0.4).

For the investigation of asphalt concrete cyclic fatigue in a laboratory, the following factors were taken: asphalt concrete water saturation as physical-climatic; aging after heating during 5 h at t 150°C as chemical; asphalt concrete composition formula as internal ones.

Cyclic fatigue under dynamic loads is tested on the specially constructed diagnostic tester in the laboratory providing load application with frequency of 868 min^{-1} in the regime of cyclic bending of the 2.5 × 4 × 16 cm size test beams at the fixed deformation amplitude. The test temperature is 20°C. The fatigue deformation amplitude was registered by the moment of destruction. The deformation amplitude while testing was 0.0021 (the test beam kink was equal to zero).

The use of shungite powder instead of standard mineral powder promotes the growth of asphalt concrete fatigue life especially in the conditions of asphalt concrete water saturation. It may be connected with the higher adhesion of shungite powder.

One of the important properties of asphalt concrete predetermining its durability (during the destruction) is its structural stability in the changing humidity regime and temperature conditions. At water saturation adsorption aqueous layers simplify the formation of new surfaces in asphalt concrete during its deformation reducing the surface energy (Rehbinder's effect). Wedge effect of aqueous layers separating mineral grains and exfoliating bitumen strengthens the destruction effect. The mechanism of accelerated fatigue destruction of water saturated asphalt concrete at surface operation in the regime of cyclic dynamic loads is conditioned by the appearance of hydrodynamic pulsations in filled with water porous. This fact was investigated by Rudensky [17].

Asphalt concrete composition	Aging (Thermal effect)	Number of destruction cycles	
		dry	Water saturated
Granit sieving – 92%, lime-stone mineral powder –8%, bitumen of road oil asphalt 60/90–7.5%	Without heating	14,756	10,416
	After 5 h of heating	17,707	11,488
Granit sieving –92%, shungite mineral powder – 8%, road oil asphalt 60/90–7,	Without heating	17,757	14,322
	After 5 h of heating	20,092	16,297

Table 4. Values of fatigue durability of asphalt concrete samples.

The assessment of asphalt concrete by the method of dynamic effect on the water saturated samples is more corresponding to the operation conditions of the material in road construction than the known methods based on statistic water action. That is why the tests of comparison fatigue of asphalt concrete durability were conducted on dry and water saturated samples. Water saturation of asphalt concrete samples was 1.65–1.66% by volume. Values of fatigue durability of asphalt concrete samples were given in **Table 4**.

The application of shungite material instead of the lime stone mineral powder promotes the growth of asphalt concrete fatigue durability especially in the conditions of water saturation. It connects with high adhesion activity of shungite carbon with shungite carbon containing material [21].

The increase of asphalt concrete fatigue durability with the application of shungite mineral powder as a result of aging (after the heating for 5 h) demonstrates the properties stability under temperature effect. It is clear that it connects with oil migration from pores in carbon containing materials and bitumen film rejuvenation during aging.

8. Asphalt concrete bindings effect

An important structure unit of asphalt concrete that is asphalt concrete binding is generated while mixing mineral filler with organic binding. The nature of asphalt concrete is determined by the quality of the asphalt concrete binding. The properties of the system "bitumen-mineral powder" depend on the properties of binding, chemical-mineral composition and mineral filler dispersion, and also these components proportions [18].

The interaction of bitumen with mineral powder is conditioned not only by the rather developed specific filler surface but also by internal grain surface generated by the branched microspores system. Porous fine dispersed fillers cause strong structuralized influence than dense ones. Micropores differ by the high adsorption potential due to which the significant part of surface active components of bitumen sorbs in them. Therefore, it is possible to suggest that porous mineral fillers will cause strong structuralized influence on bitumen than the traditional materials. It is clear that the interest to porous fillers in asphalt concrete composition is conditioned by this.

Mineral powder from shungite (Medvezhjegorskoe deposit) differs from the traditional one from dolomite by bigger porosity of the surface and high developed micropores system and also bitumen of the grade БНД 60/90 of Ryazan НПЗ was investigated. The dispersed raw under study will be correct to test for accordance with the demands [5] asphalt concrete mixes mineral powders. Characteristics of under research fillers properties are shown in **Table 5**.

Mineral powder meets the requirements. But the porosity of the material 20–25% higher than the porosity of lime stone filler and the value of specific surface at the same grain metric proportion is 60–70% higher. So the conclusion is the system of micropores of under research carbon containing raw material is more developed.

The results of the experimental researches of asphalt concrete binding agent. From our point the most informative method of studying the method "bitumen-mineral powder" is the investigation of asphalt concrete binding agent characterized by greater homogeneity of the structure and stability of the properties. Asphalt concrete binding agent was produced from according to the standard method with the before bitumen amount batching. Thus for the mineral powder from shungite the index is 16.2% which is 26% more than traditional lime stone mineral powder has. These data are correlated with the indexes of bitumen proportion in powders. The research results of the physical chemical properties of asphalt concrete binding agent with optimal amount of bitumen are given in **Table 6**.

Coefficient of composite thermal stability and its water resistance are the most interesting among all the data. Therefore, the index of heat resistance of shungite is 14.3% more than while using the traditional mineral powder from lime stone or dolomite. Thermal resistance of mixes and asphalt concrete is characterized by the coefficient of thermal resistance defined as the ratio of ultimate compression strength at t-20°C to the ultimate stress limit at t = 50°C.

As the under research, raw material is porous; the water can penetrate inside the shungite grains and decompose the asphalt concrete binding agent on the bitumen- mineral powder interface if binding agent covers mineral powder surface partially or if the adhesion bonds are weak.

Names of the characteristics	Mineral powder МП-2	Demands of the standard
	shungite	
Grain proportion, % on mass: Smaller 1.25 mm <0.315 mm <0.071 mm	100 98 80	Not less than 95 from 80 to 95 not less 60
Porosity, %	37	Not more 40
Indexes bitumen capacity, g	48	Not more 80
Humidity, % on mass	0.55	Not more 2.5

Table 5. Characteristics of under research illers properties.

Name of the index	Mineral powder МП-2	Requirements of established order
	shungite	
Compression strength, М Па: при		
20°C		
20°C in water saturated condition 50°C	4.3	
3.9		
2.7	—	
Water saturation of samples from powder and bitumen mix	0.9	Not less than 0.7
Swelling of samples from powder and bitumen mix, %	1.87	Not more than 3.0
Coefficient of heat resistance	0.63	—

Table 6. Physical chemical properties of asphalt binding agent.

9. Research of structuralized processes in asphalt concrete

The growth of the strength of the samples from shungite is 9.3% compared with the analogical indexes on lime stone [6, 15]. At less density, the growth of strength characteristics in time which attests the process of structurization of asphalt concrete on shungite mineral powder takes place.

This process can be explained from the point of active extraction of bitumen components with mineral powder from shungite. While analyzing the structure of asphalt concrete agent by microscope (C3M) NanoEducator at increase in 1000 and 2000 times (**Figure 15**), it is possible to mention the substance structure vagueness. The following figure present microstructure of asphalt concrete binding agent from (a) shungite (b) lime stone.

It is clear that an original solution of shungite occurs during amalgamation of binder and shungite. Modern image about the material under study is the precondition for such a supposition. It is known that shungite substance is not a simple amorphous carbon but a mix of various carbonic allotropes which have small lattices connected with amorphous carbon [20]. The more so various mixes of compound organic substances are a part of shungite composition. It is 97–99% of water saturated organics of shungite which escapes into the solution together with fullerenes and determine their properties.

Chemical composition of mineral powders from shungite and lime stone is shown in **Table 7**.

10. Conclusions

1. Experimental researches of physical-mechanical properties of asphalt concrete mixes of different types proved the possibility mineral powder from shungite application. So the interaction of bitumen with shungite mineral powder provides the formation of a stable structure of asphalt concrete with the predominance of closed pores.

2. The results of investigation of shungite mineral powder on scan prob. microscope with nanodementional resolution demonstrate the ability of shungite with high adsorption activity in comparison with organic binding promotes its structurization. Together with bitumen it forms the structured dispersed system acting as a binding material in concrete.

3. It is determined that at the result of shungite mineral powder application asphalt concrete has better water resistance and long water saturation properties which provides high corrosion resistance of asphalt concrete carpet in comparison with traditional powder from lime stone rock.

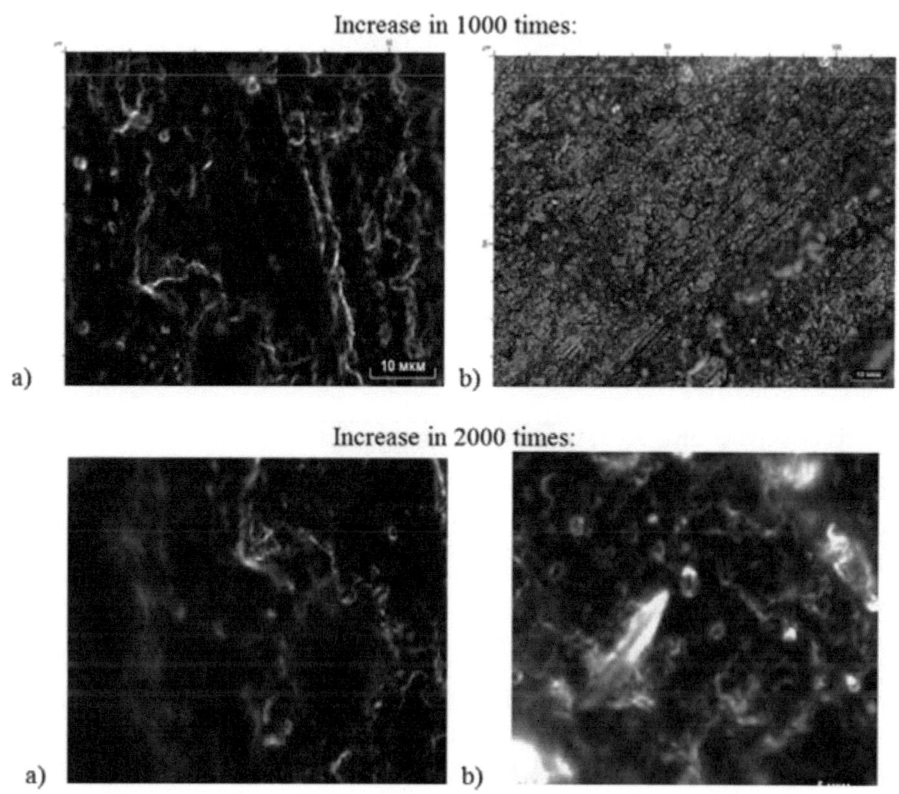

Figure 15. Microstructure of asphalt concrete binding agent from a) shungite b) from lime stone.

Sample	Mass portion, %			
	SiO_2	Al_2O_3	Fe_2O_3	Fe_2O_3
1 (shungite)	45.5	2.3	1.9	1.0
2 (lime stone)	2.1	0.1	1.2	0.8

Table 7. Chemical composition of mineral powders from shungite and lime stone.

4. The use of shungite material instead of lime-stone mineral powder increases the fatigue durability of asphalt concrete especially in the water saturation condition. It connects with the high adhesion activity of shungite carbon containing material.

5. During the process of plot arrangement it was determined that asphalt concrete mix has good placeability and compactability which allows promote designed characteristics of evenness and compacting factor while constructing road carpets.

Author details

Podolsky Vladislav Petrovich[1]*, Lukashuk Alexandr Gennadievich[1], Tyukov Evgeny Borisovich[1] and Chernousov Dmitry Ivanovich[2]

*Address all correspondence to: tatvikt@mail.ru

1 Department of Road Construction, Operation of Voronezh State Technical University, Voronezh, Russian Federation

2 Stroitelstvo e Expluatatziya Dorog, Solnechnogorsk, Moscow Region, Russian Federation

References

[1] Curl RF, Fullerenes SRE. In the world of science 1991;**12**(1):14-24

[2] Podolsky VlP, Erohin AV. Asphalt concrete resistance with application of mineral powder from carbon consisting materials. Scientific Bulletin of Voronezh State University of Architecture and Civil Engineering. 2008;(9):149-152

[3] Shcherbinina SE, Chernousov DI. Substantiation of application of mineral powder from shungite in asphalt concrete mixes. M.: News in Road Science and Engineering. 2009;**2**:34-35

[4] Podolsky VlP, Chernousov DI. Influence of mineral powder on the changing of asphalt concrete properties in time. International congress. Nauka I innovatzii v stroitelstve. Sovremennye problem stroitelnogo materialovedeniya I technologii. Book 2. Voronezh, 2008;**1**:394-399

[5] Podolsky VlP, Chernousov DI, Tyukov EB, Shiryaeva SM. Influence of different mineral powder on rheological characteristics of asphalt concrete. Bulletin of Central Department of RAACH. Voronezh. 2010;(9):173-184

[6] Podolsky VlP, Chernousov DI. Research of bitumen shungite binding agent on scanning microscope. Voronezh. Scientific bulliten of VGASU. 2010;**4**(20);93-99

[7] Podolsky VlP, Bykova AA, Trufanov EV. Influence of Carbon Consisting Mineral Powder on Operation Ability of Sand Asphalt Concrete. Kazan; 2008:26-30

[8] Podolsky VlP, Lukashuk AG, Chernousov DI. Science-practical results of application of asphalt bitumen binding agent in road construction. 2014;**17**(4):86-95

[9] Podolsky VlP, Vysotskaya MA, Kuznetsov DA, Chernousov DI. Formation of asphalt-binder substance during interaction of shungit mineral powder and bitumen. Scientific Bulletin of VGASU. Construction and Architecture. 2013;(29)75-81

[10] Boguslavsky AM, Sarkha, Efremov LG. Zavisimost dependence of asphalt concrete rheological ability on its composition. Roads. 1977;(8):22-24

[11] Rudensky AV, Radovsky BS, Konovalov SV. About the regularities of fatigue failure of road base. Works by GiprodorNII. 1973;(7):7

[12] Fatigue strength of asphalt-concrete mixtures. Moscow: Express-inform. VINITI RAAS. 1976;**41**:7-19

[13] Basics of a simplified method for calculating bitumen-mineral mixtures. Moscow: Express-inform. VINITI RAAS. 1975;**3**:1-16

[14] Podolsky VlP, Rudensky AV, Chernousov DI, Trufanov EV. Comparison Analysis of Fatigue Durability of Blankets from Asphalt Concrete Mixes Based on Lime Stone and Shungite Powder with Carbon Consisting Material. Moscow: MADI Association of Researches of Asphalt Concrete; 2011. pp. 137-143

[15] Podolsky VP, Chernousov DI, Trufanov EV. Research of fatigue life of asphalt – Concrete based on shungite mineral powder. Scientific Bulletin of VGASU. Construction and Architecture. 2011;**1**(21):80-86

[16] Rudensky AV. The Experience of Asphalt Concrete Blanket Construction in Different Climatic Conditions. Moscow: Transport; 1983. 64p

[17] Rudensky AV. Analysis of asphalt concrete structures with nonstandard performances. GiprodorNII works. 1979;(27):66-78

[18] Gezencvei LB, Boguslavsky AM. Road Asphalt Concrete. L.B. Gorelushev, Moscow: Transport. 1985. 350p

[19] Rysiev OA. Shungite is the stone of life. St. Petersburg: Publishing house "Dilya"; 2010

[20] Chernousov DI. Application of asphalt binding material with shungite in the construction of road surfaces. Voronezh; author's abstract PhD diss; 2011

[21] Podolsky VlP, Trufanov EV, Bykova AA, Kovalyov NS. Effect of Carbonaceous Mineral Powder on Operational Properties of Sand Asphalt-Concrete. Kazan. pp. 26-31

Evaluation of Structural and Thermal Properties of Rubber and HDPE for Utilization as Binder Modifier

Farzaneh Tahmoorian, Bijan Samali and
John Yeaman

Additional information is available at the end of the chapter

http://dx.doi.org/10.5772/intechopen.75535

Abstract

Today, high-performance requirements for asphalt pavements demand enhanced properties for bitumen to withstand the environmental condition and traffic demand. On the other hand, the rapid growth in population and economy results in a continuously increased material consumption, and subsequently waste generation. Among various waste materials, rubber and plastic, including high-density polyethylene (HDPE), constitute some part of the non-biodegradable solid wastes worldwide. Because of the great difficulties in managing the non-biodegradable wastes and the required volume of bitumen, the idea of using plastic and rubber as bitumen modifier in new asphalt mixtures appears to be an effective and meaningful utilization of these materials. As binder plays an important role in the final performance of the asphalt mixture, an understanding of modified binder properties is essential in designing an asphalt mixture. To this point, since compatibility of asphalt mixture with polymer is the most important factor in the blend of polymers and asphalt, the properties of the waste polymers were evaluated in this ongoing research by means of advanced thermal analysis and scanning electron microscope (SEM). This chapter presents the results of this experimental study to evaluate the properties of polymers as potential modifier for virgin bitumen in asphalt mixture.

Keywords: bitumen, modified binder, plastic, rubber, HDPE, TGA, DSC, SEM

1. Introduction

Aggregate and binder are two principal constituents of asphalt mixtures. Although the mechanical and chemical properties of aggregates can vary significantly depending on their

IntechOpen

© 2018 The Author(s). Licensee IntechOpen. This chapter is distributed under the terms of the Creative Commons Attribution License (http://creativecommons.org/licenses/by/3.0), which permits unrestricted use, distribution, and reproduction in any medium, provided the original work is properly cited. (cc) BY source, the overall durability and other performance characteristics of asphalt mixtures are generally limited by the performance of the asphalt binder. Failure of asphalt pavement due to the asphalt binder can be attributed to three main sources including:

- rutting occurs at high temperatures as asphalt softens and the elasticity of the binder decreases,
- fatigue cracking due to the repeated loading and aging of the pavement, and
- thermal cracking at low temperatures as asphalt becomes brittle.

Failure of asphalt binders is obviously undesirable. Referring to the considerable research in recent years focused on improving the binders' functional properties, the use of polymers as additives greatly improves the performance of bitumen, including elasticity increase, cohesion improvement, and temperature susceptibility reduction, which all subsequently lead to the improvement of asphalt mixture performance in terms of flexibility, cohesion, and deformation resistance at high temperatures. However, the costly nature of polymer modifiers has stimulated research into cheaper and more cost-effective modifiers produced from recycled materials. For these reasons, and as a quite effective way of disposing of the increasing volume of non-biodegradable wastes, which each year are unavoidably generated in different sectors of human activity, polymer wastes like plastic and rubber can be a reasonable potential materials for consideration as binder modifier.

In light of this, the purpose of the present work is to study the characteristics of waste polymers as potential modifier of bitumen. In that sense, as discussed in the following sections, several samples of bitumen and waste polymers were prepared, and a further thermal and microstructural characterization was carried out on these materials.

2. Bitumen

Bitumen is a construction material obtained from crude oil through distillation processes in the petroleum refinery. Bitumen is useful in road construction due to some of its properties such as adhesion, being water proof and durable. The bitumen is principally used in the road construction as a binder where it is mixed with aggregates to produce asphalt mixture. The asphalt mixture must have adequate properties in order to withstand the permanent deformation, fatigue cracking, and thermal cracking, which are the main distress modes of flexible pavements. Therefore, it should have proper stiffness and bearing capacity and must be able to spread load evenly over the pavement layers.

A large number of procedures have described different fractions in bitumen. Based on the most common procedure, bitumen is divided into four generic groups of saturates, aromatics, resins, and asphaltenes, namely SARAs. The molecular weight of these fractions increases in the order $S < A < R < A$ [1].

The structure of bitumen is mostly regarded as a colloidal system. In this system, the asphaltene particles are dispersed into the oily dispersion medium called the maltenes, which is composed of saturates, aromatics, and resins. As illustrated in **Figure 1**, a shell of resins has covered the asphaltene particles. The temperature and equilibrium between the covering part and dissolved part of resins affect the shell thickness [2–4].

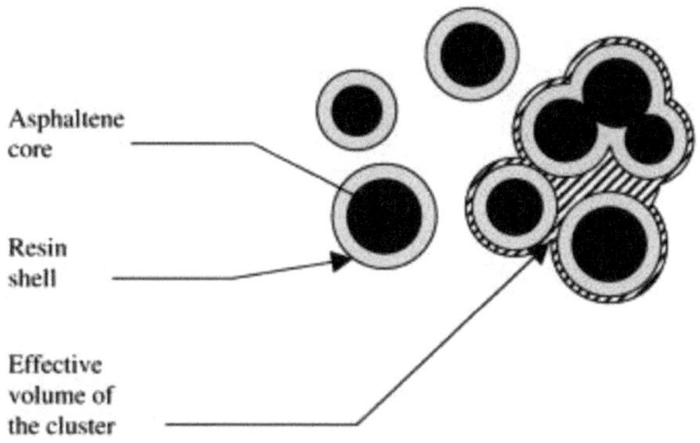

Figure 1. A simplified view of the colloidal structure of bitumen [2].

In 1987, as part of the Superior Performing Asphalt Pavement Program (Superpave), an asphalt binder specification system was developed for evaluation of asphalt performance properties and classification of binders based on its maximum and minimum service temperature. In this specification, the physical properties of an asphalt binder was matched with a performance grade (PG) based on climatic and environmental condition.

In this PG system, two numbers are assigned to each asphalt grade. For example, for asphalt grade of PG 67-22, the first number shows the maximum temperature (in °C) at which the binder can still resist permanent deformation adequately. This number is an average seven-day maximum pavement service temperature. The second number is the minimum temperature at which the binder can perform properly to resist thermal cracking, and hence it is the minimum pavement service temperature. The binders with the maximum temperature ranging from 46 to 82°C and the minimum temperature ranging from −10°C to −46°C (both in increments of 6°) are commercially available, as shown in **Figure 2**. The diagonal line that connects the asphalt grades of PG 82-10 to PG 46-46 is the border between the asphalt grades, which can be produced at refineries and are those produced only by modification (shaded areas). Useful temperature range (UTR) is a measure of the difference between upper and

Upper Specification Temperature, °C	Lower Specification Temperature, °C						
PG 82	-10	-16	-22	-28	-34		
PG 76	-10	-16	-22	-28	-34		
PG 70	-10	-16	-22	-28	-34	-40	
PG 64	-10	-16	-22	-28	-34	-40	
PG 58		-16	-22	-28	-34	-40	
PG 52	-10	-16	-22	-28	-34	-40	-46
PG 46					-34	-40	-46

Figure 2. Performance grades for commercially available binders (after Ref. [5]).

lower service temperatures. The binders produced at refineries without modification have a UTR of not more than 86°C, whereas the modified binder often have a UTR of more than 92°C. Therefore, UTR can be used as an indicator showing the degree of required modification and the cost needed for modification. As UTR increases, this cost increases accordingly [5].

3. Polymer modifiers

The binder characteristics strongly influence the mechanical properties of asphalt mixture. To this point, binder should have a certain mechanical and rheological requirements as follows in order to fulfill the pavement criteria:

- For homogeneous coating of aggregates, the bitumen should be fluid enough at mixing and construction temperatures of about 160°C.
- To resist permanent deformation, the bitumen should be stiff enough at high temperatures (about 60°C depending on the local climate)
- To resist the cracking, the bitumen should be soft enough at lower temperature that pavement experiences (approximately down to −20°C depending on the local climate).

Accordingly, it can be concluded that obtaining bitumen to work well under all aforementioned conditions can be difficult. To surmount this problem, many researchers have tried to develop the asphalt pavement performance by improving the asphalt binder behavior using different modifiers. There are a large varieties of materials, which are often used for modifying the binder, of which polymers are widely known to be easy to use and cost effective. Referring to available literature (for example, see [6–9]), polymer addition may result in both a more flexible binder at low in-service temperature and enhanced properties at high in-service temperature, which significantly prevent the pavement from being deformed. They also improve the adhesive bonding to aggregate particles [10]. Polymers can exist in two different morphologies while in a solid phase:

- Amorphous, in which molecules are randomly oriented within the polymer when the material is cooled in a relaxed state. The cooled state of amorphous materials is highly similar to their molten state. The only difference between these two states is the molecules' distance. These polymers can easily be altered in shape and generally exist in a rubbery state.
- Semicrystalline is an arrangement of ordered molecules with some amorphous regions. As the semicrystalline polymer cools, a portion of the molecular chains forms crystals by folding up into densely packed regions. The polymer is classified as semicrystalline, if more than 35% of the polymer chain forms these crystals. These polymers are stiff and exist in a glassy state.

The degree of crystallinity in a polymer is affected by different factors such as polymer type, additives, and cooling rate. The morphology and degree of crystallinity significantly influence the polymers' properties. Polymers with high degree of crystallinity have a higher glass transition temperature and higher modulus, toughness, stiffness, tensile strength, and hardness. In addition, they have more resistance to solvents but are less resistant to impact strength [11].

Today, there are a large varieties of polymers that are often used for modifying the binder. These polymers can be mainly classified into the following categories:

- Elastomers such as rubber can be stretched and then recover their shape when the stretching force is released. Elastomers contribute to the elastic component of binder. The addition of elastomers to the binder results in an increase in the binder stiffness at high temperatures and loading and subsequently will improve the resistance to permanent deformation. However, elastomers will not substantially improve the ability of asphalt mixtures in thermal resistance.
- Plastomers such as polyethylene can form tough, rigid, three-dimensional networks within the bitumen resulting in increase of the initial strength of the bitumen, and subsequently improving the ability of asphalt concrete to resist the heavy loads. Plastomers have less elasticity compared to elastomers as they do not provide an increase in the plastic component of binder while they increase the binder's stiffness at high temperature and loading [12]. Therefore, plastomers can improve the rutting resistance, but they lack the improvements in fatigue resistance, cracking resistance, and low-temperature performance [13] because of increased intermediate and low-temperature stiffness. This makes them inferior to elastomers.

Although modifying the binders will result in the enhancement of binder's properties, using virgin additives as modifier will increase the road construction cost. Therefore, in recent years, many investigations have been conducted on modifying binders using waste materials as additives. Among these waste materials, application of waste plastics and rubbers in certain amount as binder modifier can substantially enhance the stability, strength, fatigue life, and generally the asphalt performance in one hand [14], and on the other hand, it would be an ideal solution for reducing the environmental pollution associated with these non-biodegradable wastes. In light of this, according to the characteristics of elastomers and plastomers, this chapter analyses the characteristics of high-density polyethylene (HDPE) and crumb rubber as potential binder modifier. The tests used to evaluate the polymer properties were thermal analysis by thermogravimetric analysis (TGA) and differential scanning calorimetry (DSC), and microstructure analysis by scanning electron microscopy (SEM).

4. Materials

4.1. Bitumen

Bitumen of C320 was used as a base material for this research, which was kindly supplied by Boral Ltd. This bitumen corresponds to the most common bitumen used in Australia. C320 is classified and manufactured in accordance with AS 2008 (2013) and is suitable for medium to heavy asphalt applications as well as for heavy duty and hot climate seals. The typical characteristics of bitumen C320 are presented in **Table 1**.

As emphasized earlier and shown in **Figure 3**, the complex structure of bitumen is composed of unsaturated structures divided into two main groups of asphaltenes (which are insoluble in n-heptanes) and maltenes. The maltenes are further split into saturates, aromatics, and resins. The proportion of bitumen fractions and the molecular weight of each fraction is presented in **Table 2**.

Characteristics	Unit	Methods	Value
Softening point	°C	AS 2341.18	52
Penetration at 25°C	dmm	As 2341.12	Min 40
Flashpoint	°C	AS 2341.14	Min 250
Viscosity at 60°C	Pa·s	AS 2341.2	320
Viscosity at 135°C	Pa·s	As 2341.2	0.5
Specific Gravity	Kg/m³	AS 2341.7	1.03

Table 1. Characteristics of the original bitumen.

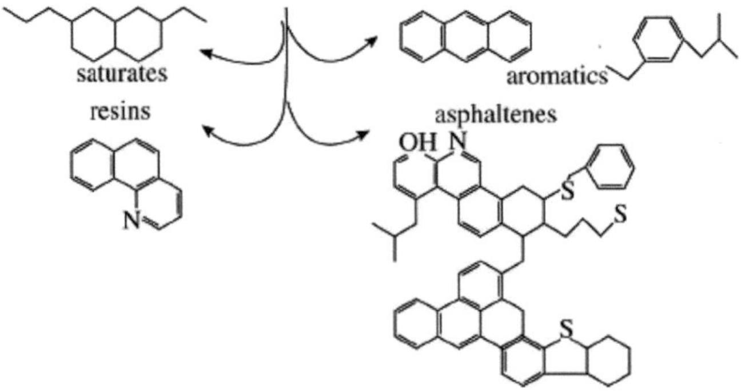

Figure 3. Main compounds in representative structures of the four bitumen fractions [15].

Fraction	Proportion of the overall bitumen	Molecular weight	Description
Asphaltenes	5–25%	600–3000	Substantial effect on bitumen rheological properties
Resins	15–25%	500–1300	Dispersing agent for asphaltenes; their proportion to asphaltenes determines the structural character of the bitumen
Aromatics	40–65%	300–800	Major dispersion medium for asphaltenes
Saturates	5–20%	300–600	Non-polar viscous oil

Table 2. Proportion and molecular weight of bitumen chemical fractions [3, 16–19].

The rheological properties of bitumen are highly affected by the asphaltene content due to its physical parameters such as glass transition and bitumen viscosity [20, 21]. An increase in the asphaltenes content will generally result in harder bitumen with a lower penetration, higher softening point, and higher viscosity [22].

4.2. HDPE

Among plastics, polyethylene (PE) forms the largest portion followed by polyethylene terephthalate (PET). To this point, this study focused on polyethylene and particularly on high-density polyethylene (HDPE). The HDPE used in this research were obtained from plastic recycling plant. As shown in **Figure 4(a)**, the HDPE is in the granular form with the particle size of 2.36 mm. HDPE like other plastics is a polymer consisting of very large molecules made up of smaller units called monomer, which are joined together in a chain by a process called polymerization. Polyethylene is semicrystalline material with a wide range of properties and appropriate resistance to chemicals and fatigue. A molecule of polyethylene has a very simple structure and is composed of a long chain of carbon atoms with two hydrogen atoms attached to them, as shown in **Figure 4(b)**. Sometimes other elements such as oxygen, nitrogen, chlorine, or fluorine are attached to these polymer molecules. These are lightweight molecules with low moisture absorption rates and good resistance to organic solvents.

HDPE is one type of thermoplastics. As most of the thermoplastics can soften at temperature ranging from 130 to 140°C with no gas emission in the temperature range of 130–180°C, they can be a potential option for blending with bitumen in asphalt production because the heating temperature for bitumen ranges from 155 to 165°C in the whole processes for asphalt pavement construction [23]. **Table 3** presents the information regarding the thermal behavior of polyethylene, which emphasizes its suitability as a binder modifier.

In order to find the relevant properties of HDPE, some tests were conducted on this material and the result of these test are presented in **Table 4**.

4.3. Crumb rubber

Crumb rubber used in this research was obtained from tyre recycling plant that processes the car tyres into crumb rubber through the ambient grinding method. The crumb rubber was provided in granule form, as shown in **Figure 5(a)**.

The particle distribution test was performed on crumb rubber. The result of sieve analysis is presented in **Table 5**. The ground tyre rubber has a particle size averaged between 8 and 50 mesh (2.36 and 0.300 mm).

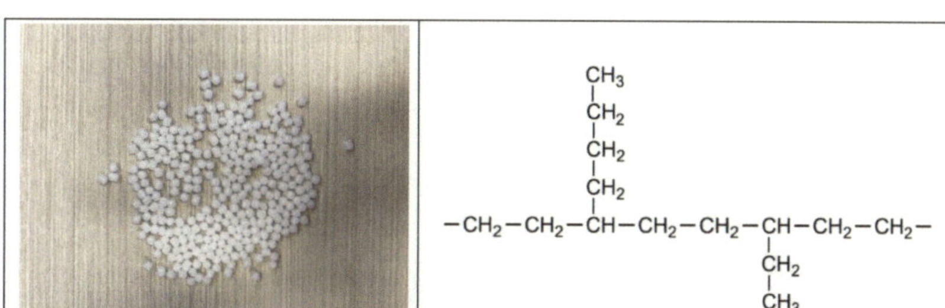

Figure 4. Analyzed material (a) HDPE and (b) chemical structure of HDPE.

Characteristics	Unit	Methods	Reported products
Solubility in water	—	Nil	—
Softening temperature	°C	100–120	No gas
Decomposition temperature	°C	270–350	CH_4, C_2H_6
Ignition temperature range	°C	> 700	CO, CO_2

Table 3. Thermal characteristics of polyethylene [23].

Characteristics	Unit	Methods	Value
Density	g/cm³	AS 1141.5	0.963
Size	mm	AS 1141.11.1	2.36
Water absorption	%	AS 1141.5	0.0

Table 4. Characteristics of high-density polyethylene (HDPE).

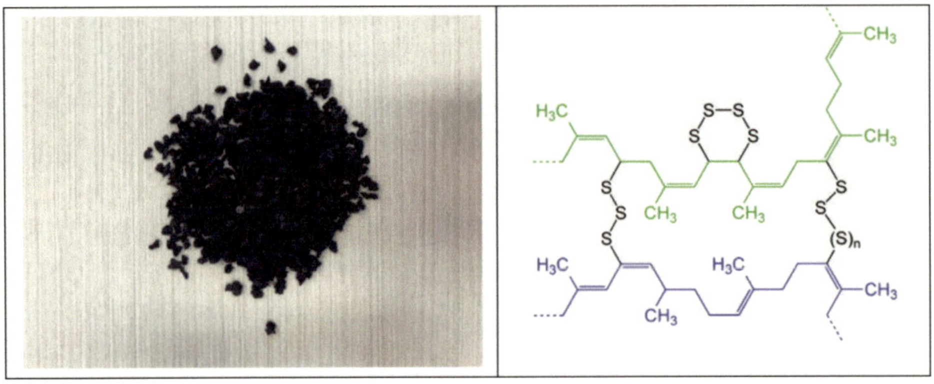

Figure 5. Analyzed material (a) crumb rubber and (b) chemical structure of rubber.

Sieve no.	Sieve size (mm)	Mass retained (%)
4	4.75	0.0
8	2.36	25.7
16	1.18	67.7
30	0.600	6.2
50	0.300	0.4
100	0.150	0.0
200	0.075	0.0

Table 5. Particle size distribution of crumb rubber.

Characteristics	Unit	Methods	Value
Density	g/cm^3	AS 1141.5	0.982
Size	mm	AS 1141.11.1	1.18–2.36
Water absorption	%	AS 1141.5	0.1

Table 6. Characteristics of crumb rubber.

The properties of crumb rubber are presented in **Table 6**, which are obtained from conducting relevant tests on crumb rubber. It should be noted that tyre rubber is typically a composition of three polymers including polyisoprene (natural rubber), polybutadiene, and polystyrene-butadiene [24]. The main compounds in rubber are shown in **Figure 5(b)**.

5. Methodology

In this research, different polymers (i.e., bitumen, HDPE, and crumb rubber) were analyzed based on their calorimetric curve, thermal transition, and their overall quality. The analysis of the materials was performed by means of thermogravimetric analysis (TGA), differential scanning calorimetry (DSC), and scanning electron microscopy (SEM).

5.1. Sample preparation

For performing analysis on polymers, the samples were prepared based on the requirements of equipment. In order to study the thermal behavior of individual polymers, in all cases, a small amount of material (5 to 10 mg) was placed in a measuring pan. To prevent pressure buildup during the test, it was advised to have a small opening in the small pan. After this preparation, the samples were placed in DSC or TGA equipment.

5.2. Analysis methodology

It is expected that the addition of polymers influences the microstructure of binder. In theory, the addition of polymers containing hard segments provides higher strength, whereas the soft segment polymers improve toughness and low-temperature cracking. Since the binder modification depends on the compatibility of bitumen and polymer as modifier, this chapter covers the study of the individual polymers to identify some of their physical and chemical properties, their thermal behavior, and their microstructure. The tests to characterize the analyzed materials were performed in the Advanced Materials Characterization Facility (AMCF) at Western Sydney University, Australia. These tests included thermal analysis, structural characterization, and microstructure analysis, and the main points of testing procedure are represented in the following sections.

5.2.1. Thermal analysis

Thermal analysis corresponds to a group of techniques used to measure the physical and chemical properties of materials as a function of temperature. The measurements can be performed in different atmospheres including inert atmosphere (nitrogen, argon, helium) or in an oxidative atmosphere (air, oxygen). The gas pressure can also selectively vary in thermal analysis. In this research, the thermal behavior of materials was investigated through DSC and TGA.

5.2.1.1. *Differential scanning calorimetry (DSC)*

Parameters such as glass transition temperature (T_g), melting point, and the degree of crystallization were monitored by DSC. It should be noted that the glass transition temperature is more important in polymer applications compared to the melting point, because it corresponds to the polymer behavior under ambient conditions.

In this research, DSC analysis was performed according to ASTM E473-85 in an NETZCH DSC 204 F1 to obtain the thermal critical points of materials. The test specimens weighing about 5 mg were heated up at different temperature ranges, depending on polymer type, in an aluminum crucible under an air flow (100 mL/min) at a rate of 10°C·min^{-1}.

For bitumen, DSC experiments were carried out on about 5 mg samples in aluminum crucibles with perforated covers. Before conducting DSC, the bitumen sample was homogenized at 130°C for about 1 hour and then placed in the DSC equipment. To conduct DSC, first, the samples were cooled from room temperature to −100°C at a heating rate of −10°C/min. The samples were maintained at the low temperature for about 15 minutes to ensure a stabilized initial reading. Then, they were heated up to 200°C at a heating rate of 10°C/min. The DSC thermograph recorded during this heating scan is considered as the first scan. On completing the first scan, the sample was maintained at 200°C for 5 min to remove thermal history and then quickly cooled from 200°C to its starting temperature (−100°C) at a cooling rate of −10°C/min and again held for about 15 min before being reheated to 200°C at a heating rate of 10°C/min. The DSC thermograph recorded during this second heating scan is considered as the second scan. The same procedure was repeated to provide the third heating scan.

For rubber, similar to bitumen, three cycles of cooling and heating were considered as the method of the experiment with the same heating rate of 10°C/min, cooling rate of −50°C/min, and temperature range of −100–200°C. For HDPE, the DSC procedure was the same as bitumen and rubber with an exception of the temperature range, which was considered from −160 to 200°C.

Glass transition and melting point were measured from DSC curves. The percentage of crystallized fraction (CF) was determined from the following equation through dividing the observed melting enthalpy (ΔH_{obs}) by the melting enthalpy of 100% crystalline material (ΔH_o).

$$CF = \frac{(\Delta H_{obs} \times 100)}{\Delta H_o} \quad (1)$$

The values of ΔH_0 depend on the material type and can be found in literature. For example, a value of 200 J/g was used by Lesueur et al. [4] and Claudy et al. [25]. Values of 180 and 121 J/g were used by Michon et al. [26] and Lu and Redelius [27], respectively.

5.2.1.2. *Thermogravimetric analysis (TGA)*

Thermogravimetric analysis (TGA) was performed to study the kinetics and to investigate the degradation process of materials at a higher temperature. In this study, the thermal decomposition was verified in 5 mg samples in an aluminum crucible under air flow (100 mL/min) heated from 30 to 600°C at a heating rate of 10°C·min^{-1}. The TGA curves and its differential (DTG) were carried out in an NETZSCH STA-449C thermogravimetric analyzer. The onset temperature of the mass loss effect (T_O) and temperature of peak rate of mass loss (T_p) were determined from TGA thermographs.

5.2.1.3. Microstructure characterization

The microstructure of the materials was investigated under scanning electron microscope (SEM). Scanning electron microscopic analysis was done in 6510LV SEM employing between 10 and 20 kW. The specimens in this study were examined with magnifications of 100–1000, and the results of this study with best magnification are presented in the following sections.

6. Results and discussion

6.1. Thermal analysis by DSC

Heating the polymers results in a number of phase changes such as the glass transition (T_g), crystallization transition (T_c), and melting point (T_m). DSC analysis is a useful technique to identify the location of these thermal parameters. In the DSC curves, the sharp peaks are related to the polymer melting and the areas under these peaks provide the heat of fusion (ΔH). Furthermore, the smaller inconsistencies at the lower temperature are most likely related to the glass transition.

In this research, DSC technique is used to investigate the transition temperatures and the crystallization degree of different polymers. Accordingly, the DSC curves were examined to evaluate the physical characteristics of individual materials. It should be noted that for DSC runs, the complete set of heating-cooling process were repeated three times for each polymer, where the first run is usually carried out to remove any impurities and moisture from the sample. In addition, in order to evaluate the transitions accurately, a temperature scan over a wide range temperature is considered for DSC analysis.

In DSC analysis, the thermal parameters for bitumen depend on the refined petroleum source as well as the petroleum refining process. **Figure 6** shows the DSC thermograph of neat bitumen and its corresponding first-derivative curve. The effects detected in the thermograph, as assigned previously (for example, see [28–30]), are described below.

An increase in the heat capacity for neat bitumen can be observed in the DSC curve by an abrupt change in the slope of the curve placed in the low-temperature region (around −30°C) corresponding to the glass transition temperature (T_g) of the bitumen. The glass transition temperature (T_g) is a material's temperature at which all molecular transitional motion is frozen; therefore, the material becomes rigid and brittle at or below this temperature. The glass transition temperature of polymers is one of the most important parameters as it is related to the average molecular weight of polymers and hence provides information about their composition. Moreover, it demonstrates the viscoelastic behavior of polymers at low temperatures [31]. Therefore, the glass transition temperature of neat bitumen is believed to be closely related to the low-temperature performance of asphalts. As shown in **Figure 6**, the middle point of the temperature range where the transition occurs is considered as the

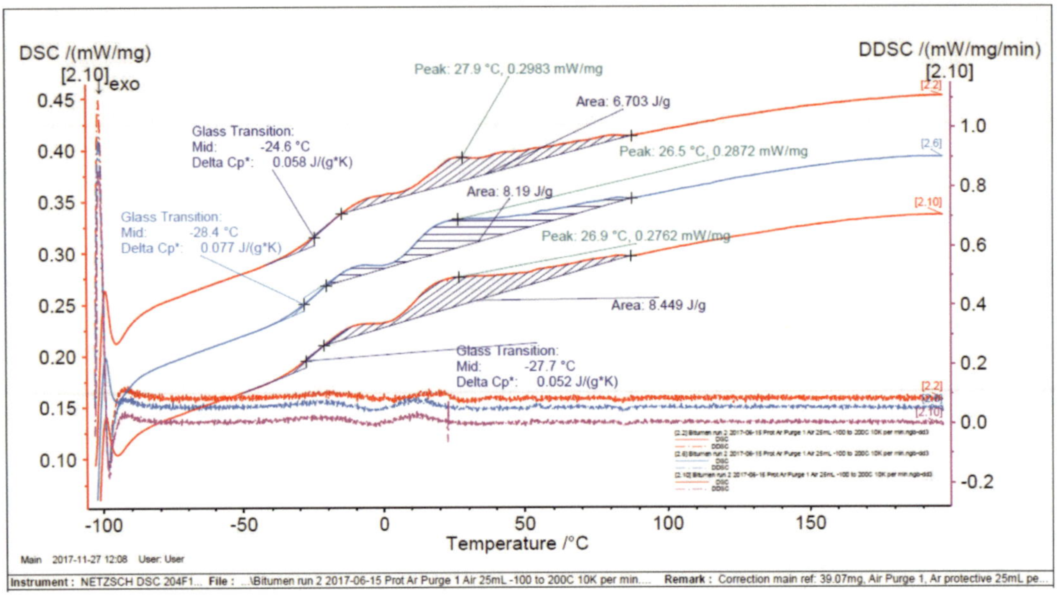

Figure 6. DSC and DDSC thermograms of bitumen.

glass transition temperature. In addition, as shown in **Figure 6**, T_{gonset} of bitumen is at about −40°C. Referring to [32], the T_{gonset} temperature is more closely related to the glass transition temperature of the saturate fraction that has the lowest T_g. The main T_g reflects the characteristics of the glass transition temperature of the majority of components.

At temperature above T_g, an exothermal peak and a broad endothermal peak from about −20 to 85°C is observed. The big exothermal peak next to the glass transition is most likely the result of crystallization of small paraffin molecules, and melting of the crystallites formed during heating or cooling is known as the main reason to produce endothermal peaks in this region [32].

Referring to [31], the exothermic effect just above the T_g in DSC thermographs is negligible, as it has been associated in previous studies (for example, see [28, 30]) with the crystallization of certain molecules, which are not crystallized during cooling.

In addition, referring to the literature (for example, see [25, 28–30, 33]), the dissolution of the crystallized fractions (CF) is the main reason of the enthalpy changes and can be calculated from the area under the peak to a reference enthalpy of dissolution. As shown in **Figure 6**, in order to calculate this parameter, a straight baseline between the end of the glass transition and the end of the endothermic effects is drawn. In this research, the reference enthalpy value of 180 J/g is used for the estimation of the amount of crystallized fraction of bitumen based on previous investigations [1, 4, 15, 25, 26, 33]. The calculation of the crystallizable fraction content shows a value of about 4%, which is considered small. The presence of wax content in bitumen is commonly responsible for the extent of crystallizable fractions, which is the main reason for the problem of pavement exudation and inappropriate thermal susceptibility [15].

To achieve DSC curve for rubber, similar to bitumen, three cycles of cooling and heating were considered as the method of the experiment with the same heating rate of 10°C/min, cooling rate of −10°C/min, and temperature range of −100–200°C.

The DSC thermograph of rubber (**Figure 7**) presented a glass transition temperature (T_g) at −55° C. However, due to amorphous nature of rubber, DSC curve does not present a well-defined melting temperature.

For HDPE DSC analysis, as shown in **Figure 8**, it can be observed that HDPE started to lose its solid form at around −100°C corresponding to the glass transition temperature (T_g) of HDPE. As the temperature increases, a strong endothermic peak average value at 134°C can be observed, which is most likely related to the melting of crystalline domains of HDPE.

The DSC curve of HDPE in second heating cycle is illustrated in **Figure 9**. As can be observed, the energy consumption for melting of crystalline domain of HDPE was 221.1 J/g that occurred between the beginning and end of melting point.

The cooling cycle involves the rate of cooling temperature of 10°C /min. As shown in **Figure 10**, the peak point for HDPE becomes totally solid at around 115°C, which means that the crystallization temperature (Tc) of HDPE, an exothermic peak, is 115°C.

To estimate the amount of crystallized fraction of HDPE, the values of ΔH_o of 287.3 J/g has been found from literature for HDPE (for example, see [34, 35]). The calculation of the crystallizable fraction content for HDPE shows a value of about 76%.

It should be noted that fully crystalline polymers do not exhibit glass transition temperature, and their structure will not change until the melting point. However, according to DSC thermographs, HDPE is considered as semicrystalline polymer.

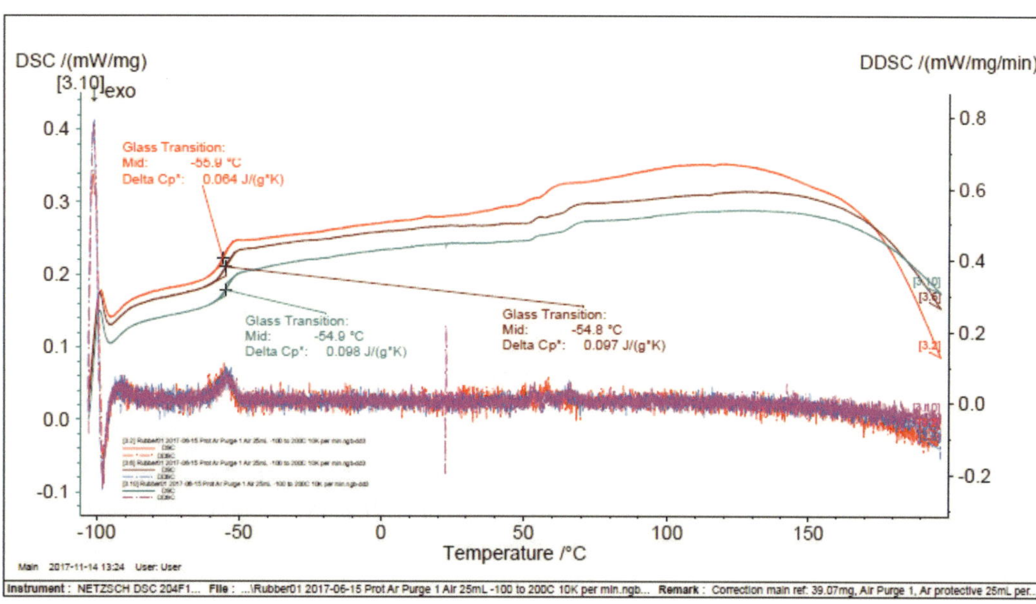

Figure 7. DSC and DDSC thermograms of rubber.

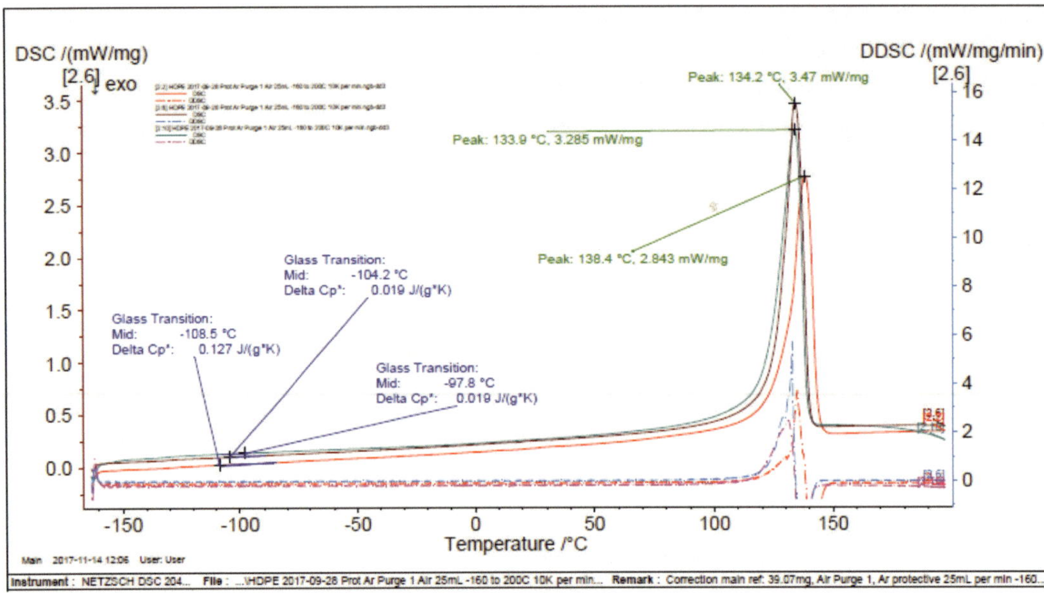

Figure 8. DSC and DDSC thermograms of HDPE for heating cycles.

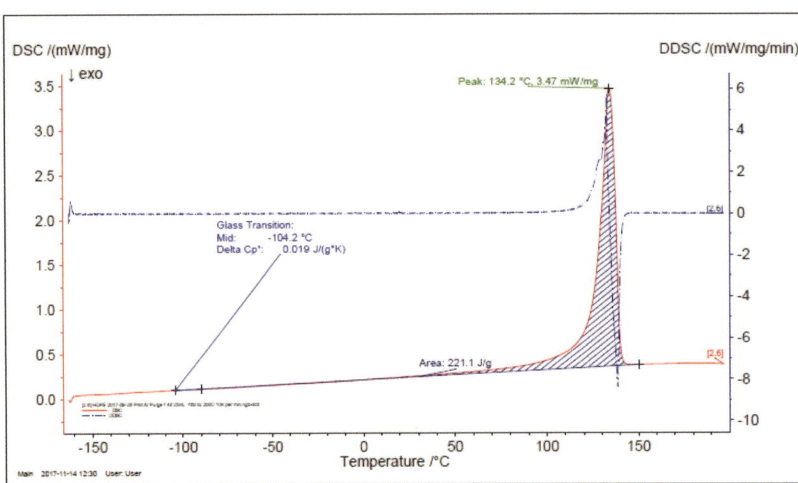

Figure 9. DSC and DDSC thermograms of HDPE for second heating cycle.

In addition, as HDPE has higher molecular weight than bitumen, the melting and crystallization temperature of HDPE is higher to provide more energy for reaching to these points. In addition, the melting and crystallization temperature of HDPE are close to each other, which can be confirmed from literature survey [36].

6.2. Thermal analysis by TGA

The thermal stability of polymers is an important property to be considered for fitting their performance to the proper final application. Thermogravimetric analysis is a good technique

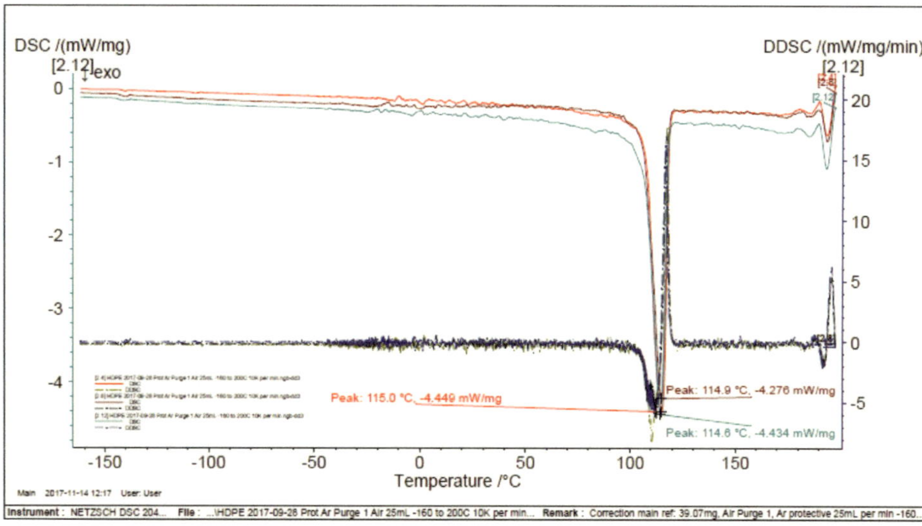

Figure 10. DSC and DDSC thermograms of HDPE for cooling cycles.

to evaluate the thermal stability of materials. Hence, in this research, the thermal stability of three polymers was studied by TGA in air and the main features of the curves including the onset temperatures of the mass loss effects (T_O) and the peak temperatures (T_p) were calculated from the TGA and DTG curves.

For bitumen, the thermogravimetric experiment results for 5 mg samples under air atmosphere over the temperature range of 30–590°C using a total purge gas flow of 100 mL/min and a heating rate of 10°C·min^{-1} show that the onset temperature of the main mass loss effect (T_O) is 370°C, as shown in **Figure 11**.

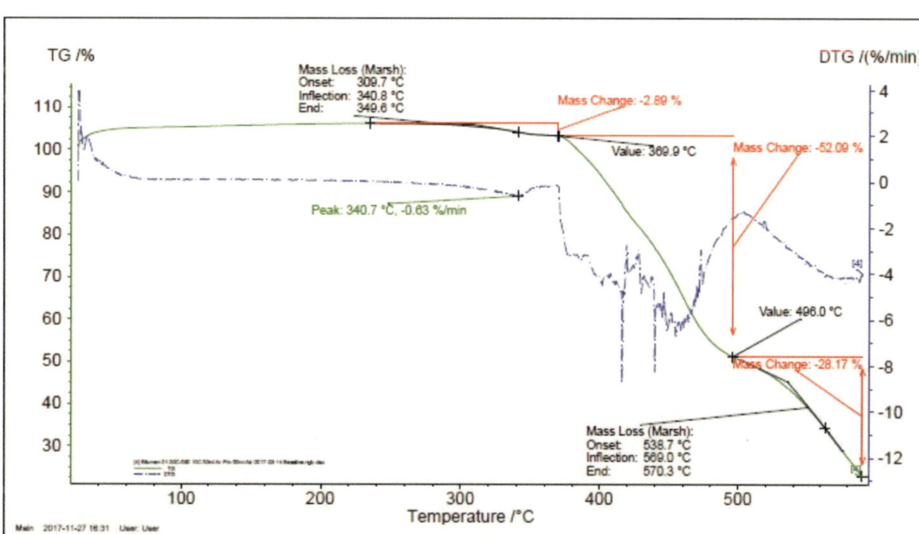

Figure 11. TGA, DTG, and D2TG thermograms of neat bitumen.

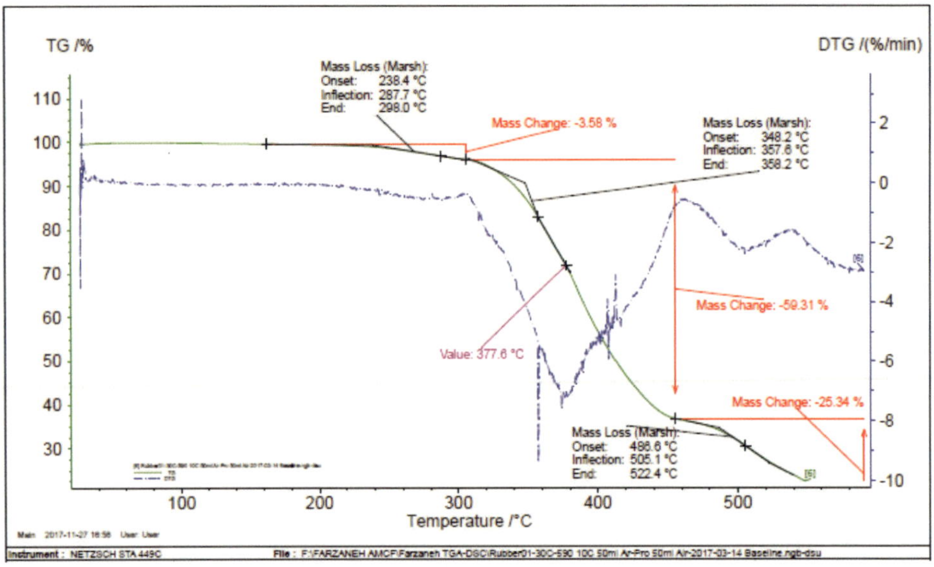

Figure 12. TGA, DTG, and D2TG thermograms of rubber.

Referring to [31], the decomposition of bitumen occurs in at least three steps, considering three temperature ranges, as shown in **Figure 11**. In the temperature range of T < 350°C, the decomposition of saturates and aromatics results in mass loss of bitumen. Over the temperature range of 350 < T < 500°C, resins and aromatics as well as asphaltenes are the main decomposed fractions, and at high temperatures of T > 500°C, the substantial mass change in bitumen occurs as a result of decomposition of asphaltenes. However, resins and aromatics are still decomposed in this range of temperature.

When selecting materials for modifying the binder, it is important that the modifier begins to degrade at a temperature above the bitumen modification temperature or the asphalt production temperature. Otherwise, it will lose its initial properties by the time the modification process is finished. In this research, TGA is used for determination of the degradation temperature of the waste materials, which are used for modifying the binder (i.e., rubber and HDPE). **Figure 12** shows the result of TGA on rubber.

As can be seen, the onset temperature of degradation for rubber is 238°C and the peak temperature of mass loss is 378°C, which can be observed as a peak in the first-derivative curve.

Similar to other polymers, TGA of the HDPE samples was done on approximately 5 mg samples over the range of room temperature to 590°C under air with 100 mL/min flow rate at a heating rate of 10°C·min^{-1}.

The onset degradation temperature and peak temperature are determined from the derivative TGA curves for HDPE, as shown in **Figure 13**. In this figure, it can be observed that HDPE remains thermally stable up to a temperature of 430°C. After this temperature, HDPE starts to degrade dramatically followed by a substantial step with maximum mass loss rates placing at

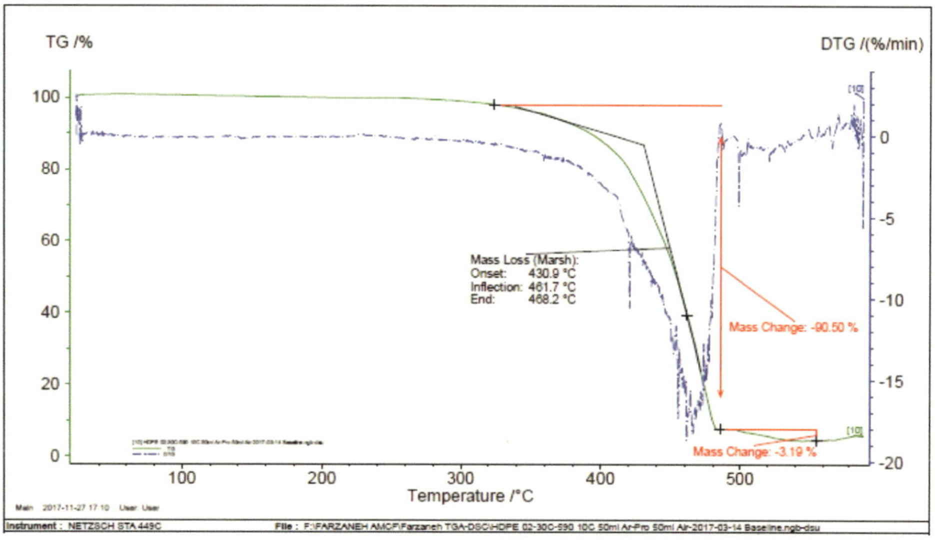

Figure 13. TGA, DTG, and D2TG thermograms of HDPE.

462°C in the DTG curve. This degradation involves a mass loss of about 91% in HDPE due to the thermal cracking of hydrocarbon chains and the production of oxygenated hydrocarbons including CO, CO_2, and H_2O [37]. The degradation ends approximately around 490°C.

6.3. Microstructure analysis by SEM

The analysis of the microstructure of polymers was performed using scanning electron microscope (SEM). The results of the microscopy of as well as the energy dispersive spectroscopy (EDS) analysis on the individual polymers are given in **Figures 14–16**.

As can be observed in **Figure 14**, the surface of bitumen appears as networks of highly entangled strings.

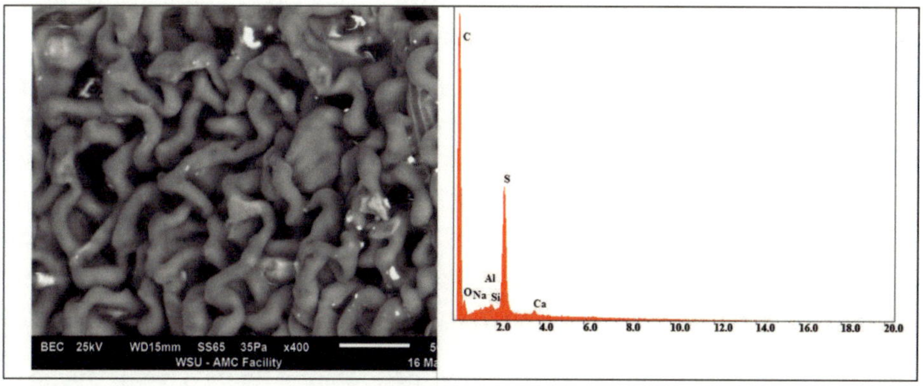

Figure 14. EDS analysis and SEM image of bitumen at 400 magnification.

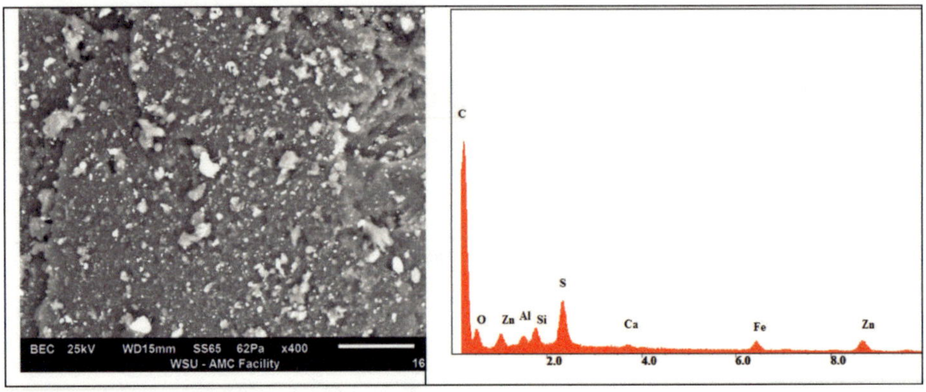

Figure 15. EDS analysis and SEM image of rubber at 400 magnification.

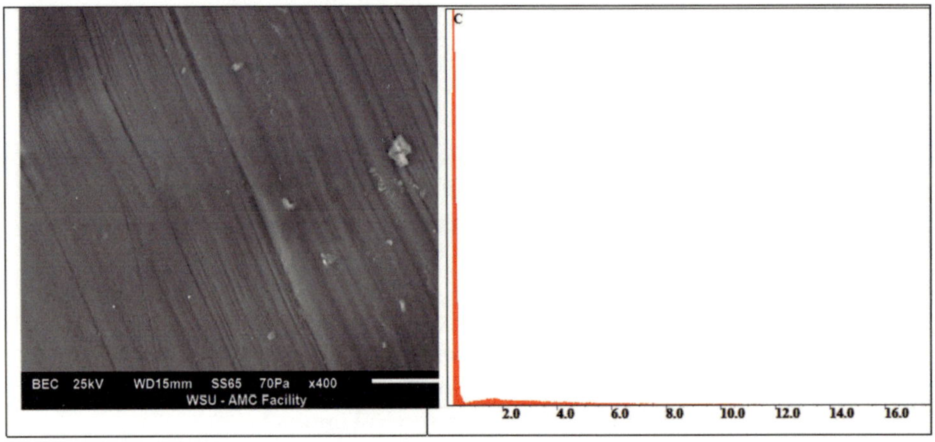

Figure 16. EDS analysis and SEM image of HDPE at 400 magnification.

Figure 15 shows the coarse texture of rubber. The irregular shape and rough texture of rubber can be attributed to its processing method, which is the ambient procedure.

Furthermore, the microstructure of HDPE is shown in **Figure 16**. It should be noted that HDPE has a higher viscosity compared to bitumen. The materials with high viscosity do not separate easily, and therefore, they present in the form of dispersed phase, as can be clearly observed in **Figure 16**.

7. Summary and conclusion

Today, pure bitumen no longer provides suitable performance for pavements due to the current traffic. Therefore, attempts have been made to maximize the effectiveness of asphalt binders selected for construction projects based on a standard asphalt binder classification system. Binder modification technique is used as an alternative to minimize the pavement failures due to poor performance of asphalt binders, as well as to increase the PG grade of the asphalt binder [38]. Based on these research studies, the utilization of polymers as modifier improves some of the bitumen's properties such as elasticity, cohesion, and temperature susceptibility, which they all subsequently lead to the improvement of asphalt mixture performance.

For these reasons, and as a quite effective way of disposing of the increasing volume of non-biodegradable wastes, which are increasingly generated in societies, plastic wastes and rubbers can be a reasonable potential materials for consideration as binder modifier.

In modification of bitumen with additives, having knowledge about the effects of modifiers on thermal stability is of high importance resulting in manufacturing more thermally stable binders. Accordingly, in this research, the thermal behavior of modifiers and bitumen was studied using TGA, DSC, and SEM facilities. The thermal parameters of T_g and T_m, enthalpy of fusion, ΔH_m, and the percentage of crystallinity, CF (%) of samples can be easily determined from the DSC curves. This information can be useful in understanding the characteristics and the composition of polymers. In addition, many researchers have proposed different equations for estimation of the glass transition temperature of mixture based on the composition and the glass transition of the components of the mixture. All these equations are basically representing the relation between the glass transition temperature of a mixture and those of its components using a basic mathematical form but with minor variations. The glass transition calculated using these equations for the blend of bitumen with 2% HDPE and 8% rubber is −28.8°C. The calculated T_g value for blend is lower than those obtained for neat bitumen, which could be attributed to several effects including certain level of miscibility between the additives (i.e., HDPE and rubber) and bitumen. Thus, using these equations, it may be possible to achieve the formulation of the desired modified binder considering the composition of components and the results of DSC analysis on each component.

Furthermore, in this research, TGA is used in determining the degradation temperature of the waste materials. A modifier that begins to degrade at a temperature below the bitumen modification temperature or the asphalt production temperature is not adequate since it will have lost its initial properties by the time the modification process is finished. In the case of analyzed waste materials, all degrade at temperatures above 200°C and therefore should be adequate for bitumen modification. The main features of TGA curves for individual polymers were discussed in previous sections. From these results, it was observed clearly that HDPE followed by bitumen have higher thermal stability than crumb rubber.

Acknowledgements

The authors would like to acknowledge the Advanced Materials Characterization Facility (AMCF) at Western Sydney University for their expert technical assistance.

Conflict of interest

The authors declare no conflict of interest.

Author details

Farzaneh Tahmoorian[1]*, Bijan Samali[1] and John Yeaman[2]

*Address all correspondence to: f.tahmoorian@westernsydney.edu.au

1 Centre for Infrastructure Engineering, Western Sydney University, Kingswood, NSW, Australia

2 Faculty of Science, Health, Education and Engineering, University of Sunshine Coast, Sippy Downs, Queensland, Australia

References

[1] Claudy P, Létoffé JM, King GN, Planche JP, Brulé B. Characterization of paving asphalts by differential scanning calorimetry. Fuel Science & Technology International. 1991;**9**:71-92

[2] Redelius P. The structure of asphaltenes in bitumen. Road Materials and Pavement Design. 2006;**7**:143-162

[3] Lesueur D. The colloidal structure of bitumen: Consequences on the rheology and on the mechanisms of bitumen modification. Advances in Colloid and Interface Science. 2009;**145**:42-82

[4] Lesueur D, Gerard JF, Claudy P, Letoffe JM, Plance JP, Martin DA. Structure related model to describe asphalt linear viscoelasticity. Journal of Rheology. 1996;**40**(5):813-836

[5] Asphalt Institute, editor. Asphalt Binder Testing MS-25. 2nd ed. USA: Asphalt Institute; 2008

[6] Yeh PH, Nien YH, Chen WC, Liu WT. Evaluation of thermal and viscoelastic properties of asphalt binders by compounding with polymer modifiers. Polym Compos. 2010;**31**:1738-1744

[7] Afroz Sultana SK, Prasad KSB. Utilization of waste plastic as a strength modifier in surface course of flexible and rigid pavements. International Journal of Engineering Resources Applications. 2012;**2**(4):1185-1191

[8] Huang Y, Bird RN, Heidrich O. A review of the use of recycled solid waste materials in asphalt pavements. Journal of Resource Conservation Recycling. 2007;**52**:58-73

[9] Casey D, McNally C, Gibney A, Gilchrist MD. Development of a recycled polymer modified binder for use in stone mastic asphalt. Journal of Resource Conservation Recycling. 2008;**52**(10):1167-1174

[10] Ghuzlan KA, Al-Khateeb GG, Qasem Y. Rheological properties of polyethylene-modified asphalt binder. Athens Journal of Technology and Engineering. 2013. Vol. X, No. Y. pp. 1-14

[11] Farukkawa T. Molecular structure, crystallinity and morphology of polyethylene/polypropylene blends studied by Raman mapping, scanning electron microscopy, wide angle X-ray diffraction, and differential scanning calorimetry. Polymer Science. 2006;**38**:1127-1136

[12] Lavin PEP. Binder Performance. Asphalt Pavements. New York, NY: Spon Press; 2003. Available from: http://www.ARRMAZ.com

[13] Vlachovicova Z, Wekumbura C, Stastna J, Zanzotto L. Creep characteristics of asphalt modified by radial styrene–butadiene-styrene copolymer. Construction Building Material Journal. 2007;**21**:567-577

[14] Verma SS. Roads from plastic waste. Science tech entrepreneur. The Indian Concrete Journal. November 2008:43-44

[15] Lucena MC, Soares SA, Soares JB. Characterization of thermal behavior of polymer-modified asphalt. Materials Research. 2004;**7**:529-534

[16] Read J, Whiteoak D. The Shell Bitumen Handbook. 5th ed. London: Thomas Telford Ltd; 2003

[17] Nellenstyen FJ. The constitution of asphalt. Journal of the Institute of Petroleum Technologists. 1924;**10**:311-325

[18] Airey GD. Chapter 23: Bitumen properties and test methods. In: ICE Manual of Construction Materials. UK: Institution of Civil Engineers; 2009

[19] Paliukaitė M, Vaitkus A, Zofka A. Influence of bitumen chemical composition and ageing on pavement performance. The Baltic Journal of Road and Bridge Engineering. 2015;**10**(1):97-104. DOI: 10.3846/bjrbe.2015.12

[20] Corbett LC. Relationship between composition and physical properties of asphalts. In: Proceedings of the Association of Asphalt Paving Technologists. 1970;**39**:481-498

[21] Halstead JC. Relation of asphalt chemistry to physical properties and specifications. In: Proceedings of the Association of Asphalt Paving Technologists. 1985;**54**:91-117

[22] Scholz TV. Durability of Bituminous Paving Mixtures [PhD thesis]. University of Nottingham; 1995

[23] Gawande A, Zamare G, Renge VC, Tayde S, Bharsakale G. An overview on waste plastic utilization in asphalting of roads. Engineering Research and Studies Journal. 2012;**3**(2):1-5

[24] Quek A, Balasubramanian R. Mathematical modeling of rubber tire pyrolysis. Journal of Analytical and Applied Pyrolysis. 2012;**95**:1-13

[25] Claudy PM, Létoffé JM, Martin D, Planche JP. Thermal behaviour of asphalt cements. Thermochimica Acta. 1998;**324**:203-213

[26] Michon LC, Netzel DA, Turner TF. A C13NMR and DSC study of the amorphous and crystalline phases in asphalts. Energy and Fuels. 1999;**13**:602-610

[27] Lu X, Redelius P. Effect of bitumen wax on asphalt mixture performance. Journal of Construction Building Material. 2007;**21**:1961-1970

[28] Claudy P, Létoffé JM, Germanaud L, King GN, Planche JP, Ramond G, Such C, Buisine JM, Joly G, Edlanio A. Thermodynamic behavior and physicochemical analysis of eight SHRP bitumens. Transportation Research Board. 1993:1-9

[29] Bosselet F, Létoffé JM, Claudy P, Valentin P. Study of the Thermal Behavior of n-Alkanes in Complex Hydrocarbon Media by Differential Calorimetric Analysis. Thermochimica Acta. 1983;**70**:19

[30] Letoffe JM, Claudy P, Kok MV, Garcin M, Volle JL. Crude oils – characterization of waxes precipitated on cooling by DSC and thermomicroscopy. Fuel. 1995;**74**:810

[31] Jiménez-Mateos JM, Quintro LC, Rial C. Characterization of petroleum bitumens and their fractions by thermogravimetric analysis and differential scanning calorimetry. Fuel. 1996;**75**:1691-1700

[32] Harrison IR, Wang G, Hsu TC. A differential scanning calorimetry study of asphalt binders. Strategic Highway Research Program, National Research Council. SHRP-A/UFR-92-612. Washington, DC; 1992

[33] Masegosa Fanego RM, Cañamero P, Sánchez Cabezudo M, Viñas Sánchez MT, Salom Coll C, Gonzáez Prolongo M, Paez A, Ayala M. Thermal behaviour of bitumen modified by sulphur addition. In: 5th Eurasphalt and Eurobitume Congress. 13-15 June 2012; Istanbul. Turkey.pp. 142-148. ISBN 978-972-8692-45-2

[34] Banat R, Fares MM. Thermo-gravimetric stability of high density polyethylene composite filled with olive shell flour. American Journal of Polymer Science. 2015;**5**(3):65-74. DOI: 10.5923/j.ajps.20150503.02

[35] Mirabella FM, Bafna A. Determination of crystallinity of polyethylene/&alpha-olefin copolymers by thermal analysis relationship of the heat of fusion of 100% polyethylene crystal and the density. American Journal of Polymer Science. Part B: Polymer Physics. 2002;**40**:1637-1643

[36] Ashraf A. Thermal analysis of polymers (LDPE, HDPE) by differential scanning calorimetry technique [thesis]. University of Qatar; 2014

[37] Purohit V, Orzel RA. Polypropylene: A literature review of the thermal decomposition products and toxicity. International Journal of Toxicology. 1998;**2**:221-242

[38] Somayaji S. Civil Engineering Materials. Upper Saddle River, NJ: Prentice Hall; 2001

AF794094

IntechOpen

Supporting open minds since 2005

Modern Applications of Electrostatics and Dielectrics
http://dx.doi.org/10.5772/intechopen.78806
Edited by Dengming Xiao and Krishnaswamy Sankaran

Contributors
Chao Wang, Katsuyuki Takahashi, Isao Hiyoshi, Yosuke Enomoto, Shinichi Yamaguchi, Hidemi Nagata, Koichi Takaki, Reza Haghighi Osgouei, Yan Yang, Guangning Wu, Muhammad Mukhlisin, Almushfi Saputra, Zion Menachem, Su Zhao, Dengming Xiao, Krishnaswamy Sankaran, Yifan Kang, Yonglin Bai

© The Editor(s) and the Author(s) 2020
The rights of the editor(s) and the author(s) have been asserted in accordance with the Copyright, Designs and Patents Act 1988. All rights to the book as a whole are reserved by INTECHOPEN LIMITED. The book as a whole (compilation) cannot be reproduced, distributed or used for commercial or non-commercial purposes without INTECHOPEN LIMITED's written permission. Enquiries concerning the use of the book should be directed to INTECHOPEN LIMITED rights and permissions department (permissions@intechopen.com).
Violations are liable to prosecution under the governing Copyright Law.

(cc) BY

Individual chapters of this publication are distributed under the terms of the Creative Commons Attribution 3.0 Unported License which permits commercial use, distribution and reproduction of the individual chapters, provided the original author(s) and source publication are appropriately acknowledged. If so indicated, certain images may not be included under the Creative Commons license. In such cases users will need to obtain permission from the license holder to reproduce the material. More details and guidelines concerning content reuse and adaptation can be found at http://www.intechopen.com/copyright-policy.html.

Notice
Statements and opinions expressed in the chapters are these of the individual contributors and not necessarily those of the editors or publisher. No responsibility is accepted for the accuracy of information contained in the published chapters. The publisher assumes no responsibility for any damage or injury to persons or property arising out of the use of any materials, instructions, methods or ideas contained in the book.

First published in London, United Kingdom, 2020 by IntechOpen
IntechOpen is the global imprint of INTECHOPEN LIMITED, registered in England and Wales, registration number: 11086078, 7th floor, 10 Lower Thames Street, London, EC3R 6AF, United Kingdom
Printed in Croatia

British Library Cataloguing-in-Publication Data
A catalogue record for this book is available from the British Library

Additional hard and PDF copies can be obtained from orders@intechopen.com

Modern Applications of Electrostatics and Dielectrics
Edited by Dengming Xiao and Krishnaswamy Sankaran
p. cm.
Print ISBN 978-1-83880-301-8
Online ISBN 978-1-83880-302-5
eBook (PDF) ISBN 978-1-83962-984-6

For EU product safety concerns:
IN TECH d.o.o., Prolaz Marije Krucifikse Kozulić 3, 51000 Rijeka, Croatia, info@intechopen.com or visit our website at intechopen.com.

We are IntechOpen, the world's leading publisher of Open Access books

Built by scientists, for scientists

4,900+
Open access books available

124,000+
International authors and editors

140M+
Downloads

151
Countries delivered to

Our authors are among the
Top 1%
most cited scientists

12.2%
Contributors from top 500 universities

CLARIVATE ANALYTICS BOOK CITATION INDEX INDEXED

WEB OF SCIENCE™

Selection of our books indexed in the Book Citation Index in Web of Science™ Core Collection (BKCI)

Interested in publishing with us?
Contact book.department@intechopen.com

Numbers displayed above are based on latest data collected.
For more information visit www.intechopen.com

Meet the editors

Dengming Xiao was born in Guangxi Province, China in 1953. He received a PhD from the School of Electrical Engineering, Xi'an Jiaotong University, China, in 1994. Currently, he is Professor in the Department of Electrical Engineering, Shanghai Jiao Tong University, Shanghai, China. He has worked on gas insulation systems, electrical equipment online monitoring, and fault diagnosis for more than 20 years. His current focus is on environmental friendly substitutes for the insulating medium SF6. He has presided over a number of projects supported by the Natural Science Foundation of China and has published more than 150 papers.

Dr. Krishnaswamy Sankaran is the Chief Executive Officer of Radical Innovations Group AB, a clean energy and circular economy infrastructure development company based in Finland. He is the winner of the 2020 Mission Innovation Champion Award under the flagship of 24 countries around the world plus European Commission for his pioneering work in Clean Energy and Circular Economy. He is an established industrialist and entrepreneur with several years of demonstrated track record in energy, utilities, infrastructure, manufacturing, and recycling industries. He has served in various operational and leadership roles in industries, the World Economic Forum, and the European Commission in more than sixteen countries and four continents. He is nominated by the Government of Finland for the prestigious 2020 Mission Innovation Champion Award, a global recognition for clean energy innovation (http://mission-innovation.net/). Dr. Sankaran received a doctorate in Engineering Sciences from the Swiss Federal Institute of Technology ETH Zurich, Switzerland, a master's degree from Karlsruhe Institute of Technology KIT, Germany, and a joint executive master's degree in Organisational Development and Leadership from Wharton School, Columbia University, INSEAD, and London Business School. He has several years of training in yoga and Advaita Vedanta, an Indian traditional system of philosophy.

Contents

Preface

Electrostatics and dielectrics have important applications in modern society. The research on electrostatics and dielectrics has been very active, and the emergence and development of modern science and technology has made a great contribution to the in-depth study of various dielectric materials and electrostatics. This research has played a great role in promoting the further production and development of science and technology. However, with the continuous development of modern society, better characteristic requirements have been put forward for dielectric materials and electrostatics. More and more equipment needs to operate at high frequency, high voltage, high temperature, and other harsh conditions. Therefore dielectric materials need high performance requirements as well as good work stability. This book contains the research progress and application status of electrostatics and dielectrics in different fields.

There are eight chapters in the book. The first chapter is the introductory chapter of modern applications of electrostatics and dielectrics. The second chapter is about the electrostatic potential modulation of atoms. The third chapter examines the development of a new electrostatic eliminator using a pulse width modulation (PWM) inverter. The fourth chapter discusses the electrostatic effect application of touch screens. The fifth chapter analyzes the variable frequency motor insulation failure mechanism and improvement measures. The sixth chapter is about the dielectric analyzing model of soil moisture content. The seventh chapter discusses dielectric materials in the study of the calculation method of the rectangular waveguide. Finally, the last chapter is about the study of synergistic effect of mixed-gas dielectrics.

Dengming Xiao
Shanghai Jiao Tong University,
China

Krishnaswamy Sankaran
Radical Innovations Group - RIG,
Finland

Chapter 1

Introductory Chapter: Modern Applications of Electrostatics and Dielectrics

Krishnaswamy Sankaran

1. Introduction

Electrostatics is one of the most basic, yet very important, branches of physics with applications in almost every domain. We observe how a plastic comb after its use on dry hair attracts small bits of paper. We are stunned by the power of lightning from dark rainy clouds during thunder storms. Time to time, we might have also experienced small electric shocks when we touch any metal after rubbing hands on certain materials. All these experiences have to do with the underlying phenomena of electrostatics.

Materials are broadly classified into two large groups based on their electrical properties. Some are called conductors and the others are termed as insulators (or dielectrics). In dielectrics, contrary to metals, all charges are attached to specific atoms and molecules. Such charges are known as bound charges. These charges can be displaced within an atom or a molecule. Compared to the dramatic rearrangements of charges in a conductor under electric field, charge displacements in dielectric materials are only microscopic in nature. However, these microscopic displacements can have cumulative effects that are responsible for several characteristic behaviours of dielectric materials.

Electrostatics-based applications are ubiquitous. Lightning conductors, surface coaters, electrostatic imagers, non-impact printers, industrial processes such as material separation, electrodialysis, static dischargers, etc., are some of the most common applications [1–5].

Dielectrics have been used for developing various devices particularly using their unique material properties. Recent advancements in material science have made it possible to develop and engineer devices for applications in elasto-optics, electro-optics, ultrasonic [6] and surface-mount electronics [7]. Dielectrics are also used to develop compact and efficient antennas [8]. There is an increase in the use of dielectrics for developing advanced defence applications [9]. Dielectric material properties can carefully engineered to develop artificial materials and coatings that can efficiently absorb or dampen incident electromagnetic waves. Such materials are both used in real-time stealth applications [10, 11] and modelling absorbing layers to truncate computational domains [12].

Accurate modelling of dielectric properties is one of the most challenging tasks. For basic applications like studying oil spills on ocean surface, one can easily model oil spills as dielectric with complex permittivity [13]. Other advanced applications require more detailed modelling of dielectric properties. For example, dielectric elastomers are actively studied for applications involving artificial muscle actuators for robots and solid-state optical devices for various electronic components [14].

 IntechOpen

These advanced applications of electrostatics and dielectrics rely on accurate modelling methods and tools. The choice of right electromagnetic tools for modelling is an important challenge for design and application engineers [15, 16]. Various advanced modelling tools are being developed for modelling such applications that includes some of the classical computational methods [17, 18], and non-mainstream computational methods [19, 20].

In this book, we have introduced seven advanced applications of electrostatics and dielectrics. These papers bring forth recent developments using electrostatics and dielectrics. In the first paper on electrostatic potential modulation of atoms by strong light field, Wang presents tunnelling ionization and single isolated attosecond light pulse application. In the second paper, Takahashi et al. discuss development of an electrostatic eliminator utilizing high-voltage AC power supply-driven by pulse width modulation (PWM) inverter. In the third paper, Osgouei shows how electrostatic friction mechanism can be used in display technology to enhance touchscreen experience. In the fourth paper, Wu and Yang investigate dielectric failure mechanism and property modification in inverter-fed motors. Mukhlisin and Saputra elaborate their dielectric analysis model for measuring soil moisture (water content) using electric capacitance volume tomography in the fifth paper. Menachem presents techniques for calculating dielectric maternal parameters for waveguide applications in the sixth paper. In the final paper of this book, Xiao and Zhao discuss research progress on synergistic effect between insulation gas mixtures.

I believe this compendium of research papers in the domain of electrostatics and dielectrics will open new opportunities for future research. I thank all the authors for their collaborative efforts and excellent contribution. Final words of gratitude are due to IntechOpen for making this work possible. Good luck!

Author details

Krishnaswamy Sankaran
Radical Innovations Group AB, Vaasa, Finland

*Address all correspondence to: krish@sankaran.org

IntechOpen

© 2020 The Author(s). Licensee IntechOpen. This chapter is distributed under the terms of the Creative Commons Attribution License (http://creativecommons.org/licenses/by/3.0), which permits unrestricted use, distribution, and reproduction in any medium, provided the original work is properly cited.

References

[1] Bright AW. Modern electrostatics. Physics Education. 1974;**9**(6):381. DOI: 10.1088/0031-9120/9/6/303

[2] Sharp KA, Honig B. Electrostatic interactions in macromolecules: Theory and applications. Annual Review of Biophysics and Biophysical Chemistry. 1990;**19**(1):301-332. DOI: 10.1146/annurev.bb.19.060190.001505

[3] Haga K, Chang JS, Kelly AJ, Crowley JM. Applications of the electrostatic separation technique. In: Handbook of Electrostatic Processes. New York: Marcel Dekker; 1995. pp. 365-386

[4] Jeon JU, Higuchi T. Electrostatic suspension of dielectrics. IEEE Transactions on Industrial Electronics. 1998;**45**(6):938-946. DOI: 10.1109/41.735338

[5] Castle GS. Industrial applications of electrostatics: The past, present and future. Journal of Electrostatics. 2001;**51**:1-7. DOI: 10.1016/S0304-3886(01)00068-7

[6] Spencer EG, Lenzo PV, Ballman AA. Dielectric materials for electrooptic, elastooptic, and ultrasonic device applications. Proceedings of the IEEE. 1967;**55**(12):2074-2108. DOI: 10.1109/PROC.1967.6087

[7] Hennings D, Klee M, Waser R. Advanced dielectrics: Bulk ceramics and thin films. Advanced Materials. 1991;**3**(7-8):334-340. DOI: 10.1002/adma.19910030703

[8] Fumeaux C, Almpanis G, Sankaran K, Baumann D, Vahldieck R. Finite-volume time-domain modeling of the mutual coupling between dielectric resonator antennas in array configurations. In: The Second European Conference on Antennas and Propagation (EuCAP). 2007. pp. 1-4. DOI: 10.1049/ic.2007.0909

[9] Sankaran K. Recent trends in computational electromagnetics for defence applications. Defence Science Journal. 2019;**69**(1):65-73. DOI: 10.14429/dsj.69.13275

[10] Musal HM, Hahn HT. Thin-layer electromagnetic absorber design. IEEE Transactions on magnetics. 1989;**25**(5):3851-3853. DOI: 10.1109/20.42454

[11] Abbas SM, Chandra M, Verma A, Chatterjee R, Goel TC. Complex permittivity and microwave absorption properties of a composite dielectric absorber. Composites Part A: Applied Science and Manufacturing. 2006;**37**(11):2148-2154. DOI: 10.1016/j.compositesa.2005.11.006

[12] Sankaran K, Kaufmann T, Fumeaux C, Vahldieck R. Different perfectly matched absorbers for conformal time-domain method: A finite-volume time-domain perspective. In: 23rd Annual Review of Progress in Applied Computational Electromagnetics. 2007. pp. 1712-1718

[13] Sankaran K, Fortuny-Guasch J. Radar remote sensing for oil spill classification (optimization for enhanced classification). In: Proceedings of the 12th IEEE Mediterranean Electrotechnical Conference MELECON. 2004. pp. 511-514. DOI: 10.1109/MELCON.2004.1346979

[14] Pelrine R, Sommer-Larsen P, Kornbluh RD, Heydt R, Kofod G, Pei Q, et al. Applications of dielectric elastomer actuators. In: Smart Structures and Materials 2001: Electroactive Polymer Actuators and Devices. Vol. 4329. International Society for Optics and Photonics; 2001. pp. 335-349. DOI: 10.1117/12.432665

[15] Sankaran K. Are you using the right tools in computational

electromagnetics? Engineering Reports. 2019;**1**(3):1-19. DOI: 10.1002/eng2.12041

[16] Aakash, Bhatt A, Sankaran K. Transcending limits: Recent trends & challenges in computational electromagnetics. In: Proceedings of IEEE-INAE Workshop on Electromagnetics–IIWE. 2018

[17] Yu W, Mittra R. A conformal finite difference time domain technique for modeling curved dielectric surfaces. IEEE Microwave and Wireless Components Letters. 2001;**11**(1):25-27. DOI: 10.1109/7260.905957

[18] Brosseau C, Beroual A. Computational electromagnetics and the rational design of new dielectric heterostructures. Progress in Materials Science. 2003;**48**(5):373-456. DOI: 10.1016/S0079-6425(02)00013-0

[19] Sankaran K. Beyond DIV, CURL and GRAD: Modelling electromagnetic problems using algebraic topology. Journal of Electromagnetic Waves and Applications. 2017;**31**(2):121-149. DOI: 10.1080/09205071.2016.1257397

[20] Aakash, Bhatt A, Sankaran K. How to model electromagnetic problems without using vector calculus and differential equations? IETE Journal of Education. 2018;**59**(2):85-92. DOI: 10.1080/09747338.2018.1554456

Chapter 2

Coulomb Potential Modulation of Atoms by Strong Light Field: Electrostatic Tunneling Ionization and Isolated Attosecond Light Pulse

Chao Wang, Yifan Kang and Yonglin Bai

Abstract

An attosecond research upsurge has been overwhelmingly rising since the establishment of novel light source—single isolated attosecond laser in extreme ultraviolet/X-ray resulted by strong field high-order harmonics generation (HHG). In this chapter, based on the electrostatic tunneling ionization from Coulomb potential modulation of atoms by strong light field, we scrutinized the intrinsic phase of high-order harmonics and analyzed qualitatively the salient dependence of two mainstream single isolated attosecond pulse generation techniques as polarization gating(PG) and amplitude gating(AG) on carrier-envelope phase (CEP) of femtosecond driving laser. The conclusion is that the optimized CEP corresponding to the highest intensity contrast between the main and sideband attosecond pulses is $\pi/2$ and 0 for polarization gating and amplitude gating, respectively. Further, an experimental implementation was given in detail to exemplify the tricks for optimum phase-matching process of HHG from the interaction of high-intensity femtosecond laser field with noble gas target. The effects of the relative location between Gaussian-shaped driving femtosecond laser field focus and the gas target source used on the HHG phase matching were studied, and the conclusion found that the expected position of gas target for optimum phase matching is always lying behind the focal point of the driving field used.

Keywords: tunneling ionization, high-order harmonics generation, single isolated attosecond pulse, carrier-envelope phase, phase matching, few-cycle femtosecond laser

1. Introduction

With the invention in general and realization of the laser in particular by Maiman in 1960, field of optics soon entered the new era of nonlinear optics. In this regime, the optical properties of materials are no longer independent of the intensity of light—as was believed for hundreds of years before—but rather change with light intensity, giving rise to a wealth of new phenomena, effects, and applications. Today, nonlinear optics has entered our everyday life in many ways and has also been the basis for numerous new developments in spectroscopy and laser technology. Indeed, from the moment of birth of nonlinear optics, laser physics and nonlinear optics have been intimately related to each other. Within "traditional" nonlinear optics, the absolute changes of the optical properties are tiny if one follows them versus time on a timescale of a cycle of light. This simple fact is the basis of many concepts and approximations of "traditional" nonlinear optics. Over the years, however, lasers have improved in many ways, especially in terms of the accessible peak intensities and in terms of the minimum pulse duration available. These days, over 50 years after the invention of the laser, the shortest optical pulses generated are about one and a half cycles of light in duration. This comes close to the ultimate limit of a single optical cycle [1].

Thanks to the technique of chirped-pulse amplification (CPA) and the availability of CEP stabilized few-cycle laser [2], amplified laser pulses with focused peak intensities in the range of 10^{22} W/cm^2 are available in some laboratories. As a result of this, today's light intensities can lead to substantial or even to extreme changes on the timescale of light, and the range of optics research is now shifting from the stage of perturbative to extreme (or nonperturbative) nonlinear optical mechanism [3, 4]. As for the latter, one remarkable achievement is the establishment of novel light source—single isolated attosecond (1 attosecond[as] = 10^{-18} s) laser pulse in extreme ultraviolet or even X-ray electromagnetic spectrum based on strong field HHG [5–7]. Since the fundamental processes of chemistry, biology, and materials science are triggered or mediated by the motion of electrons inside or between atoms, and that the atomic-scale motion of electrons typically unfolds within tens to thousands of attoseconds, breakthroughs in light source science are now opening the door to watching and controlling these hitherto inaccessible microscopic dynamics, opening up a new horizon of science by observing, controlling, and manipulating nature in a new dimension. Consequently, an attosecond research upsurge is overwhelmingly rising from physics, chemistry, and material science to even information processing and other fields [8–10]. What is worth mentioning is that transnational or transregional attosecond science research infrastructures are emerging, such as Extreme Light Infrastructure Attosecond Light Pulse Source (ELI-ALPS) in Szeged, Hungary. Its main objective is to establish a unique attosecond laser facility which provides developers and users with light sources within the THz to X-ray frequency range in the form of ultrashort pulses at high repetition rate.

Until now, the HHG based on the interaction between high-intensity few-cycle femtosecond laser and noble gases is regarded as the main process for generating attosecond laser pulse. The physical process in essence is frequency up-conversion resulted by optical field caused by atomic tunneling ionization, which means that, no matter what kind of single isolated attosecond pulse generation technique, such as amplitude gating [11] or polarization gating [12], the phase matching involved is the key issue, which affects the macroscopic response of the multiatom gas system and eventually determines the conversion efficiency. According to the three-step scenario of HHG proposed by Corkum [13], phase matching in process of HHG is microcosmically influenced by multiple factors involved: fundamental driving laser field, the gas system used, the high-order harmonics field that generated, and the plasma formed by the atomic ionization. It is very difficult and even impossible to carry out such a complete and accurate analysis on phase matching of HHG. So far, most theoretical analysis are addressing incompletely on one or two factors mentioned above, while the experimental demonstrations are just concerned about high-order harmonics yield, omitting intentionally or unintentionally the important details of phase matching optimization [14, 15]. Consequently, the experimental details to optimize HHG attosecond laser pulse generation are still not made clear.

In this chapter, based on the three-step scenario of HHG, we scrutinized the intrinsic phase of high-order harmonics from the atom tunneling ionization induced by strong laser field and analyzed qualitatively the salient dependence of two fundamental single attosecond pulse generation techniques as polarization gating and amplitude gating on CEP of femtosecond driving laser. Finally, an experimental implementation was given in great detail to show the way to realize optimum HHG phase matching during interaction of strong optical field with noble gas target.

2. Strong field tunneling ionization and HHG

The light intensities necessary to rapidly ionize an atom are on the order of $I \approx 10^{14} \sim 10^{16}\ W/cm^2$, which is well within the nonrelativistic regime. Thus, we can ignore the laser magnetic field for the moment and focus on the laser electric field [3]. **Figure 1** visualizes this kind of ionization by high-intensity few-cycle femtosecond laser pulse for an electron bound in the Coulomb potential of a nucleus (assumed to be much more massive or fixed in space). The Gaussian linearly polarized laser pulse of $t_{FWHM} = 5fs$ is characterized by its electric field $E(t) = \tilde{E}(t)\, cos\,(\omega_0 t)$ with carrier photon energy $\hbar\omega_0 = 1.5\, eV$, together with the two-dimensional scheme of the resulting electric potential experienced by an electron initially bound in an atom at three characteristic points in time. The large "tilt" along the electric field vector axis in the center of the pulse can lead to tunneling of the electron out of its binding potential through the potential barrier. If the barrier height is lowered below the binding energy, above-barrier ionization can occur. For circularly polarized light, the "tilt" stays constant, but its axis rotates in time [1].

Whether we can use the concept of electrostatic tunneling for light fields oscillating depends on the ratio between the light field period and the time the electron spends within the barrier, electron tunneling time. If the tunneling time is shorter than the period of light, the laser electric field can indeed be viewed as a static field along its polarization direction that parametrically changes its instantaneous value, which is the case of HHG process-based single attosecond pulse generation. Within the "static-field approximation," the tunneling ionization rate $\Gamma_i(t)$ depends exponentially on the instantaneous barrier width $l(t)$ because the electron wave function

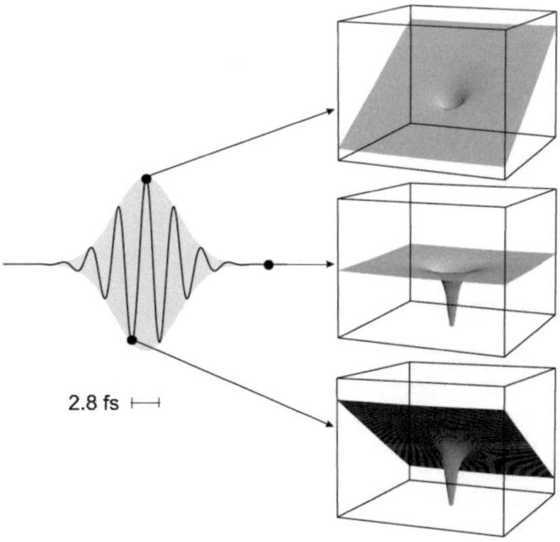

Figure 1.
Coulomb potential modulation of atoms by strong light field.

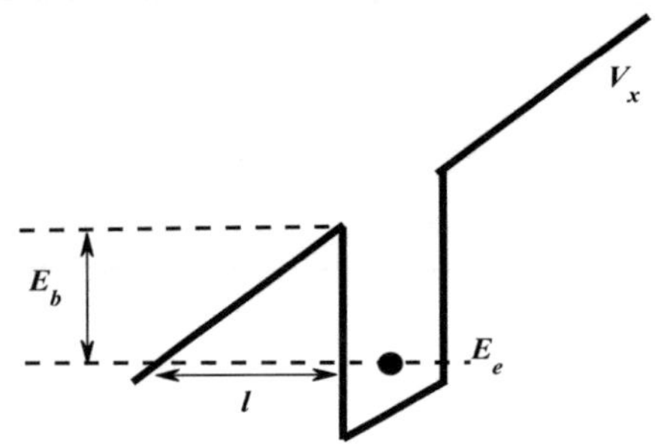

Figure 2.
Strong field-induced tunneling ionization.

is decaying exponentially in the barrier according to $\psi(x) \propto exp(-|k_x|x)$, as in **Figure 2**. The probability of tunneling through the barrier is $\propto|\psi(l)|^2$. Further, one can expect the general behavior $\Gamma_i(t)$ as

$$\Gamma_i(t) \propto e^{-\frac{E_b^{\frac{3}{2}}\sqrt{2m_e}}{\hbar e|E(t)|}l(t)} \tag{1}$$

Apparently, the ionization characteristics would definitely have a strong dependence on CEP of the exciting laser pulses.

To get the exact description of the HHG spectrum, one can in principle solve the time-dependent Schrödinger equation numerically for the ionization and compute the atomic dipole moment from the known instantaneous wave function $\Psi(\boldsymbol{r},t)$ via the expectation value $\langle\Psi(\boldsymbol{r},t)|-e\boldsymbol{r}|\Psi(\boldsymbol{r},t)\rangle$. Multiplying by the atoms' density delivers the macroscopic optical polarization $\boldsymbol{P}$, which is one of the main parameters in the Maxwell equations. Neglecting the propagation effects, the radiated electric field is proportional to the second temporal derivative of $\boldsymbol{P}$. The square modulus of its Fourier transform delivers the intensity spectrum of high-order harmonics. The most intuitive way to discuss semiclassically the extreme nonlinear interaction that leads to attosecond pulses is through the so-called three-step scenario introduced by Corkum [13], as in **Figure 3**. The first step (①) is the atom ionization via tunneling through the distorted Coulomb potential. Now the electron wave packet is released in the continuum where it gets accelerated by the external field (②). The third step (③) is the possible recombination with the nucleus and emission of a high-energy photon.

As for the second step, the movement of the electron wave packet can be described using Newton's second theorem of classical mechanics. Again, the magnetic field of the laser and the Coulomb potential bound on the electrons are ignored, and only the electric field of the laser is considered. And also we neglect the spatial dependence of the electric field since the wavelength of the laser field is much bigger than the distance the electron moves within the external field. We assume a laser field that the atom is exposed to is $E(t) = E_0 \cos(w_L t)$. The electron tunnels through the potential wall and is released into the continuum with zero velocity at position zero at birth time $t = t_i$, $v(t_i) = 0$,and $x(t_i) = 0$. When we assume that the only force experienced by the electron is $F(t) = m_e x'' = -eE(t)$, the expression for $x(t)$ follows by integrating.

$$v(t) = \frac{eE_0}{m_e w_L}[\sin(w_L t) - \sin(w_L t_i)], \tag{2}$$

$$x(t) = \frac{eE_0}{m_e {w_L}^2}[\cos(w_L t) - \cos(w_L t_i) + w_L(t - t_i)\sin(w_L t_i)]. \tag{3}$$

Now one can use Eq. (3) to calculate the return time t_r of the electron for different tunneling times t_i. Since the process is periodic in time with a periodicity of π/w_L, we use $0 \leq t_i \leq 0.5T = \pi/w_L$. Using Eq. (3), one can easily get that only those electrons freed at $0 \leq t_i \leq 0.25T$ have the chance to return to the nucleus, as illustrated in **Figure 4**. Intuitively, the return time t_r of electron with birth time t_i is the horizontal ordinate of the intersection point formed by the laser field curve and its tangent line passing the point with $t = t_i$. And two important values for analyzing HHG are free time $\tau = t_r - t_i$ that the electron spends in the continuum and the return energy W_{kin}.

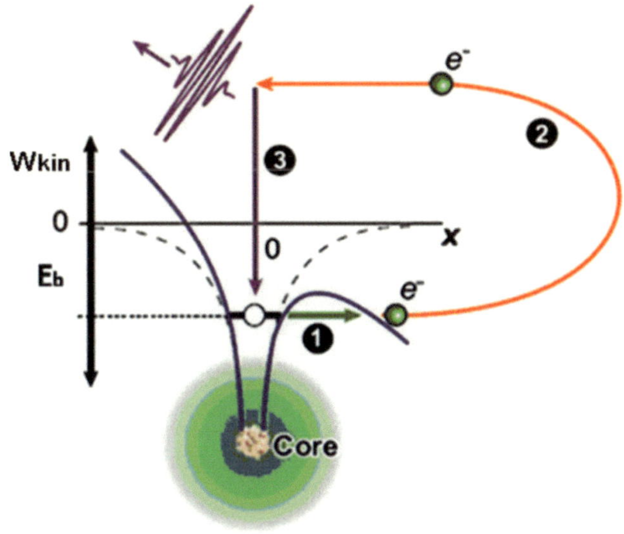

Figure 3.
Three-step scenario of high-harmonic generation.

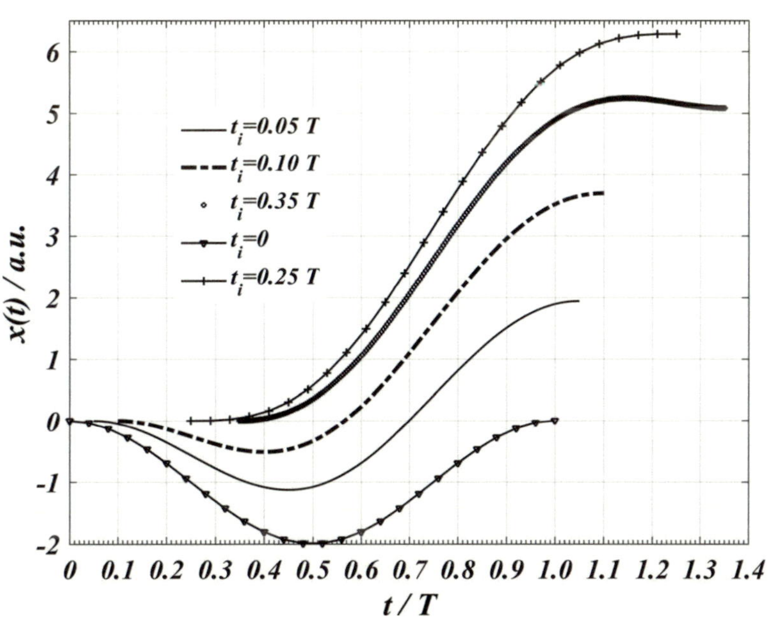

Figure 4.
Wiggling motion of electrons in external laser field.

$$W_{kin}(t_i) = 2U_p[sin\,(w_L t_r) - sin\,(w_L t_i)]^2, \quad (4)$$

$$t_i = \frac{1}{w_L} atan\left[\frac{-w_L\tau + sin\,(w_L\tau)}{1 - cos\,(w_L\tau)}\right]. \quad (5)$$

Here $U_p = e^2{E_0}^2/4m_e{w_L}^2$ is the wiggling electron's ponderomotive energy which is the period-averaged kinetic energy in external laser field $E(t) = E_0\,cos\,(w_L t)$.

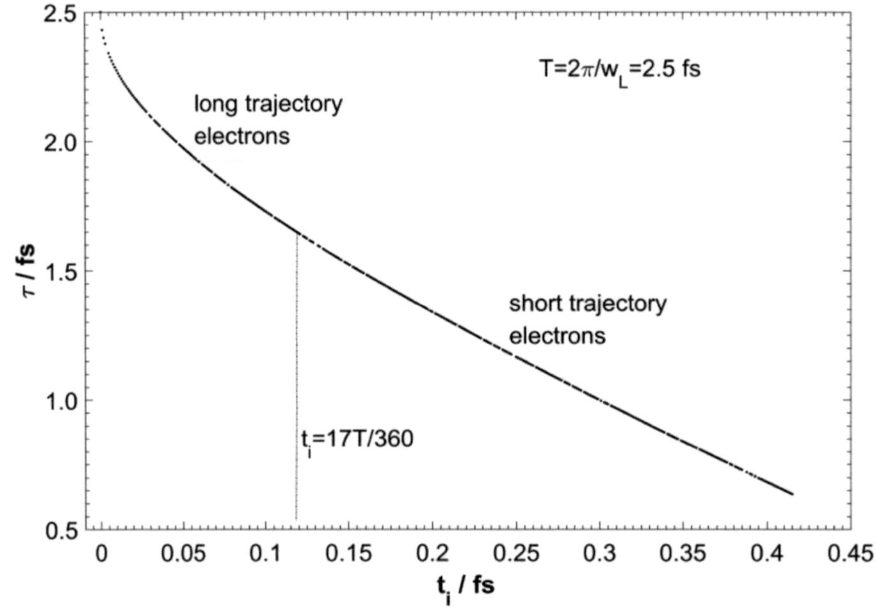

Figure 5.
Electron's free time during the process of HHG.

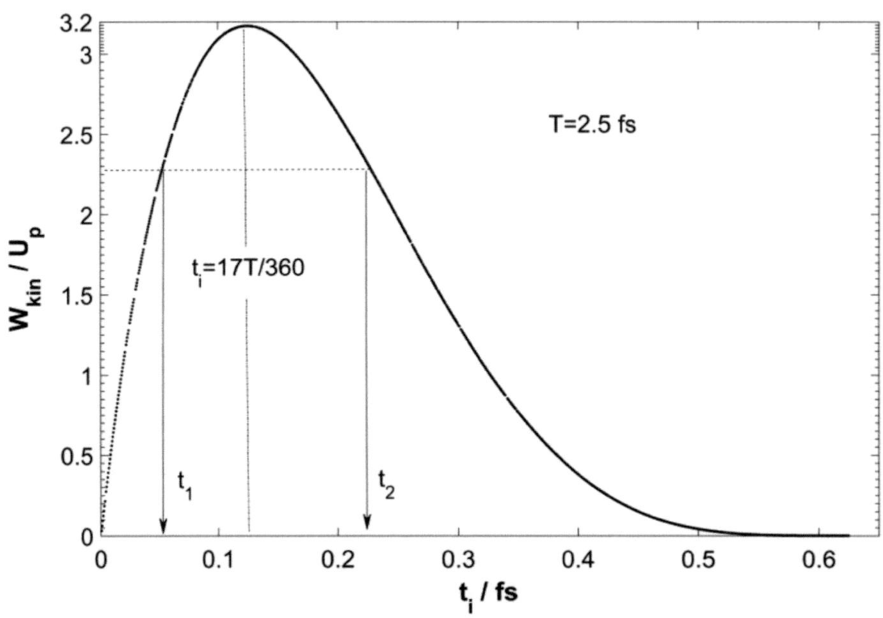

Figure 6.
Electron's return energy during the process of HHG.

The electrons with different freed time t_i are bound to have different free times τ but might have equal return kinetic energy, as shown in **Figures 5** and **6**. Taking the characteristic time $t_m = 17T/360$ as the boundary, which corresponds to the maximum return kinetic energy of 3.17 U_p, electrons freed earlier are called long trajectory electrons, while the others short trajectory electrons. The intrinsic phase of high-order harmonics resulted from the latter definitely has linear chirp, since the parameter τ linearly determining the phase has linear feature.

3. CEP dependence of single attosecond pulse generation

Single isolated attosecond pulses originated from HHG process have been demonstrated experimentally by a variety of gating techniques, which include spectral selection of half-cycle cutoffs as in amplitude gating (AG) [11, 16, 17] and ionization gating (IG) [18–21], temporal gating techniques such as polarization gating (PG) (including double optical gating) [12, 22–27], and spatiotemporal gating with the attosecond lighthouse effect [28, 29]. Based on the intuitive description of HHG as the three-step scenario, PG and AG are the two most fundamental techniques, of which the former utilizes the dependence of HHG yield on driving field ellipticity, while on the latter the essential dependence of HHG itself is on driving field intensity.

PG involves the synthesis of laser field tailored specifically with time-dependent ellipticity. The left and right circularly polarized Gaussian laser fields used are as follows:

$$\vec{E}_{LC}(t) = E_0\left[e^{-2\ln(2)\frac{(t-T_d/2)^2}{\tau_\omega^2}}\right]\left[\cos\left(\frac{2\pi}{T}t + CEP\right)\hat{x} + \sin\left(\frac{2\pi}{T}t + CEP\right)\hat{y}\right], \tag{6}$$

$$\vec{E}_{RC}(t) = E_0\left[e^{-2\ln(2)\frac{(t+T_d/2)^2}{\tau_\omega^2}}\right]\left[\cos\left(\frac{2\pi}{T}t + CEP\right)\hat{x} - \sin\left(\frac{2\pi}{T}t + CEP\right)\hat{y}\right]. \tag{7}$$

Here T_d is the time delay between the two laser pulses. So one can get the field synthesized and its ellipticity:

$$\begin{aligned}\vec{E}_C(t) = {} & E_0\left[e^{-2\ln(2)\frac{(t-T_d/2)^2}{\tau_\omega^2}} + e^{-2\ln(2)\frac{(t+T_d/2)^2}{\tau_\omega^2}}\right]\cos\left(\frac{2\pi}{T}t + CEP\right)\hat{x} \\ & + E_0\left[e^{-2\ln(2)\frac{(t-T_d/2)^2}{\tau_\omega^2}} - e^{-2\ln(2)\frac{(t+T_d/2)^2}{\tau_\omega^2}}\right]\sin\left(\frac{2\pi}{T}t + CEP\right)\hat{y},\end{aligned} \tag{8}$$

$$\xi(t) = \frac{\left|1 - e^{-4\ln(2)\left(T_d/\tau_\omega^2\right)t}\right|}{\left|1 + e^{-4\ln(2)(T_d/\tau_\omega^2)t}\right|}. \tag{9}$$

The field in Eq. (8) can be decomposed into two linearly polarized fields, the driving field and gating field for HHG, respectively. Taking $\tau_\omega = 2T = 5fs$, $T_d = 5fs$, and the critical field ellipticity $\xi_c = 0.2$ [30], one can have the gating zone with notable HHG effect as $T_G = 1.5fs$. As shown in **Figure 7**, the two channels for HHG are appearing different evolutionary characteristics corresponding to the CEP changing from 0 to $\pi/2$, with channel ① depressed and channel ② enhanced. The intensity contrast between the HHG that resulted reaches the maximum at CEP $= \pi/2$, meaning the generation of single isolated attosecond pulse from one driving laser pulse. Consequently the optimized phase setting for driving field is CEP $= \pi/2$.

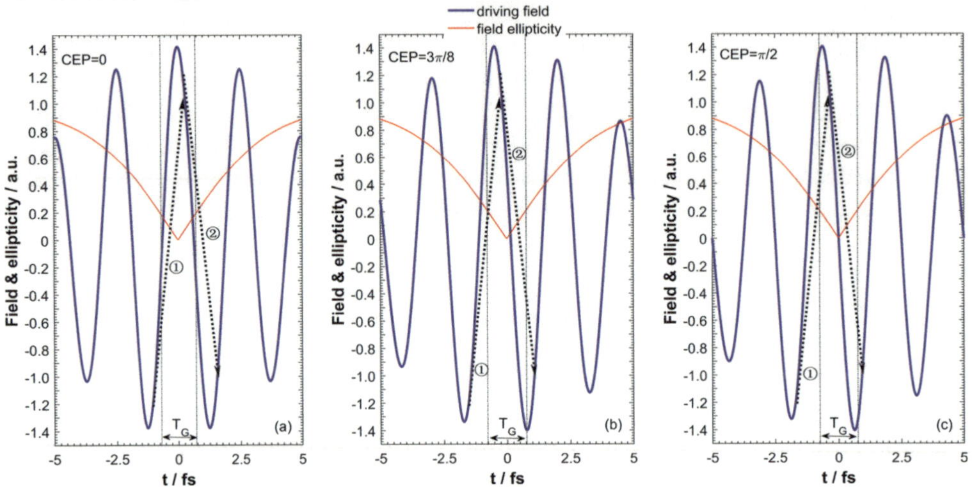

Figure 7.
CEP dependence of HHG for PG.

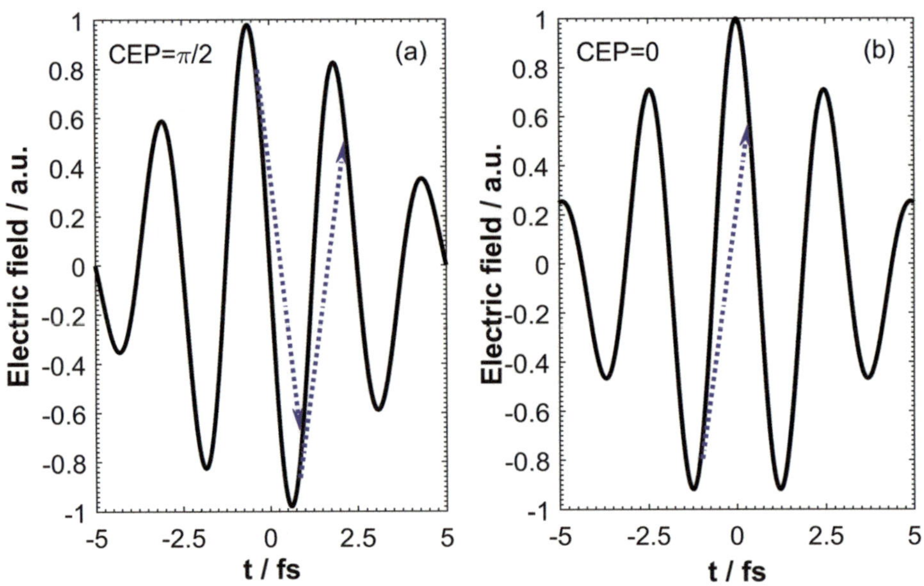

Figure 8.
CEP dependence of HHG for AG.

As for AG, the essence is spectral selection of half-cycle HHG cutoffs, so the linearly polarized driving field with CEP = 0 could guarantee that the cutoff spectrum only comes from one channel and thus is supercontinuum supporting single isolated attosecond pulse, as shown in **Figure 8**. So the optimized phase setting for AG comparatively is CEP = 0, which has been proven by the experimental results in Ref. [16].

4. Phase matching for high-energy attosecond pulse

The device used for optimizing HHG phase matching is schematically shown in **Figure 9**. The CEP stabilized few-cycle femtosecond laser (ladled as 1) is used as the driving optical field for HHG, with the inert gas (ladled as 5) in the nickel tube (ladled as 4) as the target source. The metal filter (ladled as 8), e.g., aluminum or zirconium, is used to filter the spectrum of residual driving laser from the HHG generated. Acrylic cover (ladled as 12) is specifically adopted for easier monitoring of the driving laser condition when the system is in vacuum. The procedure for optimizing high-order harmonics phase matching includes three steps. The details are deliberated in order as follows.

The first step is done in air. A Ti:sapphire oscillator (Rainbow, Femtolasers GmbH) produces sub-10-fs seed pulses with 2.4-nJ energy at 76-MHz. The oscillator was pumped by a continuous wave laser (Coherent Verdi) at 532-nm and 3.10-watt. The oscillator output pulses are then stretched before coupling them into the amplifier. The output of the nine-pass amplifier with three-mirror configuration was more than 1.1-mJ with hundreds of picosecond duration and 3-kHz repetition rate. After amplification, the beam is sent into a grating compressor consisting of two high-efficiency transmission gratings and a vertical retro-reflector to get 25-fs laser pulse of 0.95-mJ. The beam from this laser system was focused into a fused silica hollow-core fiber filled with neon gas with a length of 1.2-m (250-μm inner diameter and 750-μm outer diameter) [31]. The broad optical spectrum ranging from 420-nm to around 950-nm obtained by self-phase modulation effect in hollow-core fiber, after traveling through a set of finely designed chirped-mirror compressor, gives laser pulse width of 4.6-fs which is about 1.7 optical cycles for the center wavelength of 803.5-nm, as shown in **Figure 10**. With focus spot size of 50-μm, the peak intensity reaches above the level of 1.5×10^{14} W/cm^2, which is enough to initiate the HHG in inert gas. The high-intensity femtosecond laser pulse will bring about air breakdown plasma near its focus, the color of which is blue and purple, as shown in **Figure 11**. Under the dark atmosphere, the longitudinal mid-point of the plasma can be roughly regarded as the focus point of the driving pulse. What is worth mentioning is that, before sending the drive pulse to the vacuum chamber, both the nickel tube and the metal filter plates must not be in the propagation path of the driving pulse.

Subsequent experimental steps are done using the focus point determined as a benchmark. Firstly, keep the driving femtosecond laser in its experimental settings, and roughly place the nickel tube 3 $\sim$ 5-mm after the focus point (yet outside the driving laser path by now to avoid damage). The nickel tube is a hollow-core cylinder with an outer diameter of 2.5-mm and inner diameter of 2.0-mm, the top of which is fully sealed to keep the inert gas system. Secondly, lower the power of the driving laser and then move slowly the nickel tube into the laser path so that the

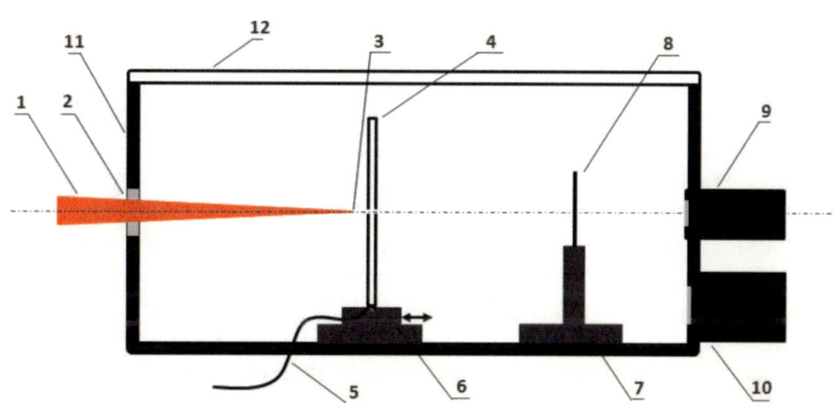

Figure 9.
Schematic diagram of the HHG experimental illustration.

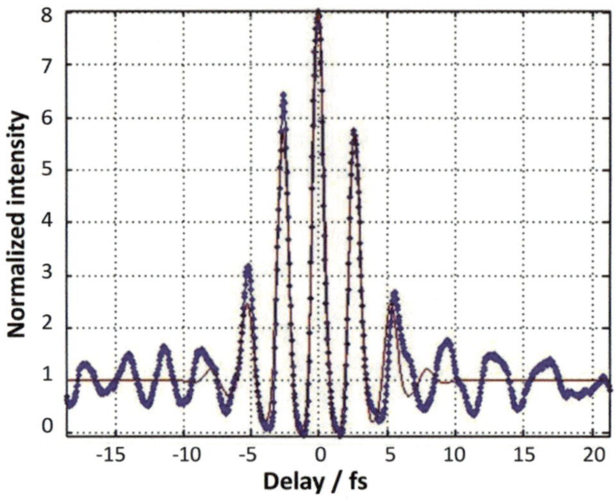

Figure 10.
Autocorrelation trace of laser pulse width measurement.

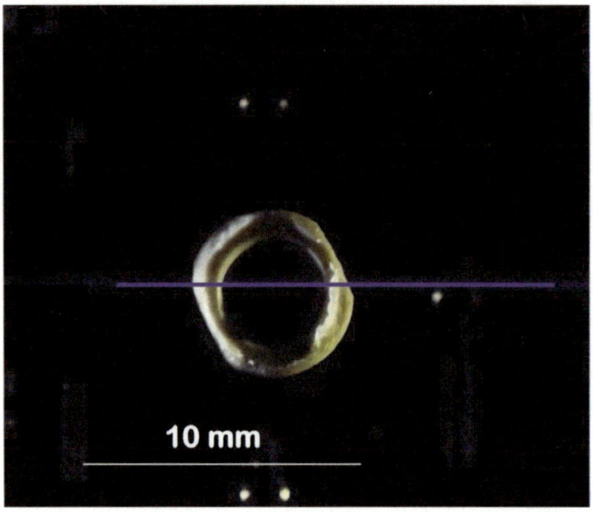

Figure 11.
Strong laser-driven air breakdown plasma.

laser beam spot is just located at the transverse center of the gas tube. At this time the laser power should be too low to make any ablation of the nickel tube. Finally, recover the driving laser power gradually to its experimental value to drill the nickel tube. To reduce the impact of the surrounding air flow on the driving laser beam pointing stability, it is necessary to put the acrylic cover back to get nice tube drilling. It takes at least 30 minutes to finish the drilling process, and the quality can be checked through the pinhole far-field diffraction image, as shown in **Figure 12**.

This part illustrates the details to achieve the optimum phase matching for the most efficient harmonic conversion output. The zirconium filter of thickness of 0.15-μm is used, and its spectral transmittance characteristics are shown in **Figure 13**. When the HHG chamber reaches the required degree of vacuum, 2×10^{-4} Pa or even lower, send the driving femtosecond laser and open the neon gas pipeline control valve to generate high-order harmonics. The high-order harmonics signal is collected by an X-ray CCD. For the given driving laser settings and nickel tube position, the maximum high-order harmonics yield can be found by finely tuning the backing pressure of the incoming neon gas. By changing the nickel tube position, one can obtain the dependency of high-order harmonics yield on the relative position between the driving laser focus and the target source.

The results are shown in **Figure 14**, in which the horizontal ordinate of 0 indicates the focus point of the driving laser traveling along the positive direction. The optimum phase matching can be arrived when the nickel tube gas target is placed at about 4-mm after the driving laser focus, indicated by the maximum high-order harmonics yield there. The harmonics yield is showing significant asymmetry and is extremely low for the target position before the driving laser focus, which is different from the result of reference [9]. The beam profiles of the high-order harmonics detected by CCD and the driving field at the gas target are shown in **Figures 15** and **16**, respectively. It shows that, under the condition of optimum harmonics phase matching with gas target positioned about 4-mm after the driving laser focus, the harmonics generated have similar intensity spatial distribution with the driving laser field, providing an experimental evidence for the commonly used

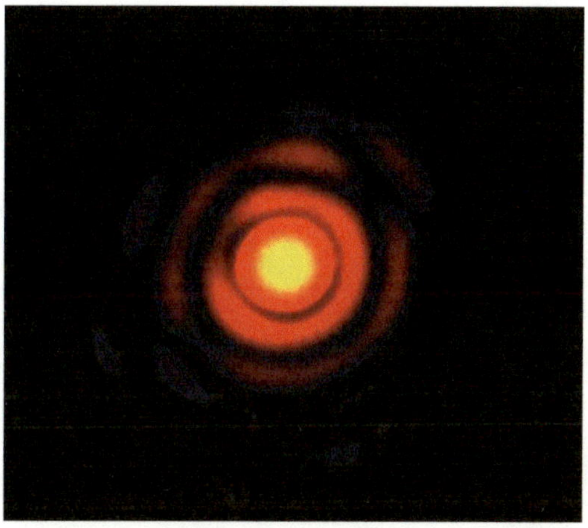

Figure 12.
Far-field pinhole diffraction image of the laser-drilled nickel tube.

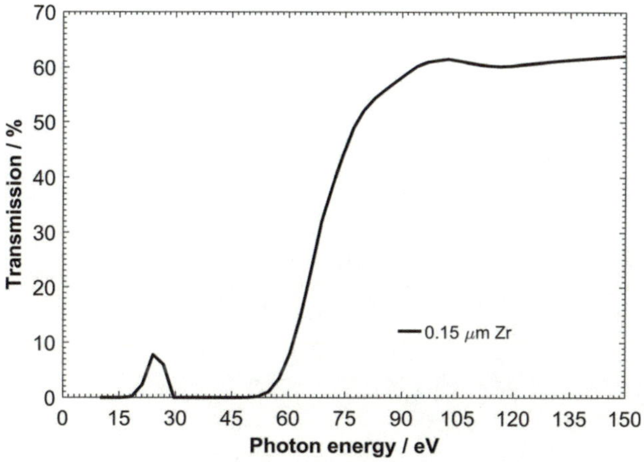

Figure 13.
Transmittance characteristics of 0.15-μm zirconium filter [32].

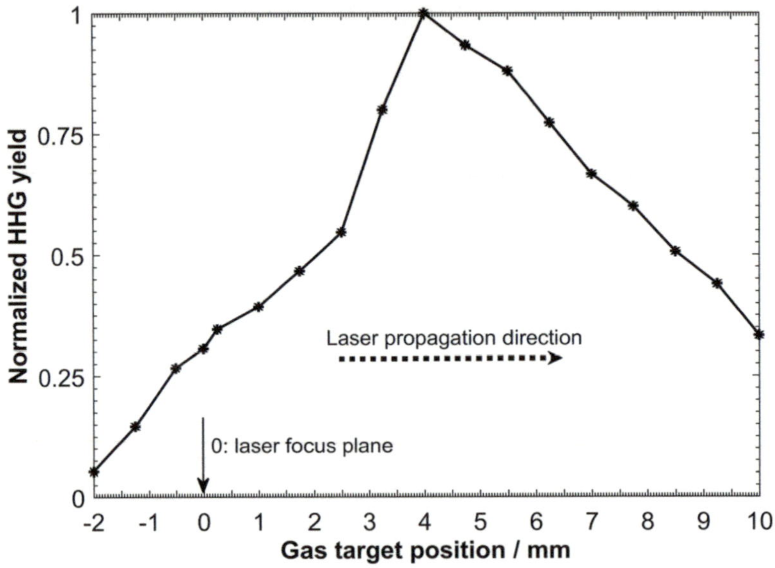

Figure 14.
Dependence of high-order harmonics yield on gas target position.

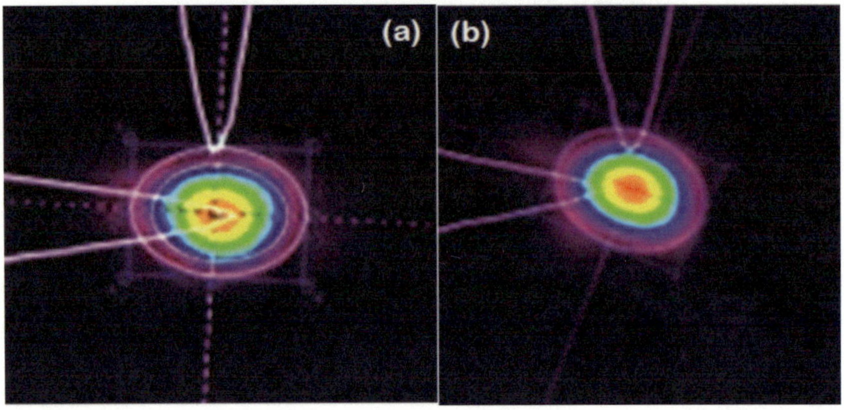

Figure 15.
Driving laser profiles at (a) 4-mm and (b) 2.5-mm after its focus.

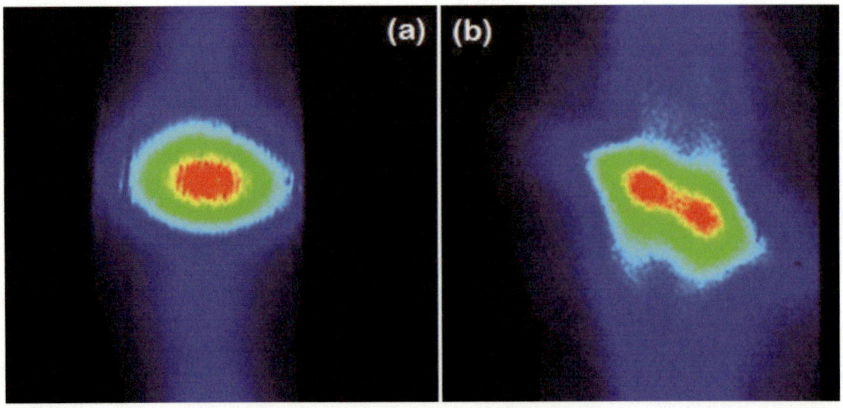

Figure 16.
Far-field profiles of the high-order harmonics with gas target positioned at (a) 4-mm and (b) 2.5-mm after the driving laser focus.

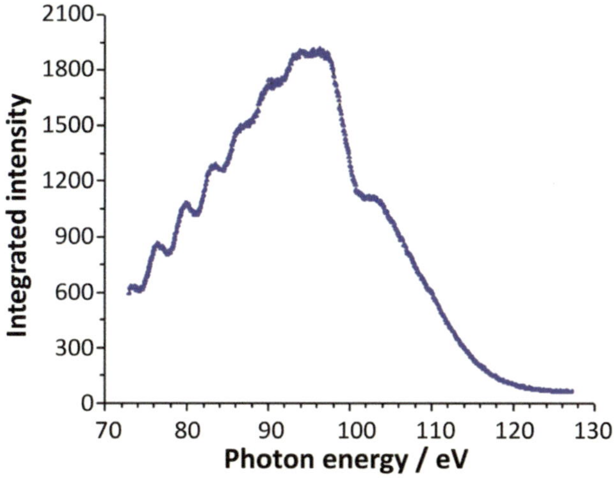

Figure 17.
High-order harmonics spectrum.

assumption that the single isolated attosecond pulse generated by a Gaussian femtosecond laser is also a Gaussian beam. But for the case of worse phase matching as in **Figure 15(b)**, a Gaussian driving laser does not necessarily generate high-order harmonics with similar profile. Using a grazing incidence flat-field spectrometer, the HHG spectrum is analyzed and shown in **Figure 17**. The spectrum in the cutoff region is continuum, indicating that only one channel gave its contribution for the HHG as in **Figure 8(b)**.

It is worth mentioning that other nickel tubes with different inner diameter sizes were also tried. The findings showed that optimum phase matching of HHG existing after the driving laser focus keeps the same for all these experimental situations, though the gas target position for best harmonic phase matching, the harmonics yield, and the backing pressure of gas target used vary to some extent. So we can safely say that the conclusion is universal for Gaussian-type driving laser-based HHG process.

5. Conclusions

The establishment of novel attosecond light source gave rise to an attosecond research upsurge from physics, chemistry, and material science to information processing, among which high-energy single isolated attosecond pulse resulted by interaction between strong optical field and noble gases which has been attracting much attention. Phase matching between the fundamental driving laser field and the high-order harmonics resulted is the key issue to such frequency conversion. In this chapter we scrutinized the intrinsic phase of high-order harmonics resulted from the atom tunneling ionization induced by strong laser field and analyzed qualitatively the salient dependence of polarization gating and amplitude gating on CEP of femtosecond driving laser. The conclusion is that the optimized CEP of driving femtosecond laser for generating single attosecond pulse is $\pi/2$ and 0 for techniques of polarization gating and amplitude gating, respectively. Meanwhile, an implementation was presented experimentally to exemplify the details for optimum HHG phase matching in the interaction of Gaussian-shaped high-intensity few-cycle femtosecond laser with inert gas target. We studied the dependence of the harmonics phase matching on the relative position between the gas target source and driving laser field focus and found that the optimum gas target position for HHG phase matching is always lying behind the focus of the driving field.

Acknowledgements

The work was supported by the National Natural Science Foundation of China (Grant No. 11675258, 11505289, 61690222).

Author details

Chao Wang[1]*, Yifan Kang[2] and Yonglin Bai[1]

1 Key Laboratory of Ultra-fast Photoelectric Diagnostics Technology, Xi'an Institute of Optics and Precision Mechanics, Chinese Academy of Sciences, Xi'an, People's Republic of China

2 School of Science, Air Force Engineering University, Xi'an, People's Republic of China

*Address all correspondence to: igodwang@163.com

IntechOpen

© 2020 The Author(s). Licensee IntechOpen. This chapter is distributed under the terms of the Creative Commons Attribution License (http://creativecommons.org/licenses/by/3.0), which permits unrestricted use, distribution, and reproduction in any medium, provided the original work is properly cited.

References

[1] Wegener M. Extreme Nonlinear Optics. Berlin Heidelberg: Springer-Verlag; 2005. p. 224. DOI: 10.1007/b137953

[2] Nisoli M, De SS, Svelto O, et al. Compression of high-energy laser pulses below 5 fs. Optics Letters. 1997;**22**:522-524. DOI: 10.1364/OL.22.000522

[3] Wegener M. Extreme Nonlinear Optics. 1st ed. Heidelberg: Springer-Verlag Berlin Heidelberg; 2005. p. 223. DOI: 10.1007/b137953

[4] Chang Z. Fundamentals of Attosecond Optics. 1st ed. Florida: CRC Press; 2011. p. 536

[5] Hentschel M, Kienberger R, Ch S, et al. Attosecond metrology. Nature. 2001;**414**:509-513. DOI: 10.1038/35107000

[6] Krausz F, Ivanov M. Attosecond physics. Reviews of Modern Physics. 2009;**81**:163-234. DOI: 10.1103/RevModPhys.81.163

[7] Chang Z, Corkum P. Attosecond photon sources: The first decade and beyond [invited]. Journal of the Optical Society of America B: Optical Physics. 2010;**27**:B9-B17. DOI: 10.1364/JOSAB.27.0000B9

[8] Calegari F, Ayuso D, Trabattoni A, et al. Ultrafast electron dynamics in phenylalanine initiated by attosecond pulses. Science. 2014;**346**:336-339. DOI: 10.1126/science.1254061

[9] Despre V, Marciniak A, Loriot V, et al. Attosecond hole migration in benzene molecules surviving nuclear motion. The Journal of Physical Chemistry Letters. 2015;**6**:426-431. DOI: 10.1021/jz502493j

[10] Hassan M. Attomicroscopy: From femtosecond to attosecond electron microscopy. Journal of Physics B: Atomic, Molecular and Optical Physics. 2018;**51**:1-30. DOI: 10.1088/1361-6455/aaa183

[11] Goulielmakis E, Schultze M, Hofstetter M, et al. Single-cycle nonlinear optics. Science. 2008;**320**:1614-1617. DOI: 10.1126/science.1157846

[12] Sansone G, Benedetti E, Calegari F, et al. Isolated single-cycle attosecond pulses. Science. 2006;**314**:443-446. DOI: 10.1126/science.1132838

[13] Corkum P. Plasma perspective on strong-field multiphoton ionization. Physical Review Letters. 1993;**71**:1994-1997. DOI: 10.1103/PhysRevLett.71.1994

[14] Salieres P, L'Huilier A, Lewenstein M. Coherence control of high-order harmonics. Physical Review Letters. 1995;**74**:3776-3779. DOI: 10.1103/PhysRevLett.74.3776

[15] Gragossian A, Seletskiy DV, Sheik-Bahae M. Classical trajectories in polor-asymmetric laser fields: Synchronous THz and XUV emission. Scientific Reports. 2016;**6**:34973/1-34973/8. DOI: 10.1038/srep34973

[16] Kienberger R, Goulielmakis E, Uiberacker M, et al. Atomic transient recorder. Nature. 2004;**427**:817-821. DOI: 10.1038/nature02277

[17] Witting T, Frank F, Okell WA, et al. Sub-4-fs laser pulse characterization by spatially resolved spectral shearing interferometry and attosecond streaking. Journal of Physics B: Atomic, Molecular and Optical Physics. 2012;**45**:074014/1-074014/6. DOI: 10.1088/0953-4075/45/7/074014

[18] Jullien A, Pfeifer T, Abel MJ, et al. Ionization phase-match gating for wavelength-tunable isolated attosecond pulse generation. Applied Physics B. 2008;**93**:433-442. DOI: 10.1007/s00340-008-3187-z

[19] Abel MJ, Pfeifer T, Nagel PM, et al. Isolated attosecond pulses from ionization gating of high-harmonic emission. Chemical Physics. 2009;**366**: 9-14. DOI: 10.1016/j.chemphys.2009.09.016

[20] Thomann I, Bahabad A, Liu X, et al. Characterizing isolated attosecond pulses from hollow-core waveguides using multi-cycle driving pulses. Optics Express. 2009;**17**:4611-4633. DOI: 10.1364/OE.17.004611

[21] Ferrari F, Calegari F, Lucchini M, et al. High-energy isolated attosecond pulses generated by above-saturation few-cycle fields. Nature Photonics. 2010;**4**:875-879. DOI: 10.1038/nphoton.2010.250

[22] Corkum PB, Burnett NH, Ivanov MY. Subfemtosecond pulses. Optics Letters. 1994;**19**:1870-1872. DOI: 10.1364/OL.19.001870

[23] Sola IJ, Mevel E, Elouga L, et al. Controlling attosecond electron dynamics by phase-stabilized polarization gating. Nature Physics. 2006;**2**:319-322. DOI: 10.1038/nphys281

[24] Zhao K, Zhang Q, Chini M, et al. Tailoring a 67 attosecond pulse through advantageous phase-mismatch. Optics Letters. 2012;**37**:3891-3893. DOI: 10.1364/OL.37.003891

[25] Mashiko H, Gilbertson S, Li C, et al. Double optical gating of high-order harmonic generation with carrier-envelope phase stabilized lasers. Physical Review Letters. 2008;**100**: 103906/1-103906/4. DOI: 10.1103/PhysRevLett.100.103906

[26] Feng X, Gilbertson S, Mashiko H, et al. Generation of isolated attosecond pulses with 20 to 28 femtosecond lasers. Physical Review Letters. 2009;**103**: 183901/1-183901/4. DOI: 10.1103/PhysRevLett.103.183901

[27] Mashiko H, Bell MJ, Beck AR, et al. Tunable frequency-controlled isolated attosecond pulses characterized by either 750 nm or 400 nm wavelength streak fields. Optics Express. 2010;**18**: 25887-25895. DOI: 10.1364/QELS.2011.QMG4

[28] Vincenti H, Quéré F. Attosecond lighthouses: How to use spatiotemporally coupled light fields to generate isolated attosecond pulses. Physical Review Letters. 2012;**108**: 113904/1-113904/5. DOI: 10.1103/PhysRevLett.108.113904

[29] Kim KT, Zhang C, Ruchou T, et al. Photonic streaking of attosecond pulse trains. Nature Photonics. 2013;**7**: 651-656. DOI: 10.1038/NPHOTON.2013.170

[30] Antoine P, L'Huillier A, Lewenstein M, et al. Theory of high-order harmonic generation by an elliptically polarized laser field. Physical Review A. 1996;**53**:1725-1745. DOI: 10.1103/PhysRevA.53.1725

[31] An J, Kim DE. Compact in-line autocorrelator using double wedge. Optics Express. 2012;**20**:3325-3330. DOI: 10.1364/OE.20.003325

[32] Available at: http://henke.lbl.gov/optical_constants/filter2.html [Accessed: 20 May 2018]

Chapter 3

Development of a Corona Discharge Ionizer Utilizing High-Voltage AC Power Supply Driven by PWM Inverter for Highly Efficient Electrostatic Elimination

Katsuyuki Takahashi, Koichi Takaki, Isao Hiyoshi, Yosuke Enomoto, Shinichi Yamaguchi and Hidemi Nagata

Abstract

The corona discharge ionizer has been widely used to eliminate electrostatic charges on insulators in a variety of manufacturing industries for the prevention of electrostatic discharge (ESD) problems. High-speed electrostatic elimination is conventionally required for ionizer performance. Because of the high sensitivity of recent electronic devices to ESD damage, an extremely low-offset voltage (ion balance) is required for the performance of electrostatic eliminators. Long-term performance stability is required to maintain the quality of the products, but the short cleaning interval of the unit increases the operating cost. The efficiency is also affected by the waveform of the applied voltage. The optimization of the applied voltage is an important factor in achieving long-term performance stability. In this study, an intermittent pulse voltage AC power supply was developed to achieve a highly efficient electrostatic elimination with long-term stability high-speed electrostatic elimination and an excellent ion balance.

Keywords: electrostatic, ionizer, corona discharge, ion balance, high voltage

1. Introduction

Elimination of electrostatic charges on insulators for the prevention of electrostatic discharge (ESD) problems in manufacturing industries such as semiconductor and electric device is very important for the enhancement of productivity and reliability [1–5]. The simplest and the most important method to prevent ESD is eliminating charges by keeping conductive materials connected to the ground. In work spaces, wearing anti-ESD wrist straps and conductive clothes and shoes and keeping conductivity of tables and floors high are required for ESD control. In the case of insulators, charges are not removed by connecting them to the ground. Therefore, generally, the charges accumulated on insulators are decayed by decreasing the surface resistance by increasing humidity. Although these methods are effective for the prevention of ESD problems, they are restricted by environment, materials, and conditions of objects. When these methods are unavailable, the methods to forcibly eliminate the charges by neutralizing using ions are required.

The corona discharge ionizer has been widely used for elimination of electrostatic charges because of its good operability. **Figure 1** shows schematic illustration of elimination of electrostatic charge accumulated on an insulator using an ionizer. The corona discharges are produced at high-voltage electrodes in the ionizer, which enable the production of positive and negative ions such as H_3O^+, CO_3^-, NO_3^-, and their cluster ions [6–9, 10]. The ions produced by the corona discharges are transported by the electric field or blown air to insulators, and the charges on insulators are neutralized by the ions. The time required for the electrostatic elimination depends on the amount of ions transported to the insulators. For the ionizer performance, not only the short electrostatic elimination time but also the good ion balance (a low-offset voltage), the balance between the amounts of ions of both polarities (**Figure 2**), is required

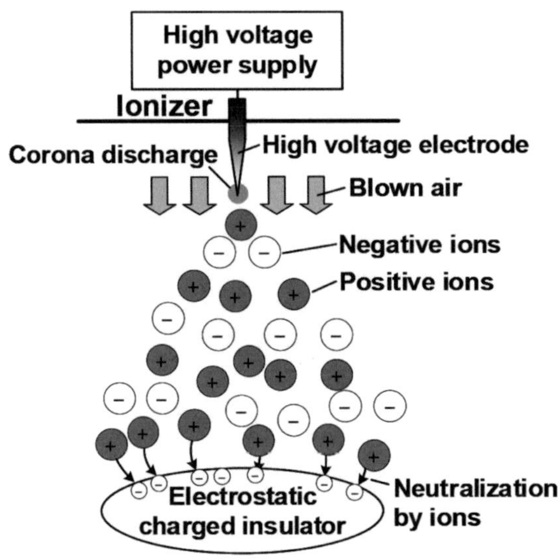

Figure 1.
Schematic illustration of elimination of electrostatic charge accumulated on an insulator using an ionizer.

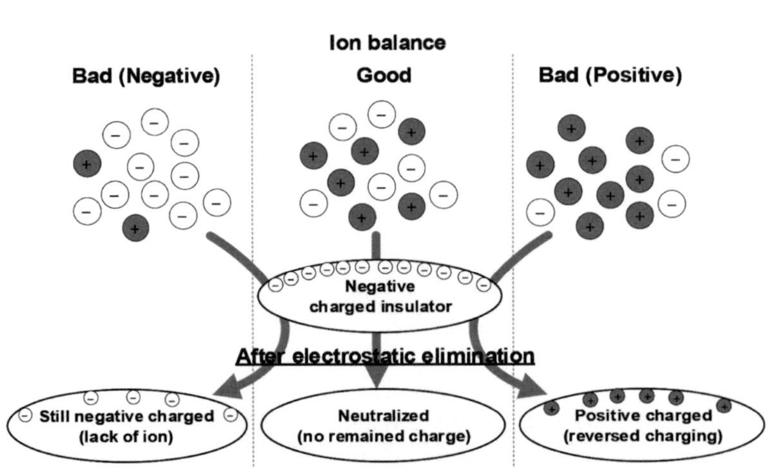

Figure 2.
Relationships between ion balance and electrostatic charges on insulator after electrostatic elimination.

	DC	Commercialized frequency AC (50/60 Hz)	Rectangular AC (< 100 Hz)*	Intermittent bipolar pulse (several hundred Hz)	High frequency AC (several tens kHz)
Electrostatic elminination without blown air	Excellent	Average	Good	Below Average	Inferior
Ion balance unevenness in deionzing space	Inferior	Average	Below Average	Good	Excellent
Ion balance fluctuation	Average	Inferior	Inferior	Good	Excellent
Performance stability	Inferior	Average	Average	Excellent	Inferior
Size and weight	Average	Inferior	Below Average	Good	Excellent

*Pulsed AC type ionizer and pulsed DC type ionzer

Table 1.
Characteristics of typical corona discharge ionizers for the various types of voltage supplies.

[11–14]. The bad ion balance causes ESD problem on the latest electronic devices, which have high sensitivity to ESD damage. For example, the ion balance is limited less than 25–100 V in the manufacturing process of electric devices, and only several volts in the process of hard disc heads have a very high sensitivity for ESD [15–18].

The polarity of the ions produced by the corona discharges is the same as the polarity of the applied voltage to the needles [19, 20]. **Table 1** shows the characteristics of the typical corona discharge ionizers for the various types of voltage supplies. The corona discharge ionizers are mainly characterized by the frequency of the applied voltage to the electrodes. There are advantages and disadvantages with each frequency, which require an optimum selection for each situation.

The charge decay time, the time required to eliminate electrostatic charge on a charged object, is significantly affected by the amplitude of applied voltage to the needles, the voltage waveform, and its timing and frequency. Usually, the ions are transported to the charged objects by blown air using a fan or compressed air. When the blown air is limited, the ions must be transported by Coulomb force along the electric field from the needle and by self-diffusion [21]. In this case, DC and low-frequency and commercialized AC are suitable for electrostatic elimination because the polarity change of the electric field is slow and ions are efficiently transported to the object; on the other hand, the fluctuation width of the ion balance becomes large with low-frequency AC [21]. The fluctuation causes ion balance unevenness in the deionizing space [20, 22, 23]. Although a high-frequency AC requires blown air for electrostatic elimination due to its rapid polarity change, the well-mixed positive and negative ions suppress the fluctuation of the ion balance.

Long-term performance stability is required to maintain the quality of the products, but the short cleaning interval of the unit increases the operating cost. The performance of the electrostatic eliminator declines with the deterioration of the high-voltage electrode because of dust attraction and abrasion from the corona discharge. The amount of ions produced by corona increases with discharge current which depends on amplitude of applied voltage and geometry of electrode [24, 25], which decreases the charge decay time. On the other hand, the efficiency of electrostatic elimination decreases with increasing discharge current that does not contribute to eliminating electrostatic charge, which causes electrode deterioration. The efficiency is significantly affected by the waveform of the applied voltage. DC-type ionizers have two polarity electrodes. Since the abrasion of the electrode depends on the polarity, the balance of the positive and negative ions collapses. In the high-frequency AC ionizers, the ions are easily lost by the ion-ion recombination reaction, and the efficiency for ion transport to the charged object is very low. The optimization of the applied voltage is an important factor in achieving long-term performance stability.

In this study, an intermittent pulse voltage AC power supply with a small and lightweight transformer was developed for highly efficient electrostatic elimination. A corona ionizer driven by a power supply offers long-term performance stability in low-offset voltage. The high stability enables the installation of an electrostatic eliminator in the manufacturing processes of high electrostatic-sensitive devices required for extremely low-offset voltage. In this chapter, the specifications of the power supply and the electrostatic eliminator are outlined with quantitative data.

2. Bipolar pulse voltage power supply driven by PWM converter

Figure 3 shows a schematic diagram of the bipolar pulse voltage power supply driven by a full-bridge PWM converter [26]. The primary current of the transformer is controlled by a full-bridge converter circuit consisting of four MOSFETs with a DC voltage supply (V_{DD}). The value of V_{DD} depends on the turn ratio of the transformer and is 24 V for this example. The MOSFETs are driven by PWM input signals (Q_1–Q_4) generated by a microcomputer through the FET driver circuits. The ON signal and PWM signal are outputted to Q_1 and Q_4, respectively, in a positive phase; the ON signal and PWM signal are outputted to Q_3 and Q_2, respectively, in a negative phase. **Figure 4a** shows the relationship between the output voltage and the control signals. The amplitude of the pulse voltage is adjusted by the duty ratio of the PWM signals at the rising edge of the pulse. The PWM signals at the falling edge of the pulse suppress the ringing induced by the resonance, which contributes to generating a pulse shape close to an ideal condition. **Figure 5** shows the typical waveform of the output voltage controlled by the PWM signals. The amplitude, interval, and generation timing of the pulses can be arbitrarily set according to the PWM signals as shown in **Figure 4b**, which realizes the generation of not only intermitted pulse voltage but also AC sinewave voltage.

The frequency and pulse width of the output voltage are determined by the characteristics of the transformer such as the material and the cross section of the core and the number of turns of the coil. Since the characteristics of the

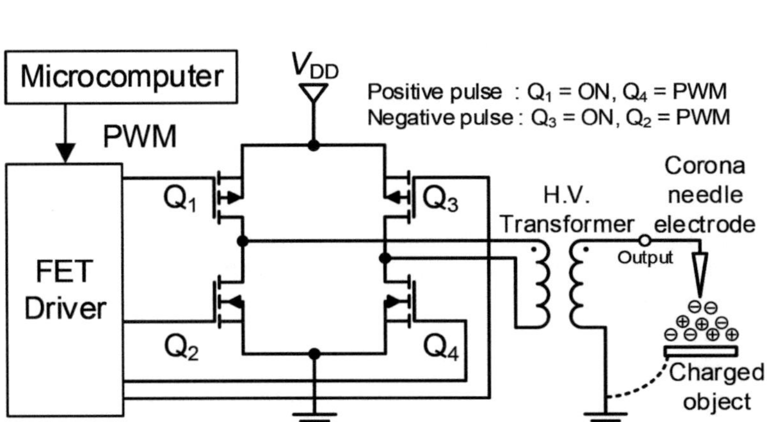

Figure 3.
Schematic diagram of the bipolar pulse voltage power supply driven by a full-bridge PWM converter [26].

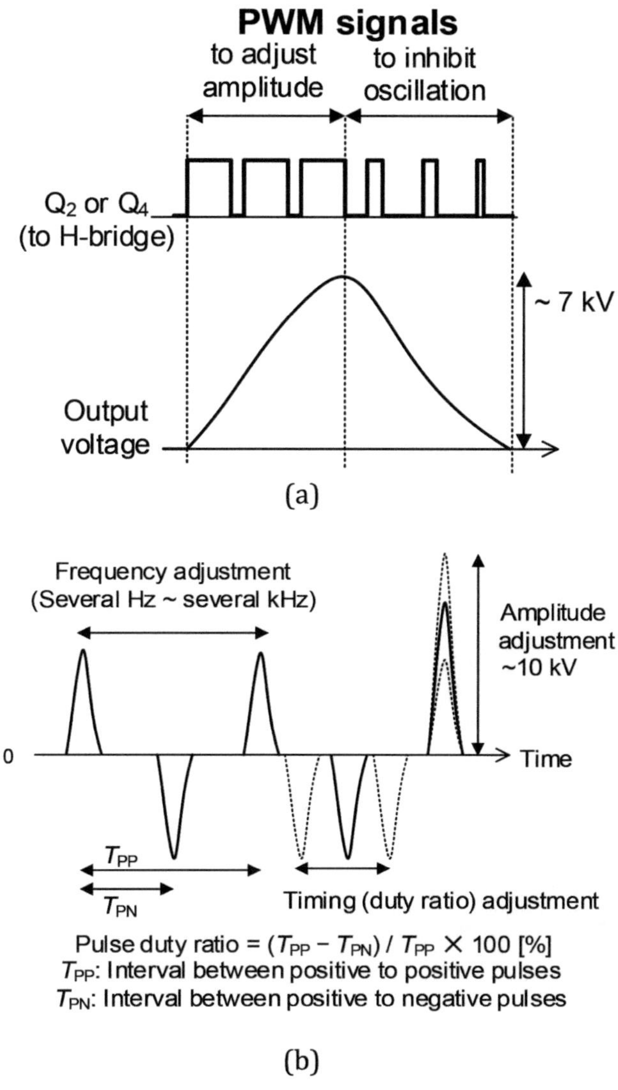

Figure 4.
Schematic diagram of PWM control [26]. (a) Relationship between the output voltage and the control signals, (b) output voltage.

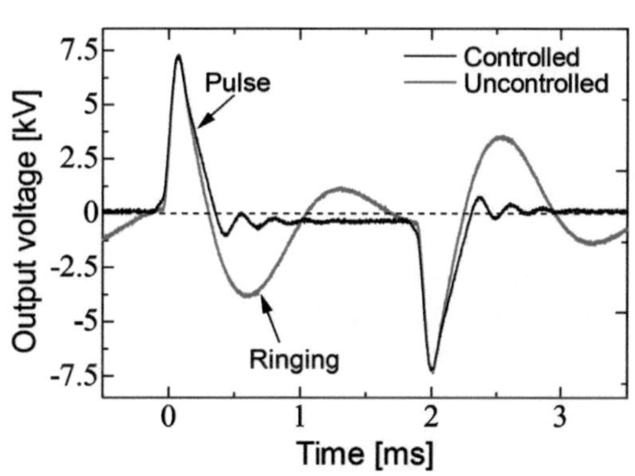

Figure 5.
Typical waveform of output voltage [26].

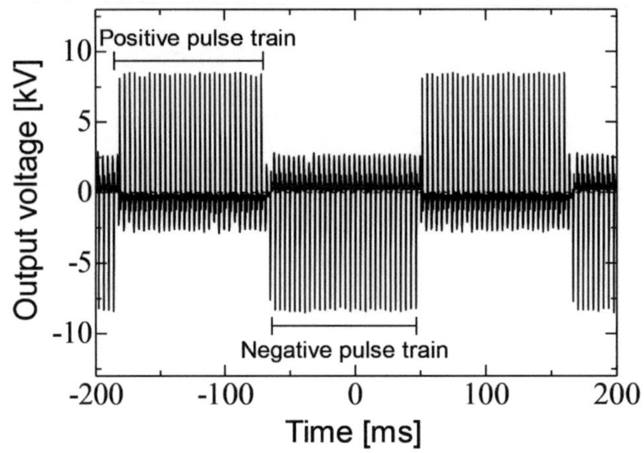

Figure 6.
Typical waveform of continuously generated pulse train [26].

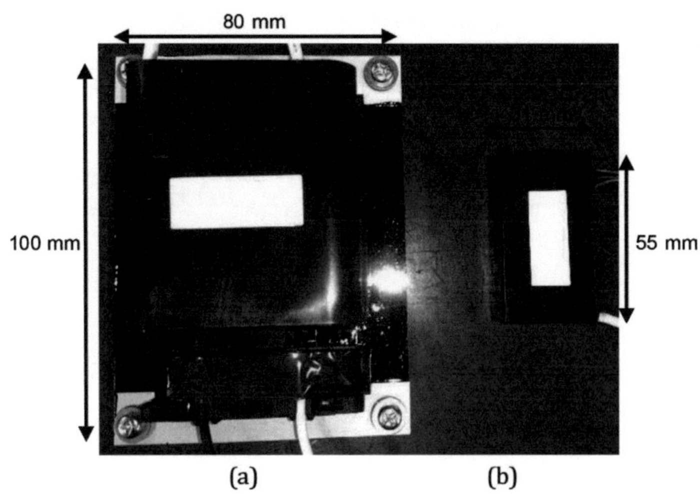

Figure 7.
Photograph of two transformers [27]. (a) Commercial frequency, (b) bipolar intermittent pulse.

electrostatic eliminator depend on the frequency as mentioned above, the various types of transformers are prepared for various applications to generate the optimized frequency. Generally, the magnetic flux inside the core required to magnetically couple the primary and secondary windings increases with decreasing the frequency, which increases the size and weight of the transformer. To generate the low-frequency voltage with a small and lightweight transformer, the same polarity pulses are continuously generated in a phase as a pulse train by the PWM converter, and the polarity is switched with a period, as shown in **Figure 6**. The apparent frequency decreases with increasing number of pulses in the period. The unipolar pulse train can continuously generate unipolar ions in a phase and transport them to the charged object by the electric field and self-diffusion, which contributes to improving the electrostatic elimination speed without blown air, as described later.

The bipolar pulse voltage power supply driven by a full-bridge PWM converter provides a certain degree of freedom in generating the pulse voltages. **Figure 7** shows a photograph of a transformer designed for commercial frequency (50 or 60 Hz) and a transformer designed for the bipolar intermittent pulse with a frequency of 250 Hz [27]. The turn ratios are 70 and 360. The cores are both made using a silicon steel sheet. Although the transformer designed for the bipolar intermittent pulse is over 10 times smaller and lighter than the transformer designed for commercial frequency, the performance for electrostatic elimination is almost the same in many situations with the optimum voltage waveforms controlled by PWM control. The optimization of voltage waveforms in various applications realizes a design of small and lightweight systems and a long-term performance stability.

3. Evaluation method of electrostatic elimination performance of ionizer

The charge plate monitor (CPM) is usually used for evaluation of electrostatic elimination performance of ionizer. CPM consists of a metal plate with a plate-to-ground capacitance of 20 ± 2 pF [28]. The plate is charge to higher than ±1 kV [28], and the voltage is reduced by the ionic current caused by ion flow in the plate. The electrostatic elimination time is defined as time required for decreasing the plate voltage from ±1 kV to ±100 V. The ionic current flow to the plate is determined by the temporal change of the plate voltage according to an equation, $i = \mathrm{d}q/\mathrm{d}t = C\,\mathrm{d}v/\mathrm{d}t$. The ion current measuring device such as CPM is required for a high insulation resistance, a low capacitance, and a high-voltage charge because the ionic current into the insulators is on the order of 10^{-9} A [29]. The frequency response of the CPM is limited by these requirements, which causes an incorrect measurement. For example, the frequency response of the conventional CPMs is lower than 100 Hz. Furthermore, the displacement current induced by the high voltage applied to the corona discharge electrodes interferes with the measurement of the ionic current.

Figure 8 shows a schematic diagram of the ion measuring system developed for a precise evaluation of the ionizer with a high-frequency component [30]. The system consists of the ion-trapping metal plate, capacitors C_{I1} consisted of three series capacitors (WIMA, FKP2-33/1000/10, 33 pF) and C_{I2} (KEMET, PHE448SB4100JR06, 1 nF), and a buffer circuit op-amp A_1 (Texas instruments, OPA454). The ion-trapping plate voltage (v_I), usually higher than ±1 kV, can be determined by measuring the voltage across C_{I2} using the buffer circuit as a low voltage because the ion-trapping plate voltage (v_I) is divided by C_{I1} and C_{I2}. The buffer circuit does not interfere with measuring v_I because of its very high input impedance ($10^{13}\Omega$) and low-bias current (1.4 pA). The ratio between v_I and the output voltage of the buffer circuit (v_{A1}) is determined by the capacitances of C_{I1} and C_{I2}. The op-amp

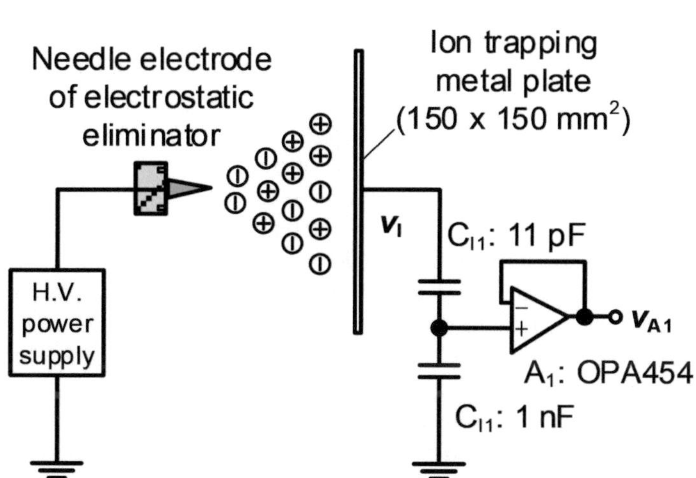

Figure 8.
Schematic diagram of the ion measuring system.

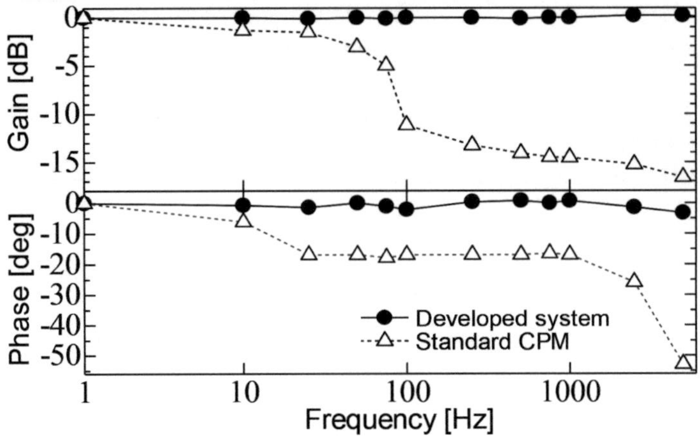

Figure 9.
Bode plots for the developed system and the standard CPM.

allows to apply the input voltage less than ±50 V. Therefore, in the system shown in **Figure 6**, the maximum voltage that can be applied to the plate is 4.25 kV. The total capacitance between the ion-trapping plate and ground is 21.2 pF with the plate area of 150 × 150 mm^2, which is based on a standard established by ESD association [28].

To compare the developed system with a conventional ion evaluation and ion measuring device, a standard CPM (Trek, Model 158) is employed. **Figure 9** shows the bode plots for the developed system and the standard CPM [30]. The gain and phase shift have constant values in the case of the developed system with the frequencies lower than 10 kHz. Although the gain of the standard CPM drastically decreases with the frequency higher than approximately 60 Hz, that of the developed system has a constant gain without phase shift. The voltage of the plate is maintained with the charge decay time of 83 minutes. The insulation resistance estimated by the decay time according to an equation is $R = t/C$ and is $0.23 \times 10^9 \Omega$. The time required for decaying 5% of initial charging voltage is 243 seconds, which is satisfied for the evaluation of the performance of the corona discharge ionizer [28].

Figure 10 shows the typical waveform of the charged voltage of the developed CPM in the case of evaluating the charge decay time of an ionizer. The initial charge voltage to the plate is ±1.1 kV and is decayed exponentially by the ions generated by the ionizer. Generally, the charged decay time is defined as the time required to decay to ±0.1 kV from ±1.0 kV. When the plate voltage assumes a steady-state condition after electrostatic elimination, the average value of the plate voltage is defined as the ion balance or the offset voltage.

Figure 11 shows the equivalent circuit model of the electrostatic elimination system using a corona discharge ionizer. The ionizer and the charged object are connected by the deionizing space. The v_O is the output voltage of the high-voltage power supply applied to the electrode of electrostatic eliminator. C_A is the stray capacitance and is on the order of 10^{-12} F. R_{ion} is resistance expressed by the ion flow and is on the order of 10^8–10^{10} Ω [27]. C_{obj} and R_{obj} are the capacitance and the leakage resistance of the charged object, respectively. Since the high-voltage supply (ionizer) and the object (CPM) are coupled by C_A, the voltage synchronized to v_O is induced on C_{obj}. In the electrostatic elimination system shown in **Figure 8**, the voltage is induced on the ion-trapping plate of the CPM and is induced by the electric field emitted from the ionizer in the same manner. Additionally, the ions generated by the corona discharges flow into the plate electrode through R_{ion}. The polarity of the ion flows into the plate is the same as the phase of v_O. Therefore,

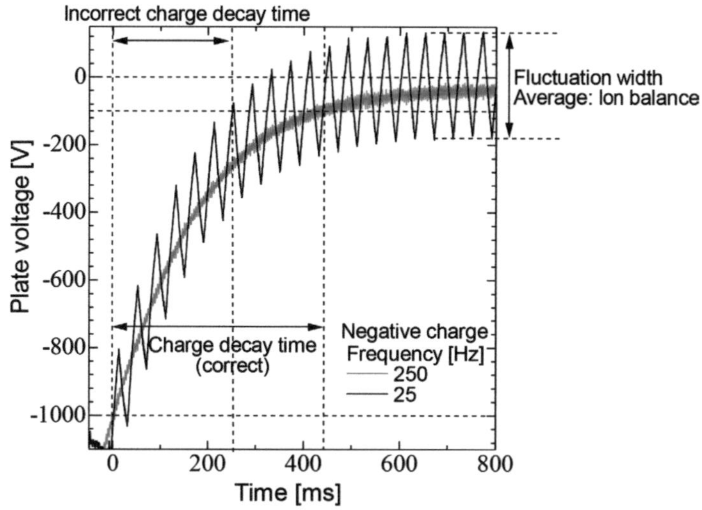

Figure 10.
Typical waveform of the plate voltage in the case of negative charged electrostatic elimination for two frequencies.

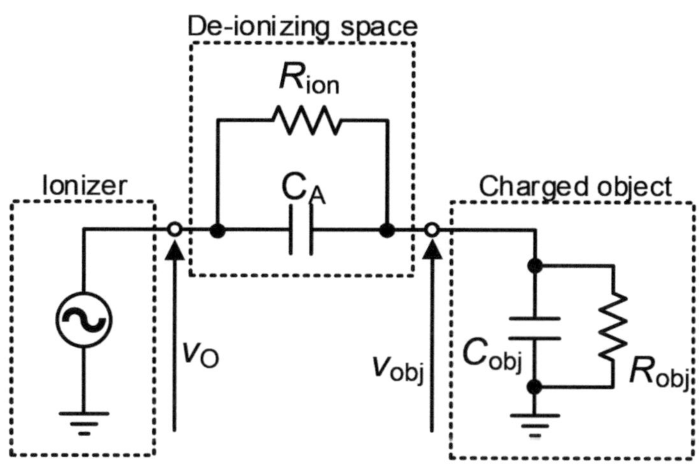

Figure 11.
Equivalent circuit model of the electrostatic elimination system using a corona discharge ionizer.

the plate voltage fluctuates during the steady-state condition after electrostatic elimination. **Figure 12** shows the example waveforms of a high voltage applied to an electrode of an electrostatic eliminator, the plate voltage, an induced voltage by high voltage, and a voltage generated by an ionic flow current to the plate. A rectangular waveform with a frequency of 1 kHz is applied to the electrostatic eliminator. The induced voltage is estimated by the electric field from the electrostatic eliminator using a probe [30]. The voltage due to the ionic current is estimated by subtracting the induced voltage from the plate voltage. The plate voltage is induced by the electric field and the ionic current, which causes a fluctuation of the voltage potential of the object.

This fluctuation is submerged when the CPM's frequency characteristics are not sufficiently high for the frequency of the voltage applied to the electrostatic eliminator, which could lead to unexpected ESD problems. Additionally, the fluctuation could lead to underestimation of the charge decay time because some CPMs define

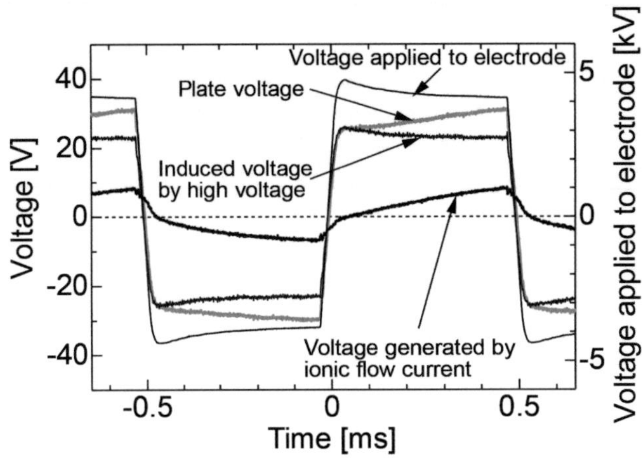

Figure 12.
Example waveforms of a high voltage applied to an electrode of an electrostatic eliminator, the plate voltage, an induced voltage by the high voltage, and a voltage generated by an ionic flow current to the plate.

the charge decay time when the plate voltage reaches to ±0.1 kV at once, although it should be measured with the time averaged voltage as shown in **Figure 8**. The precise evaluation of the voltage potential of the object is vitally important for high-level ESD control.

4. Characteristics of electrostatic elimination using bipolar pulse voltage with blown air

Figure 13 shows the charge decay time as a function of air velocity of the blown air using two intermittent bipolar pulsed voltages, uncontrolled with the ringing and controlled close to an ideal condition, as shown in **Figure 5** [26]. A bar-type corona discharge ionizer (Shishido electrostatic, CABX-350L) is used as the ionizer

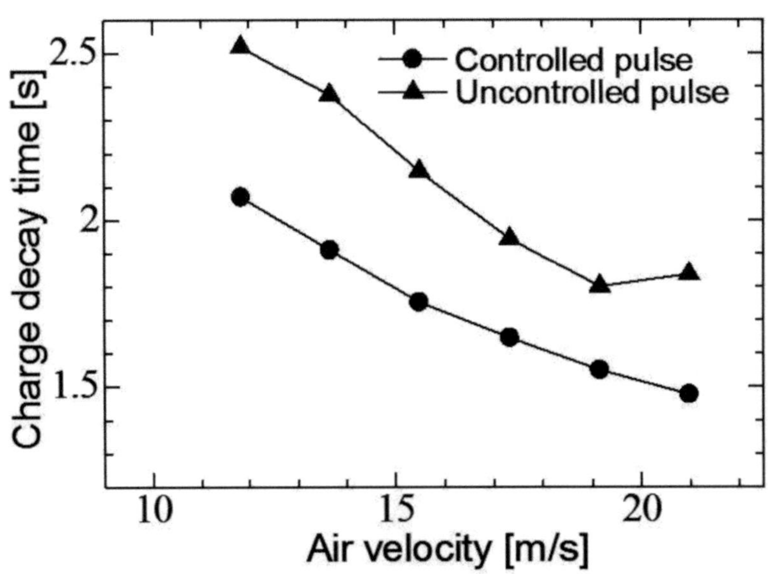

Figure 13.
Charge decay time as a function of air velocity using two intermitted bipolar pulsed voltages, controlled and uncontrolled pulses to inhibit ringing [26].

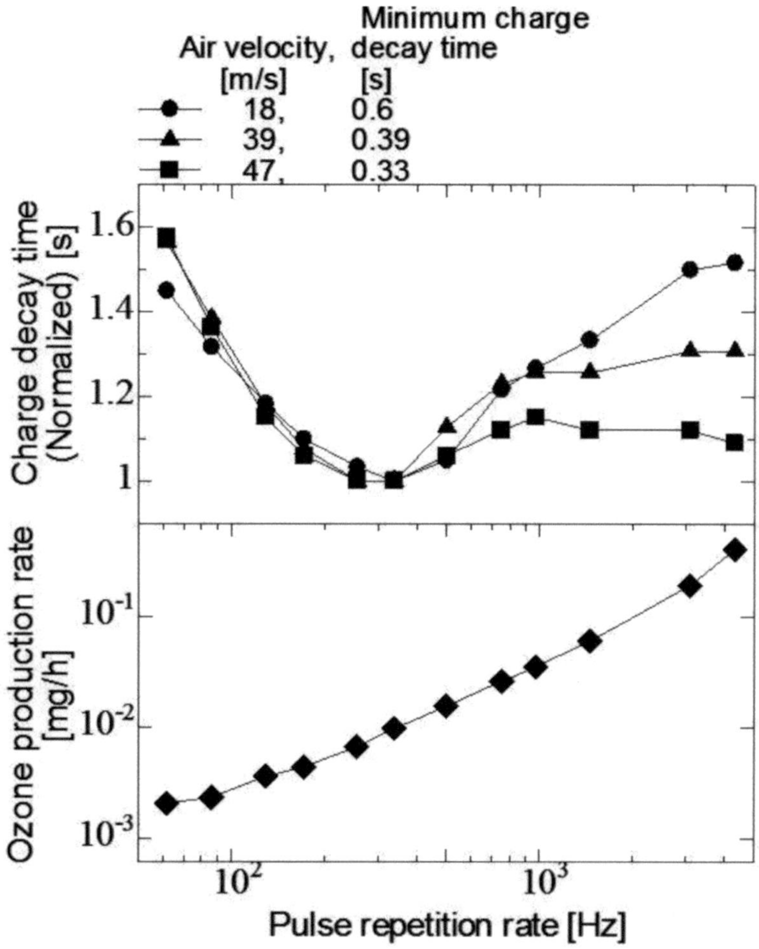

Figure 14.
Charge decay time and ozone production rate as a function of pulse repetition rate [26].

with the air blown by compressed air. The air velocity is measured at 10 mm in front of the needles. The amplitude and pulse repetition rate of the pulsed voltage are 7.5 kV and 250 Hz, respectively. The charge decay time in the case of controlled pulsed voltage is much shorter than that of uncontrolled pulsed voltage.

Figure 14 shows the charge decay time and the ozone production rate as a function of the pulse repetition rate. The pulse width is fixed at 0.2 ms. The charge decay time is normalized by the minimum charge decay time for each air velocity. The ozone concentration is determined by an ozone monitor (Ebara, EG-2001). The charge decay time has a minimum at the pulse repetition rate of 250 Hz. The ozone concentration increases with increasing pulse repetition rate. Since ozone is produced by the ionization reactions in the corona discharges, the amount of ions produced by the corona discharges increases with the ozone concentration [31]. The charge decay time depends on the ion flux into the ion-trapping plate and decreases with increasing ion concentration. The number of corona discharges per unit time decreases with decreasing pulse repetition rate, which decreases the ion concentration and increases the charge decay time. On the other hand, with increasing pulse repetition rate, the ions dragged toward the needles by Coulomb force decrease, and the amount of ions lost by the recombination reactions at the vicinity of the needle increases. Therefore, the charge decay time has a minimum value at the pulse repetition rate of 250 Hz. Because the time of flight of the ions decreases with increasing the air velocity, the recombination reactions of ions in the deionizing space decrease. Thus, the increase in the charge decay time is suppressed with the higher air velocity, as shown in **Figure 14**. The uncontrolled pulse waveform with ringing, as shown in **Figure 5**, has a higher-frequency component than the controlled pulse waveform. This is one of the reasons that the charge decay time in the case of an uncontrolled pulse is higher than that of a controlled pulse as shown in **Figure 13**.

The corona needle electrodes are degraded with a dust attachment to the electrode, a crystallization of a neutral gas molecule, such as siloxane, at the tip of the electrode, and an erosion of the electrode by an etching effect of corona discharge [32–35]. The performance of the electrostatic eliminator is degraded with needle degradation, which leads to an increase in the charge decay time due to the decrease in ion production and the collapse of the ion balance with unbalanced ion production. The increase in the ions that do not contribute to the electrostatic elimination, such as ions lost by recombination reactions, causes the degradation of the needle electrode, which is strongly affected by the frequency and the waveform of the applied voltage to the electrode. The optimization of frequency and waveform is important for improving the efficiency of electrostatic elimination and maintaining performance during long-term operation.

The ion balance can be controlled by the timing of pulse generation. **Figure 15** shows the ion balance as a function of the pulse duty ratio. The pulse duty ratio, a timing of pulse generation, is defined as shown in **Figure 5b**. The ion balance linearly increases with increasing pulse duty ratio with a controlled pulse, which enables the easy control of ion balance. Generally, the ion balance has been controlled with the amplitude of the voltages in the cases of DC and AC type ionizers. **Figure 16** shows the time change of the plate voltage using three different types of high-voltage power supply: DC, a commercialized frequency (50 Hz), and an intermittent pulse (250 Hz). The amplitude of the voltage is fixed at 7 kV in every type of high-voltage power supply. The fluctuation of the plate voltage is suppressed using the intermittent pulse type in comparison with the others because the corona discharge and ion production are stable. The ion balance control by the duty ratio of the intermittent pulse enables an easy and rapid control of the ion balance, which also contributes to long-term stability.

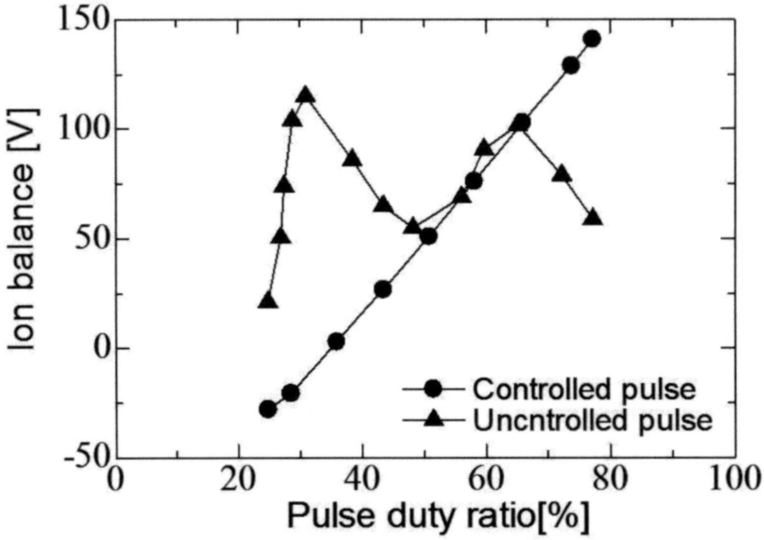

Figure 15.
Ion balance as a function of pulse duty ratio [26].

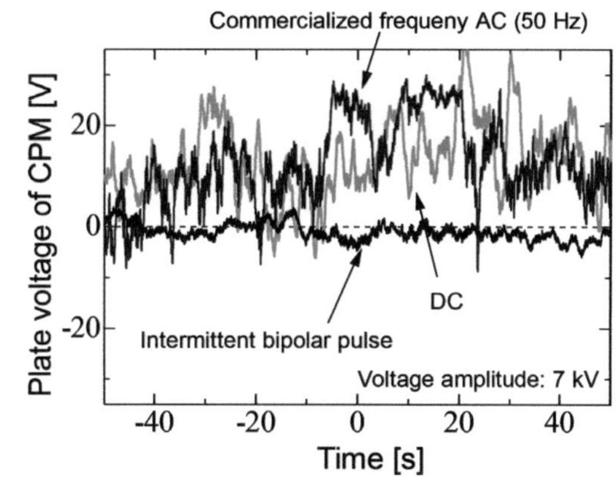

Figure 16.
Time change of the plate voltage using three different types of high-voltage power supply.

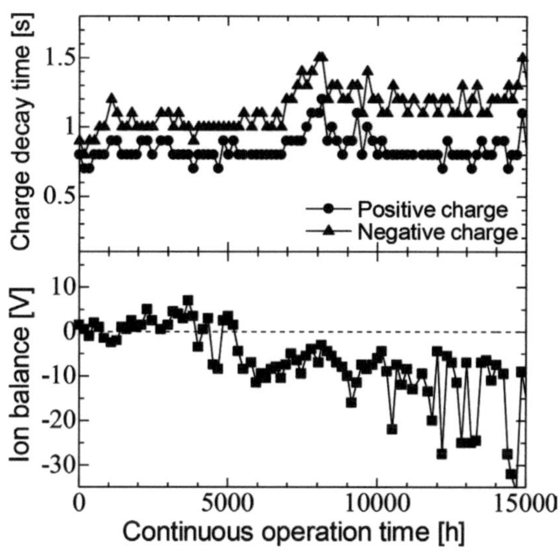

Figure 17.
Time change of the charge decay time and ion balance of a corona discharge electrostatic eliminator driven by the bipolar intermittent pulse voltage.

Figure 17 shows the time change of the charge decay time and the ion balance of a corona discharge electrostatic eliminator driven by the bipolar intermittent pulse voltage. The performance is maintained for 15,000 hours, several times longer than conventional commercial frequency AC and DC types. Recently, a perfectly balanced fan-type ionizer utilizing an intermittent pulse AC voltage power supply has been developed [36]. The short-term fluctuation range of the offset voltage is smaller than ±2 V without a sensor feedback system, which has maintained 2500 hours of continuous operation. The charge decay times are also maintained. The electric field expanded from this electrostatic eliminator has an insignificant effect on the voltage potential of the object. This ionizer is very promising for electrostatic elimination situations requiring an extremely low-offset voltage (ion balance), such as a magnetic head slider used in hard disc drives where the ion balance is restricted to a maximum of 2–10 V in the manufacturing process [15, 16, 17].

5. Electrostatic elimination without blown air

Generally, the ions produced by the corona discharges are transported to the electrostatic charged objects by blown air using a fan or a compressed air. The charge decay time decreases with increasing the air velocity as shown in **Figure 13** and is mainly dominated by the air velocity [13, 37–41]. However, blown air is limited in many situations, such as the electrostatic elimination of light weight objects and those in a clean room. In these situations, the ions are mainly transported to the objects by the Coulomb force with the electric field between the electrodes and the objects. In the cases of AC and bipolar pulse voltages, the ions are not efficiently accelerated with the rapid polarization change, which causes the decrease of the electrostatic elimination efficiency due to the increase of ion recombination loss. Therefore, a low-frequency voltage waveform is suitable for electrostatic elimination without blown air. However, as mentioned above, in a high-voltage power supply using a transformer, the magnetic flux inside the core required to magnetically couple the primary and secondary windings increases with decreasing frequency, which increases in the size and weight of the transformer. To generate a low-frequency voltage with a small and lightweight transformer, the same polarity pulse train is continuously generated in a phase by the PWM converter, as shown in **Figure 6**. **Figure 18** shows the typical waveforms of the applied voltage and the plate voltage of the CPM using the continuously generated pulses. The plate voltage continuously increases in the phase in which the same polarity pulses are continuously generated. The results show that the ions flowing into the plate and the phase have the same polarity.

When the same polarity current flows to the primary side of the transformer, the magnetic core is magnetically saturated, which decreases the amplitude and pulse width of the output voltage and causes ringing. To prevent magnetic saturation, a counter pulse with a low voltage is generated before the main pulse voltage. **Figure 19** shows the typical waveforms of pulse voltage with and without the counter pulses and the bipolar intermittent pulse. The counter pulse prevents magnetic saturation, which inhibits ringing after the main pulse and decreases the amplitude and pulse width of the pulse voltage. **Figure 20** shows the charge decay time with and without the counter pulses. Without blown air, the electrostatic charge is efficiently eliminated by the continuously generated pulses. The charge decay time

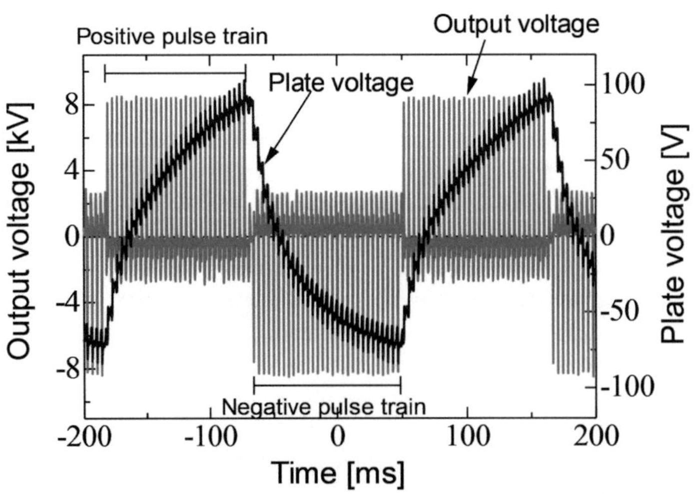

Figure 18.
Typical waveforms of the applied voltage and the plate voltage of CPM.

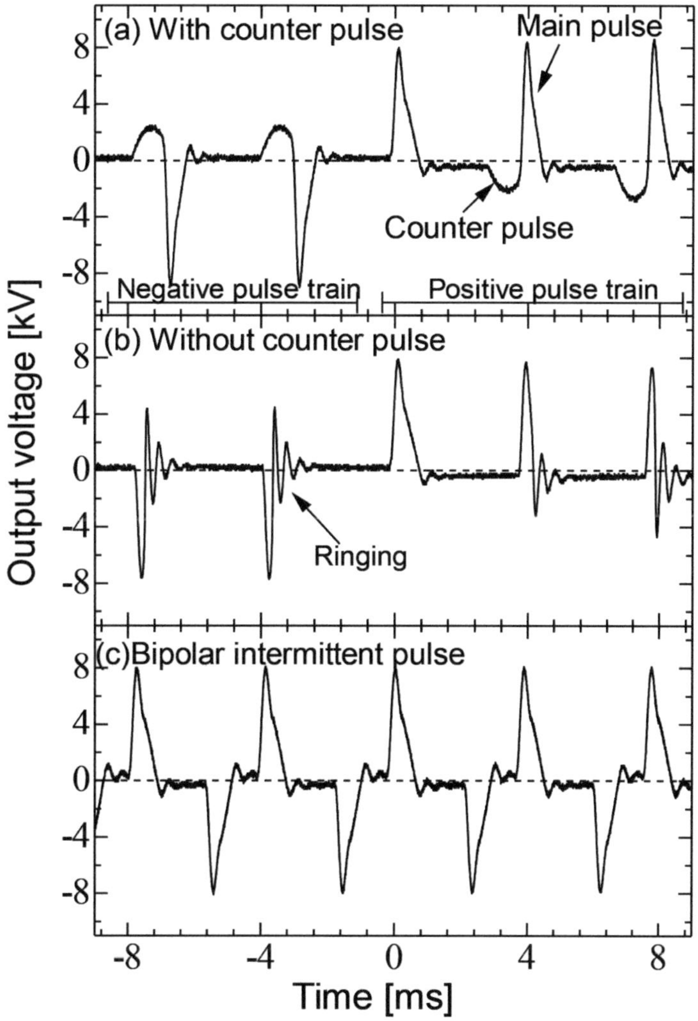

Figure 19.
Typical waveforms of pulse voltage with and without the counter pulses and the bipolar intermittent pulse [26].

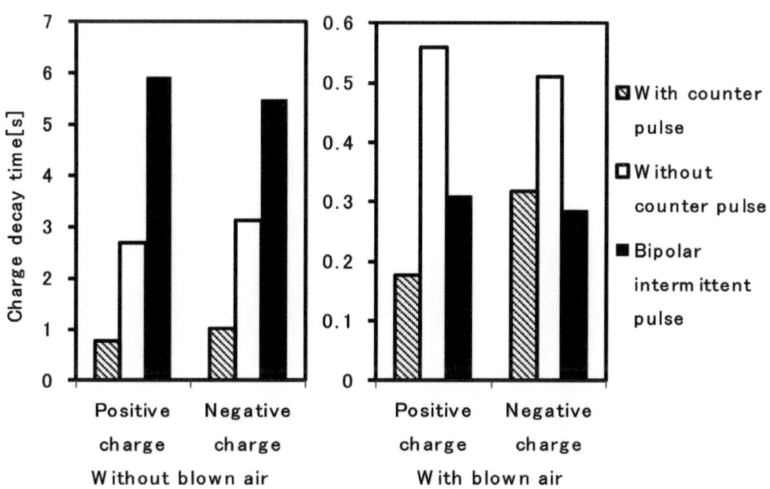

Figure 20.
Charge decay time with and without the counter pulses [26].

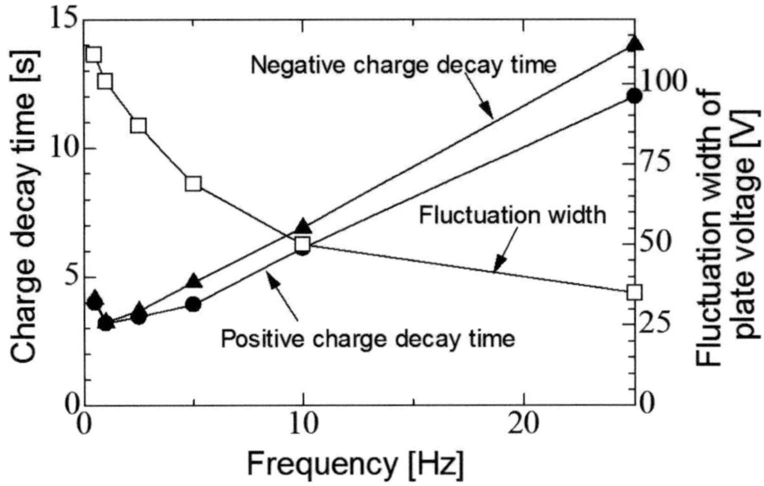

Figure 21.
Charge decay time and the fluctuation width of the plate voltage as a function of the frequency of the voltage phase [26].

with the counter pulses is shorter than that without the counter pulses. In the case of the bipolar intermittent pulse, the charge decay time is over six times longer than that of the continuously generated pulses. The ozone concentration produced by the corona discharges in the case of this method is 0.011 mg-O_3/h and is lower than that with a conventional commercial frequency AC voltage, at 0.017 mg-O_3/h. The results show that the efficiency of electrostatic elimination is much higher than conventional methods. With blown air **Figure 20b**, the charge decay time with the counter pulse is shorter than that without the counter pulses. The decrease of the amplitude and pulse width decreases the ion production and the efficiency of ion transportation as shown in **Figure 20a**.

As shown in **Figures 12** and **18**, the same polarity ions as the applied voltage continuously flow into the plate at the same polarity phase, and the plate voltage increases according to the polarity. **Figure 21** shows the charge decay time and the fluctuation width of the plate voltage as a function of the frequency of the voltage phase. The fluctuation width is measured during the steady-state condition as shown in **Figure 10**. The charge decay time and the fluctuation width increase with decreasing frequency. Without blown air, the lower frequency is suitable for the electrostatic elimination; however, it causes a large fluctuation of the voltage potential on the objects. When the capacitance and the limitation of the voltage potential of the objects are low, suppressing the fluctuation of the voltage potential is required. The developed method can generate pulse voltages with wide-range frequencies ranging from several Hz to several kHz using the same transformer and circuit, which can adjust to various situations.

6. Conclusion

An intermittent pulse voltage AC power supply controlled by a PWM inverter with a small and lightweight transformer has been developed for highly efficient electrostatic elimination, and its specifications as an electrostatic eliminator have been highlighted. The PWM control can provide a flexible control of the frequency, pulse width, and amplitude of the voltage waveform applied to the corona discharge electrode. A certain degree of freedom in generating the pulse voltages and the waveforms can be optimized in various applications, which enables the design of small and lightweight systems. The charge decay time, the time required to eliminate electrostatic charge on a charged object, is evaluated using the charge plate monitor with a high-frequency characteristic for accurate measurements. The efficiency for electrostatic elimination increases with increasing velocity of blow air and has a maximum value with pulse repetition rate of around 250 Hz in the case of intermittent bipolar pulse voltages. The high-efficiency electrostatic elimination can be realized with the optimization of the waveforms, which contributes to a long-term performance stability. The intermittent pulse voltage AC power supply can contribute to the various processes such as the processes which require very good ion balance.

Conflict of interest

The authors declare no conflict of interest.

Author details

Katsuyuki Takahashi[1,2]*, Koichi Takaki[1,2], Isao Hiyoshi[3], Yosuke Enomoto[3], Shinichi Yamaguchi[1,3] and Hidemi Nagata[3]

1 Faculty of Science and Engineering, Iwate University, Morioka, Iwate, Japan

2 Agri-Innovation Center, Iwate University, Morioka Iwate, Japan

3 Shishido Electrostatic, Ltd., Yokohama, Japan

*Address all correspondence to: ktaka@iwate-u.ac.jp

IntechOpen

© 2019 The Author(s). Licensee IntechOpen. This chapter is distributed under the terms of the Creative Commons Attribution License (http://creativecommons.org/licenses/by/3.0), which permits unrestricted use, distribution, and reproduction in any medium, provided the original work is properly cited.

References

[1] Greason WD, Castle GSP. The effects of electrostatic discharge on microelectronic devices a review. IEEE Transactions on Industry Applications. 1984;**IA-20**:247-252. DOI: 10.1109/TIA.1984.4504404

[2] Lerner A. A new additive for electrostatic discharge control in foams and elastomers. Journal of Cellular Plastics. 1985;**21**:31. DOI: 10.1177/0021955X8502100103

[3] Vinson JE, Liou JJ. Electrostatic discharge in semiconductor devices: An overview. Proceedings of the IEEE. 1998;**86**:399-418. DOI: 10.1109/5.659493

[4] Vinson JE, Liou JJ. Electrostatic discharge in semiconductor devices: Protection techniques. Proceedings of the IEEE. 2000;**88**:1878-1900. DOI: 10.1109/ICMEL.2000.840579

[5] Greason WD, Castle GSP. Review of the effect of electrostatic discharge and protection techniques for electronic systems. IEEE Transactions on Industry Applications. 1987;**IA-23**:205-216. DOI: 10.1109/TIA.1987.4504895

[6] Nagato K, Matsui Y, Miyata T, Yamauchi T. An analysis of the evolution of negative ions produced by a corona ionizer in air. International Journal of Mass Spectrometry. 2006;**248**:142-147. DOI: 10.1016/j.ijms.2005.12.001

[7] Nagato K, Nakauchi M. Experimental study of particle formation by ion-ion recombination. The Journal of Chemical Physics. 2014;**141**:164309. DOI: 10.1063/1.4898376

[8] Inaba H, Ohmi T, Morita M, Nakamura M. Neutralization of wafer charging in nitrogen gas. IEEE Transactions on Semiconductor Manufacturing. 1992;**5**:359-367. DOI: 10.1109/66.175368

[9] Inaba H, Sakata S, Yoshida T, Okada T, Ohmi T. Antistatic protection in wafer drying process by spin-drying. IEEE Transactions on Semiconductor Manufacturing. 1992;**5**:234-240. DOI: 10.1109/66.149814

[10] Manninen HE, Franchin A, Schobesberger S, Hirsikko A, Hakala J, Skromulis A, et al. Characterisation of corona-generated ions used in a neutral cluster and air ion spectrometer (NAIS). Atmospheric Measurement Techniques. 2011;**4**:2767-2776. DOI: 10.5194/amt-4-2767-2011

[11] Kraz V. Notes on maintaining Sub-1V balance of an ionizer. In: Proceedings of the 2004 Electrical Overstress/Electrostatic Discharge Symposium; 19-23 September 2004; Grapevine, TX, USA; pp. 1-6. DOI: 10.1109/EOSESD.2004.5272608

[12] Levit L, Wallash A. Measurement of the effects of ionizer imbalance and proximity to ground in MR head handling. Journal of Electrostatics. 1999;**47**:375-382. DOI: 10.1016/S0304-3886(99)00040-6

[13] Chang JS. Neutralization theory of static surface charges by an ionizer under wide gas pressure environments. IEEE Transactions on Industry Applications. 2012;**37**:1641-1645. DOI: 10.1109/28.968172

[14] Takahashi K, Sato M, Ohkubo T, Fujiwara T, Takaki K. 2-D measurement of charged particles diffusing from a double DC corona discharge ionizer. IEEE Transactions on Plasma Science. 2013;**41**:1863-1868. DOI: 10.1109/TPS.2013.2252610

[15] Mizoh Y, Nakano T, Tagashira K, Nakamura K, Suzuki T. Soft ESD phenomena in GMR heads in the HDD manufacturing process. Journal of Electrostatics. 2006;**64**:72-79. DOI: 10.1016/j.elstat.2005.03.090

[16] Wallash A, Smith D. Electromagnetic interference (EMI) damage to giant magnetoresistive (GMR) recording heads. In: Proceedings of Electrical Overstress/Electrostatic Discharge Symposium; 6-8 October 1998; Reno, NV, USA; pp. 368-374. DOI: 10.1109/EOSESD.1998.737058

[17] Baril L, Nichols M, Wallash A. Degradation of GMR and TMR recording heads using very-short-duration ESD transients. IEEE Transactions on Magnetics. 2002;**38**:2283-2285. DOI: 10.1109/TMAG.2002.802803

[18] Paasi J. Assessment of ESD threats to electronic components. Journal of Electrostatics. 2005;**63**:589-596. DOI: 10.1016/j.elstat.2005.03.021

[19] Asano K, Fukada Y, Yasukawa T. Measurement of AC ion current from a corona ionizer using a Faraday cage. Journal of Electrostatics. 2008;**66**:275-282. DOI: 10.1016/j.elstat.2008.01.006

[20] Ohsawa A, Nomura N. Continuously balanced pulse-DC ioniser to minimise the offset voltage. Journal of Electrostatics. 2016;**79**:16-19. DOI: 10.1016/j.elstat.2015.11.005

[21] Jones JE, Davies M, Goldman A, Goldman B. A simple analytic alternative to Warburgs law. Journal of Physics D: Applied Physics. 1990;**23**:542-552. DOI: 10.1088/0022-3727/23/5/012

[22] Ohsawa A. Modeling of charge neutralization by ionizer. Journal of Electrostatics. 2005;**63**:767-773. DOI: 10.1016/j.elstat.2005.03.043

[23] Ohsawa A. Computational comparison of charge neutralisations of conductors and insulators with corona ionisers. Journal of Electrostatics. 2004;**62**:219-230. DOI: 10.1016/j.elstat.2004.05.009

[24] Dordizadeh P, Adamiak K, Castle GSP. Experimental study of the characteristics of Trichel pulses in the needle-plane negative corona discharge in atmospheric air. Journal of Electrostatics. 2017;**88**:49-54. DOI: 10.1016/j.elstat.2016.12.013

[25] Frylladitakis ED, Moronis AX, Theodoridis MP. A mathematical model for determining an electrohydrodynamic accelerator's monopolar flow limit during positive corona discharge. IEEE Transactions on Plasma Science. 2017;**45**:432-440. DOI: 10.1109/TPS.2017.2663778

[26] Takahashi K, Hiyoshi I, Enomoto Y, Nagata H. Development of an electrostatic eliminator utilizing high voltage AC power supply driven by PWM inverter. Journal of International Electrostatics Japan. 2014;**38**:124-129

[27] Takahashi K, Goto A, Nagata H. Development of an electrostatic eliminator utilizing compact high-voltage AC power supply. Journal of International Electrostatics Japan. 2015;**39**:238-241

[28] ANSI/ESD STM3.1-2015. ESD association standard test method for the protection of electrostatic discharge susceptible items—Ionization: ESD Association; 2015

[29] Crowley JM, Leri D, Dahlhoff G, Levit L. Equivalent circuits for air ionizers used in static control. Journal of Electrostatics. 2004;**61**:71-83. DOI: 10.1016/j.elstat.2003.11.006

[30] Takahashi K, Kaga H, Kubo K, Takaki K, Yamaguchi S, Nagata H. Development of an ion measuring system for AC corona discharge. IEEJ Transactions on Fundamentals and Materials. 2018;**138**:551-552. DOI: 10.1541/ieejfms.138.551

[31] Chen J, Davidson JH. Electron density and energy distributions in the

positive DC corona: Interpretation for corona-enhanced chemical reactions. Plasma Chemistry and Plasma Processing. 2002;**22**(2):199-224. DOI: 10.1023/A:1014851908545

[32] Chen J, Davidson JH. Ozone production in the negative DC corona: The dependence of discharge polarity. Plasma Chemistry and Plasma Processing. 2003;**23**:501-518. DOI: 10.1023/A:1023235032455

[33] Noll CG, Lawless PA. Comparison of germanium and silicon needles as emitter electrodes for air ionizers. Journal of Electrostatics. 1998;**44**:221-238. DOI: 10.1016/S0304-3886(98)00015-1

[34] Petrov AA, Amirov RH, Samoylov IS. On the nature of copper cathode erosion in negative corona discharge. IEEE Transactions on Plasma Science. 2009;**37**:1146-1149. DOI: 10.1109/TPS.2009.2018561

[35] Weissler GL. Positive and negative point-to-plane corona in pure and impure hydrogen, nitrogen and argon. Physical Review. 1943;**63**:96-107. DOI: 10.1103/PhysRev.63.96

[36] Takahashi K, Goto A, Yamaguchi S, Saito T, Sakamoto K, Nagata H. Development of a perfectly balanced electrostatic eliminator utilizing an intermittent pulse AC voltage power supply. In: Proceedings of 2015 37th Electrical Overstress/Electrostatic Discharge Symposium (EOS/ESD), 27 September-2 October 2015; Reno, NV, USA; DOI: 10.1109/EOSESD.2015.7314753

[37] Ohsawa A. Efficient charge neutralization with an ac corona ionizer. Journal of Electrostatics. 2007;**65**:598-606. DOI: 10.1016/j.elstat.2007.01.003

[38] Osawa A. 2-D electrohydrodynamic simulations towards zero offset voltage with corona ionisers. Journal of Electrostatics. 2013;**71**:116-124. DOI: 10.1016/j.elstat.2012.12.021

[39] Noll C. Charge-carrier extraction from air and nitrogen gas streams that entrain charge from DC corona ionizers. Journal of Electrostatics. 2002;**54**:271-282. DOI: 10.1016/S0304-3886(01)00170-X

[40] Intra P, Tippayawong N. Design and evaluation of a high concentration high penetration unipolar corona ionizer for electrostatic discharge and aerosol charging. Journal of Electrical Engineering and Technology. 2013;**8**:1175-1181. DOI: 10.5370/JEET.2013.8.5.1175

[41] Intra P, Tippayawong N. Effect of needle cone angle and air flow rate on electrostatic discharge characteristics of a corona-needle ionizer. Journal of Electrostatics. 2010;**68**:254-260. DOI: 10.1016/j.elstat.2010.01.008

Chapter 4

Electrostatic Friction Displays to Enhance Touchscreen Experience

Reza Haghighi Osgouei

Abstract

Touchscreens are versatile devices that can display visual content and receive touch input, but they lack the ability to provide programmable tactile feedback. This limitation has been addressed by a few approaches generally called *surface haptics* technology. This technology modulates the friction between a user's fingertip and a touchscreen surface to create different tactile sensations when the finger explores the touchscreen. This functionality enables the user to see and feel digital content simultaneously, leading to improved usability and user experiences. One major approach in surface haptics relies on the electrostatic force induced between the finger and an insulating surface on the touchscreen by supplying high AC voltage. The use of AC also induces a vibrational sensation called *electrovibration* to the user. Electrostatic friction displays require only electrical components and provide uniform friction over the screen. This tactile feedback technology not only allows easy and lightweight integration into touchscreen devices but also provides dynamic, rich, and satisfactory user interfaces. In this chapter, we review the fundamental operation of the electrovibration technology as well as applications have been built upon.

Keywords: electrostatic display, variable friction display, electrovibration technology, surface haptics, tactile rendering, texture rendering

1. Introduction

Among the five senses, touch is the most fundamental one we are equipped from the moment we enter this world. Even newborn babies know how to utilize their sense of touch to interact with their surrounding environment. Many of the typical tasks around us require touch which without it even a very basic task would be challenging to accomplish. Just imagine how difficult it can be to grab any object if you cannot feel its shape and weight or determine the amount of force you need to apply to hold it. Touch is very important to human being, and we rely on our touch sense more than we think we do [1].

Modern technologies in this digital era added new interactive agents around us which require our touch input. Touchscreen consumer electronics such as smartphones and tablet devices are among them. They are a versatile device that displays visual content and takes touch input simultaneously. More specifically, smartphones are an inevitable part of our daily life. Users spend a significant amount of time interacting with the digital contents on their mobile phones. So, equipping such devices with functionality to provide some sort of touch feedback was inevitable and seemed to be a natural course of technological development.

IntechOpen

However, despite technological advances, these devices lack the ability to provide programmed tactile feedback, which can be essential for more natural and intuitive interaction. At best, they provide some simple monotonic vibration patterns in response to the user's touch input. This is neither appealing nor satisfactory given the expectations users have from such modern devices [1].

With the introduction of variable friction displays, this limitation has been addressed by technologies collectively called *surface haptics*. These technologies modulate friction between a user's fingertip and a touchscreen surface in order to create a variety of tactile sensations when the finger explores on the touchscreen. This functionality allows the user to see and feel the digital content simultaneously with richer haptic information, leading to improved user experience and/or usability. There exist two major approaches in surface haptics: electrovibration and ultrasonic vibration. Whereas the former increases the surface friction by modulating attractive electrostatic force, the latter decreases the friction by vibrating the surface at an ultrasonic frequency and creating an air gap. Such electrovibration displays have the advantages that they require only electrical components and that the friction can be controlled uniformly on the screen, which are particularly attractive for mobile devices with a provision of adequate amplifiers [2].

The rest of the chapter is organized as following. In the next section, a brief overview of the fundamental operation of the electrovibration technology is given. Next, the literature has been reviewed for the studies, and applications have been built upon. In the final section, conclusions and future remarks are provided.

2. Electrovibration technology

The earliest known observation of electrical attraction between the human skin and a charged surface was made by Gray in 1875 [3, 4]. Forgotten for a while, a similar phenomenon was rediscovered later and called *electroadhesion* by Johnsen and Rahbek in 1923 [4, 5]. In 1953, Mallinckrodt et al. again reported a rubber-like sensation when a coated metallic surface connected to a 110-V power line was touched by a grounded finger [4, 6]. This phenomenon is called *electrovibration* by Grimnes in 1983, explaining its principle of operation based on Coulomb's electrostatic force [7]. Electrovibration is due to the electrostatic attraction force between two conductive plates separated by a dielectric. When the finger scans an insulated electrode, a condenser is formed between the electrode and the conductive substance under the skin [7] (**Figure 1**). Exciting the electrode using a periodic voltage

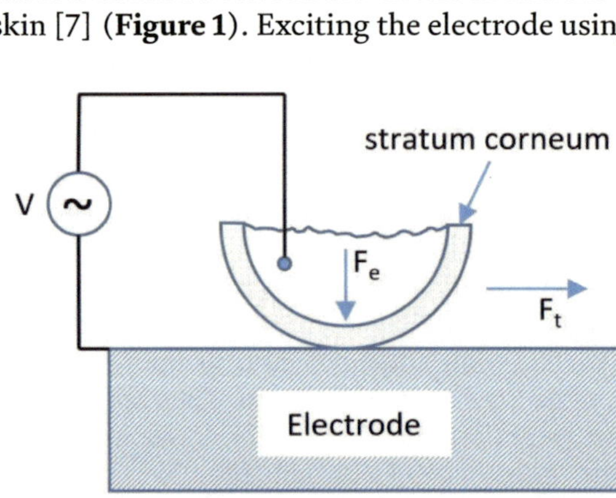

Figure 1.
Interaction between the finger, the isolating part of the skin (stratum corneum), and the conductive plate.

induces electrostatic attraction, and this increases the friction force between the surface and the moving finger.

The induced friction is perceived by the mechanoreceptors in the fingertip skin. In general, mechanoreceptors are responsible to perceive sensations such as pressure, vibration, and texture, and there are four types of them in hairless skin, Merkel discs, Meissner's corpuscles, Ruffini corpuscles, and Pacinian corpuscles, as shown in **Figure 2**. They are categorized into fast-adapting (Pacinian and Meissner) and slow-adapting (Merkel and Ruffini) receptors. The former ones detect small and fast changes such as surface roughness, while the latter ones detect static perception such as pressure. It has been shown that the electrovibration is primarily perceived through the Pacinian channel [9].

Nevertheless, when a potential is applied, the electrostatic force, $F_e = \frac{\epsilon A V^2}{2d^2}$, compresses the stratum corneum, where A is contact area, d is thickness of stratum corneum, V is instantaneous potential difference, and ϵ is dielectric constant. Because there are no nerve endings in the stratum corneum, the compression will not be sensed. By moving the skin along the metal electrode, another force perpendicular to the compressive force will arise. This frictional tangential force is given by $F_t = \mu(F_e + F_n)$, where μ is coefficient of friction and F_n is contact pressure (normal force) exerted by human body. μ is therefore an appreciable amount of transfer from the perpendicular compressional force to the tangential frictional force.

This electrostatic stimulation was introduced into a tactile display by Strong et al. [10]. They developed the first electrostatic display using a stimulator array consisting of a large number of small electrodes. They reported that the intensity of the perceived vibration was mainly due to the peak applied voltage. Later, a polyimide-on-silicon electrostatic fingertip tactile display was fabricated with 49 electrodes arranged in a square array [11]. They conducted experiments to assess the intensity and spatial resolution of the tactile percepts. In a following study, its application to present various spatial tactile patterns such as line, triangle, square, and circle to the visually impaired users is investigated [12]. In all these works, the dryness of fingertip is emphasized to be the key factor maintaining the percept, reporting that a small amount of sweat could cause the percept to fade or disappear. The direct method has difficulty in stable stimulation because of finger perspiration. Indirect stimulation was suggested as a solution. Yamamoto et al. built a display with a thin slider film between electrostatic stator electrodes and fingertip

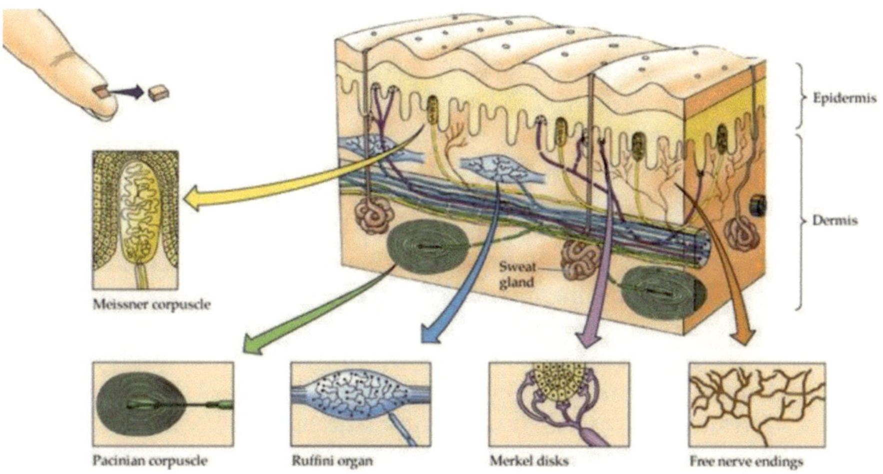

Figure 2.
Touch mechanoreceptors in the hairless (glabrous) skin of the human fingertip [8].

for presenting surface roughness [13]. In another work, multiple contact pads are used for multi-finger interaction with a large electrostatic display [14]. This was mainly to address finger perspiration during direct interaction and also to create larger force by applying higher voltage. This also enables multi-finger interaction.

Electrovibration regained attention in 2010 after a collaboration between Disney Research and Carnegie Mellon University yielded to a system for rendering 3D textures onto an electrovibration touchscreen. Called TeslaTouch [15], the developed system could deliver variable friction to user's sliding finger by modifying amplitude and frequency of the excitation signal. Implemented on top of a tablet computer, a user could perceive real-time tactile feedback correspond to the displayed digital content. Different tactile effects could be generated mimicking surface geometry such as bumps and ridges or surface texture such as frictional patterns to enhance user experience interacting with the objects in the scene.

The core of TeslaTouch is a transparent capacitive touch panel (Microtouch, 3 M, USA) driven by a high-voltage signal to modulate friction on a sliding finger. The panel is made of a thick glass layer on the bottom, a transparent electrode (indium tin oxide; ITO) in the middle, and a thin insulator layer on the top. In the usual setup, the electrode is excited by high AC voltage, and the human body is grounded electrically. The big advantage of TeslaTouch is that the capacitive panel is a commercial off-the-shelf product which requires only an additional high-voltage amplifier for proper operation. The same panel has been used in electrovibration displays by other groups [16–22]. Radivojevic et al. at Nokia introduced a flexible and bendable version by replacing indium tin oxide (ITO) with graphene [23].

While TeslaTouch was mainly designed for desktop applications, a company in Finland, Senseg, developed Tixel [24], a transparent electrostatic film targeting handheld devices. The touch panel is made of transparent electrodes on a glass plate coated with an insulating layer. By applying a periodic voltage to the electrodes via connections used for sensing a finger's position on the screen, the researchers were able to effectively induce a charge in a finger dragged along the surface. By changing the amplitude and frequency of the applied voltage, the surface can be made to feel as though it is bumpy, rough, sticky, or vibrating. The major difference is the specially designed control circuit that produces the sensations.

The tactile experience comes from two components: a coating layered atop touchscreen and electronics that modulate the electrostatic field and produce textures. Senseg's Tixel is the means by which Senseg's technology transmits electrovibration stimulus. It is an ultrathin durable coating on the touch interface that outputs tactile effects. The hardware inside a device modulates the signal for varied intensities of tactile sensation and types of tactile effects and provides accurate spatial resolution over the entire Tixel surface area.

Senseg later introduced a short-lived commercial product called Feelscreen, a 7″ Android tablet overlaid with Tixel, into the market between 2014 and 2016. Feelscreen has been used in several projects such as 3D shape rendering [25], texture gradients [26], and visual and haptic latency [27]. At the moment, Tanvas [28], a startup company in the USA, is commercializing similar products but on a larger 10″ tablet with some improvements such as generating stronger friction forces and not requiring an external power supply.

Some other researchers developed their own electrovibration display not using the 3 M capacitive touch panel. Pyo et al. built a tactile display that provides both electrovibration and mechanical vibration on a large surface [29]. They fabricated an insulated ITO electrode on top of an electrostatic parallel plate actuator, both operating based on the electrostatic principle. A nontransparent electrostatic friction display was also developed in [30, 31] using an aluminum plate covered with a thin plastic insulator film.

These displays do not support multi-touch or localized friction modulation, and all fingers in contact with the surface experience the same sensation. This issue was addressed by several prototypes presenting local stimulation. For example, a display panel was developed with multiple horizontal and vertical ITO electrodes in a grid enabling localized stimulation at the region where the vertical and horizontal electrodes cross each other [32]. In [14], a multi-finger electrostatic display was developed consisting of a transparent electrode and multiple contact pads on which users place their fingers. Applying different voltages to the pads and electrically grounding the transparent electrode induce different frictional stimuli to the multiple fingers.

The relationship between input signal and output friction in electrostatic friction displays is not clearly understood, and a number of studies have shown great interest in defining such relationship. Researchers have worked on this topic either by measuring friction forces using a tribometer [16, 31] or by estimating perceived intensities in psychophysical experiments [17, 33]. For instance, Meyer et al. [16] developed a tribometer to make precise measurements of finger friction and confirmed the expected square law of frictional force to driving voltage. They also showed a linear mapping between friction and normal force, confirming the Coulombic model of dry friction. Conducting a six-value effect strength subjective index rating, Wijekoon et al. showed a significant correlation (0.8) between signal amplitude and perceived intensity but no correlation between frequency and perceived intensity [33]. In [17], participants assigned a number between 0 and 100 to the subjective friction intensity. A linear fit in log-log scale was observed in the normalized results relating applied voltage amplitude to perceived friction force intensity.

As well as fabrication, various properties of electrovibration have been investigated too. The polarity effect of the actuation signal is studied in [34], reporting that tactile sensation is more sensitive to negative than positive pulses. Meyer et al. showed an expected square law dependence of frictional force, measured by a tribometer, on actuation voltage [16]. A similar approach is taken by Vezzoli et al. to develop a model for electrovibration effect considering frequency dependence [31]. Kim et al. proposed a current control method to provide more uniform perceived intensity of electrovibration [19]. In another work and by comparing two actuation signals, it is reported that square waves are more detectable than sine waves at frequencies lower than 60 Hz while they are same at higher frequencies [35]. Testing three methods, amplitude modulation, adding DC offset, and their combination, Kang et al. investigated low-voltage operation of electrovibration display [22]. They showed all methods increased dynamic friction force, while only DC offset increased static friction force.

3. Applications

To perceive the friction force generated on an electrovibration display, one requires to drag or slide their finger over the surface. While this type of interaction is natural and intuitive for most of handheld touchscreen devices, however, it limits the range of applications can benefit from this functionality. It is worth to recall that the two key attributes of real and simulated objects are shape (surface geometry) and texture (simply surface frictional properties) [36]. Addressing these two attributes separately, in this section we review the relevant work in the literature.

3.1 Rendering surface geometry

Rendering 3D objects on a flat surface, either using a haptic interface or a variable friction display, has not been addressed much in the literature. In an early work regarding haptic perception of curvature, Gordon and Morison showed that the gradient is an effective stimulus for curvature perception and humans rely on local curvature when perceiving surface [37]. Later, Minsky et al. demonstrated that tangential force alone can be sufficient for rendering surface texture assuming it is made of little bumps [38]. They introduced *gradient technique* to create the illusion of bumps and valleys using a 2D force-feedback joystick. As the user moves the joystick in a direction which is *up* a bump, his motion is opposed by a spring force proportional to the height of the bump. This gives the sense that it is very difficult to move to the top of the bump (springs resist being stretched) and easy to fall off the bump back into a lower region of the simulated surface (springs like to revert to a short length). For fine-grained surfaces, joystick spring forces can be computed based on a local gradient. As the user moves the joystick on the virtual surface, the change in height in the direction of motion is noted. We create virtual springs opposing the motion "up" the sides of each tiny bump. Thus, the spring forces applied to the hand are computed from local gradients of the height of the surface.

Based on the *gradient technique*, an early attempt to create the haptic illusion of a non-flat shape on a nominally flat surface was introduced in [39] using a force-shading algorithm. Later continuing their earlier work [40], Robles-De-La-Torre and Hayward demonstrated that in active exploration of a physical shape, lateral force applied to the sliding finger plays the main role in the perception of shape [41]. They investigated the accuracy of physical shape recognition using a one-degree of freedom (DoF) force-feedback device without visual cues. Different combinations of physical and virtual geometries (bump, hole, and flat surface), e.g., a virtual bump laid on a physical flat surface, were presented to participants. The virtual shapes were rendered using lateral force only. Participants could accurately identify the virtual shapes in all conditions.

This study was foundational to the gradient-based algorithm of Kim et al. [17] for rendering 3D features on a touchscreen using electrovibration. In their work, a psychophysical perceptual model, subjectively relating the perceived friction to the applied voltage, was formulated. The model was a straight line in log–log scale, fitted over average users' ratings of the perceived friction intensity in a scale of 0–100. The model then utilized to modulate friction and render three lateral force profiles: height, slope, and rectangular. They compared users' preference for three types of force profile for a visual bump displayed on the screen. Results indicated that the slope profile best matched the visual bump. They generalized this finding to a 2D gradient-based rendering algorithm for 3D features and applied the algorithm to many user interface examples.

In Ref. [25], the authors presented an effective rendering method for improving the recognition of 3D features rendered on a touchscreen using an electrostatic friction display. First, a formative user study is carried out using a basic gradient-based algorithm adapted from [41] in order to assess users' ability of recognizing primitive 3D shapes based on lateral force feedback provided by an electrostatic tablet and a force-feedback interface. Experimental results demonstrated that users are not able to associate electrovibration patterns with geometric shapes in an absolute manner without contextual information. However, when such guidance was given, participants achieved moderate recognition. Then, they extended the basic algorithm to support general 3D mesh objects. The generalized algorithm computes the frictional rendering force by estimating the gradient at the touch point and also emphasizes sharp edges on the surface by rendering perceptually salient friction effects. Lastly, they conducted a summative user study to evaluate the effectiveness of their proposed shape rendering algorithm in reducing the visual uncertainty in 3D shape perception. They found that when frictional feedback was provided, the correct recognition performance was notably increased in comparison with when only visual rendering was presented.

3.2 Rendering surface texture

Compared to the problem of rendering 3D geometries on a flat electrostatic display, rendering surface textures seems more feasible and intuitive on such displays. As mentioned earlier, depending on the actuation signal, an electrovibration display generates different textural patterns. A simple illustration is given in **Figure 3**. On one hand, a sinusoid actuation signal creates a smooth bumpiness underneath of sliding finger. On the other hand, a square wave signal generates a rough and edgy feeling. A more complicated texture can be re-created using a proper complex signal.

Therefore, the type of input signal, its waveform, its amplitude, and its frequency components play a significant role on the generated textural patterns. Hence looking at the problem from a systematic standpoint, knowing the input–output relationship of the display is vital for this problem. As stated earlier, several efforts have been made modeling the display and drawing a relationship between the input actuation voltage and the output friction force. However, aside from the fact that the output force is somehow proportional to the squared input voltage, there exists no reliable general model covering all type of input signals across a wide range of frequencies. This suggests an alternative method, the so called data-driven texture rendering. Data-driven, or measurement-based, haptic rendering is a general approach that uses recordings from real objects to generate realistic haptic feedback in virtual environments [42, 43]. It can be either parametric- and physics-based, to optimize parameters of a predefined model, or nonparametric and generic. It is usually accompanied by a generic interpolation scheme to handle the data sets not being measured. It provides a unified framework to capture and display a diverse range of physical phenomena, while not requiring simulations of complex contact dynamics. This data-driven approach enables researchers to bypass the complex step of hand tuning a dynamic simulation of the target interaction to try to match a haptic sensation. Instead, the goal of the modeling process is to capture the output response of the system (e.g., force and acceleration) given some set of user inputs (e.g., position, velocity, and force). Such methods shift the focus from reproducing the physics of the interaction to reproducing the real sensations felt by the user, and thus they have been largely successful at realistic haptic simulation [44].

While the problem of data-driven haptic texture rendering has been fairly addressed in the literature using conventional or customized haptic interfaces

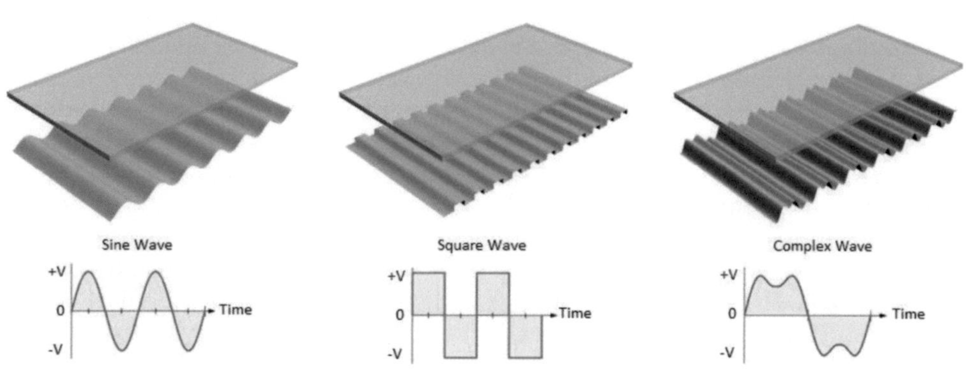

Figure 3.
How the input actuation signal makes the perceived friction different.

[45–51], little work has been done on variable friction displays and particularly using electrovibration attraction.

An electrostatic friction display creates clearly perceptible stimuli when the surface is laterally scanned, but not when the finger is stationary. This fundamental limitation has confined the application of electrostatic friction displays mostly to texture rendering. In the only relevant work [18], Ilkhani et al. proposed a data-driven texture rendering method by recording accelerations from three real materials and playing them back on an electrovibration display. Their automated data collection is done under single constraint condition (contact force 0.35 N and scanning velocity 0.74 m/s) using a servomotor controlled by an Arduino Uno. They conducted a user study to compare the perceived surface roughness generated with their data-driven signals and with that of square wave signals. The frequency of each square wave is set based on the main frequency of the corresponding acceleration. Using a visual indicator, they made the user to keep a constant scanning velocity, but not equal to the data collection velocity and presumably very slower than that. In addition, there is no mention of contact force status during experimentation. Nevertheless, they reported higher percentage of similarity between data-driven textures and real ones than square wave patterns. In their extended work [52], they applied the same approach on the data from Penn Haptic Texture Toolkit [53] and performed MDS analysis to create a perceptual space and to extract underlying dimensions of the textures. Their results showed roughness and stickiness as the primary dimensions of texture perception.

In ref. [54], a data-driven neural network for realistic texture rendering on an electrovibration display is proposed. First, a motorized linear tribometer is developed to collect lateral frictional forces from the textured surfaces under various scanning velocities and normal forces. Then an inverse dynamics model of the display is created to describe its output-input relationship using nonlinear autoregressive with external input (NARX) neural networks. Forces resulting from applying a full-band pseudorandom binary signal (PRBS) to the display are used to train each network under the given experimental condition. A comparison between the real and virtual forces in frequency domain shows promising results and reveals the capabilities and limitations of the proposed technique.

4. Conclusions

In this chapter, we have introduced the concept behind electrostatic friction displays (also called electrovibration displays) and their potential applications for shape and texture rendering. The potential uses for the technique are exciting. Electrovibration could make interactive textbooks more engaging on tablets, allowing students to explore the three-dimensional features of an object directly on each page. Software for iOS or Android could be augmented with unique haptic feedback for button presses and swipe gestures. Games could incorporate electrovibration to add a new layer of interactivity to touch controls. With some smart design, it could really improve the functionality of touchscreens used in other fields, as well. For instance, the use of touchscreens in automobiles to navigate the map or control the music playback persuades drivers to avert their eyes from the road. Possibly, with an appropriate design, the same control functionalities could be delivered using a variable touch-based feedback without the need to take our eyes off the road. Given the commonness of capacitive touchscreens, the addition of richer tactile feedback through electrovibration promises to enhance almost all of our interactions with digital contents.

While the technique has a lot of potential, the form factor remains a primary barrier to adoption. Implementing the alternating voltage results in a much bulkier device than with an ordinary capacitive touchscreen. As the technology sees more frequent use, however, there may be technological developments that allow more smartphone and tablet manufacturers to feature electrovibration without sacrificing the compactness of their designs.

Acknowledgements

I would like to thank Prof. Seungmoon Choi for his extensive personal and professional guidance teaching me a great deal about both scientific research and life in general. As my teacher and mentor, he has taught me more than I could ever give him credit for here. He has shown me, by his example, what a good scientist (and person) should be. I also would like to thank Dr. Jin Ryung Kim for his constant support and encouragement.

Conflict of interest

I certify that I have no affiliations with or involvement in any organization or entity with any financial interest (such as honoraria; educational grants; participation in speakers' bureaus; membership, employment, consultancies, stock ownership, or other equity interest; and expert testimony or patent-licensing arrangements) or nonfinancial interest (such as personal or professional relationships, affiliations, knowledge, or beliefs) in the subject matter or materials discussed in this manuscript.

Author details

Reza Haghighi Osgouei
Imperial College London, London, UK

*Address all correspondence to: r.haghighi-osgouei@imperial.ac.uk

IntechOpen

© 2020 The Author(s). Licensee IntechOpen. This chapter is distributed under the terms of the Creative Commons Attribution License (http://creativecommons.org/licenses/by/3.0), which permits unrestricted use, distribution, and reproduction in any medium, provided the original work is properly cited. CC BY

References

[1] Haghighi Osgouei R. Haptic Rendering of Surface 3D Curvature and Texture on Electrostatic Friction Display. 2017

[2] Haghighi Osgouei R, Kim JR, Choi S. Identification of primitive geometrical shapes rendered using electrostatic friction display. In: IEEE Haptics Symposium (HAPTICS). 2016. pp. 198-204

[3] Gray E, inventor. Improvement in electric telegraphs for transmitting musical, tones. United States patent US US166,096; 1875

[4] Haghighi Osgouei R, Kim JR, Choi S. Data-driven texture modeling and rendering on electrovibration display. IEEE Transactions on Haptics. 2019. DOI: 10.1109/TOH.2019.2932990

[5] Johnsen A, Rahbek K. A physical phenomenon and its applications to telegraphy, telephony, etc. Journal of the Institution of Electrical Engineers. 1923;**61**(320):713-725

[6] Mallinckrodt E, Hughes AL, Sleator W Jr. Perception by the skin of electrically induced vibrations. Science. 1953;**118**:277-278. Available from: https://doi.org/10.1126/science.118.3062.277

[7] Grimnes S. Electrovibration, cutaneous sensation of microampere current. Acta Physiologica Scandinavica. 1983;**118**(1):19-25

[8] Darian-Smith I. The sense of touch: Performance and peripheral neural processes. In: Comprehensive Physiology. Wiley [Online Library]. 2011. pp. 739-788

[9] Vardar Y, Güçlü B, Basdogan C. Effect of waveform on tactile perception by electrovibration displayed on touch screens. IEEE Transactions on Haptics. 2017;**10**(4):488-499

[10] Strong RM, Troxel DE. An electrotactile display. IEEE Transactions on Man-Machine Systems. 1970;**11**(1): 72-79

[11] Beebe DJ, Hymel CM, Kaczmarek KA, Tyler ME. A polyimide-on-silicon electrostatic fingertip tactile display. In: Engineering in Medicine and Biology Society, 1995, IEEE 17th Annual Conference. Vol. 2. 1995. pp. 1545-1546

[12] Tang H, Beebe DJ. A microfabricated electrostatic haptic display for persons with visual impairments. IEEE Transactions on Rehabilitation Engineering. 1998;**6**(3): 241-248

[13] Yamamoto A, Ishii T, Higuchi T. Electrostatic tactile display for presenting surface roughness sensation. In: IEEE International Conference on Industrial Technology. Vol. 2. 2003. pp. 680-684

[14] Nakamura T, Yamamoto A. Multi-finger electrostatic passive haptic feedback on a visual display. In: World Haptics Conference (WHC). 2013. pp. 37-42

[15] Bau O, Poupyrev I, Israr A, Harrison C. TeslaTouch: Electrovibration for touch surfaces. In: Proceedings of the 23nd Annual ACM Symposium on User Interface Software and Technology. 2010. pp. 283-292

[16] Meyer DJ, Peshkin MA, Colgate JE. Fingertip friction modulation due to electrostatic attraction. In: World Haptics Conference (WHC). 2013. pp. 43-48

[17] Kim SC, Israr A, Poupyrev I. Tactile rendering of 3D features on touch surfaces. In: Proceedings of the 26th Annual ACM Symposium on User Interface Software and Technology. 2013. pp. 531-538

[18] Ilkhani G, Aziziaghdam M, Samur E. Data-driven texture rendering with electrostatic attraction. In: International Conference on Human Haptic Sensing and Touch Enabled Computer Applications. 2014. pp. 496-504

[19] Kim H, Kang J, Kim KD, Lim KM, Ryu J. Method for providing electrovibration with uniform intensity. IEEE Transactions on Haptics. 2015;**8**(4):492-496

[20] Zhang Y, Harrison C. Quantifying the targeting performance benefit of electrostatic haptic feedback on touchscreens. In: Proceedings of the International Conference on Interactive Tabletops & Surfaces. 2015. pp. 43-46

[21] Wang Q, Ren X, Sarcar S, Sun X. EV-pen: Leveraging Electrovibration haptic feedback in pen interaction. In: Proceedings of the ACM on Interactive Surfaces and Spaces. 2016. pp. 57-66

[22] Kang J, Kim H, Choi S, Kim KD, Ryu J. Investigation on low voltage operation of electrovibration display. IEEE Transactions on Haptics. 2017;**10**(3):371-381

[23] Radivojevic Z, Beecher P, Bower C, Haque S, Andrew P, Hasan T, et al. Electrotactile touch surface by using transparent graphene. In: Proceedings of the Virtual Reality International Conference. 2012. p. 16

[24] Linjama J, Mäkinen V. E-sense screen: Novel haptic display with capacitive electrosensory interface. In: HAID, 4th Workshop for Haptic and Audio Interaction Design. 2009

[25] Haghighi Osgouei R, Kim JR, Choi S. Improving 3D shape recognition with electrostatic friction display. IEEE Transactions on Haptics. 2017;**10**(4):533-544

[26] Klatzky RL, Adkins S, Bodas P, Haghighi Osgouei R, Choi S, Tan HZ. Perceiving texture gradients on an electrostatic friction display. In: World Haptics Conference (WHC). 2017. pp. 154-158

[27] Kim JR, Haghighi Osgouei R, Choi S. Effects of visual and haptic latency on touchscreen interaction: A case study using painting task. In: World Haptics Conference (WHC). 2017. pp. 159-164

[28] TanvasTouch®. Available from: https://tanvas.co [Accessed: 02 September 2020]

[29] Pyo D, Ryu S, Kim SC, Kwon DS. A new surface display for 3D haptic rendering. In: International Conference on Human Haptic Sensing and Touch Enabled Computer Applications. 2014. pp. 487-495

[30] Giraud F, Amberg M, Lemaire-Semail B. Merging two tactile stimulation principles: Electrovibration and squeeze film effect. In World Haptics Conference (WHC). 2013. pp. 199-203

[31] Vezzoli E, Amberg M, Giraud F, Lemaire-Semail B. Electrovibration modeling analysis. In International Conference on Human Haptic Sensing and Touch Enabled Computer Applications. 2014. pp. 369-376

[32] Haga H, Yoshinaga K, Yanase J, Sugimoto D, Takatori K, Asada H. Electrostatic tactile display using beat phenomenon of voltage waveforms. In: SID Symposium Digest of Technical Papers. Vol. 45. No. 1. 2014. pp. 623-626

[33] Wijekoon D, Cecchinato ME, Hoggan E, Linjama J. Electrostatic modulated friction as tactile feedback: Intensity perception. In: International Conference on Human Haptic Sensing and Touch Enabled Computer Applications. 2012. pp. 613-624

[34] Kaczmarek KA, Nammi K, Agarwal AK, Tyler ME, Haase SJ,

Beebe DJ. Polarity effect in electrovibration for tactile display. IEEE Transactions on Biomedical Engineering. 2006;**53**(10):2047-2054

[35] Vardar Y, Güçlü B, Basdogan C. Effect of waveform in haptic perception of electrovibration on touchscreens. In: International Conference on Human Haptic Sensing and Touch Enabled Computer Applications. 2016. pp. 190-203

[36] Campion G, Hayward V. Fundamental limits in the rendering of virtual haptic textures. In: First Joint Eurohaptics Conference and Symposium on Haptic Interfaces for Virtual Environment and Teleoperator Systems. World Haptics Conference. Pisa, Italy; 2005. pp. 263-270

[37] Gordon IE, Morison V. The haptic perception of curvature. Perception & Psychophysics. 1982;**31**(5):446-450

[38] Minsky M, Ming OY, Steele O, Brooks Jr FP, Behensky M. Feeling and seeing: Issues in force display. In: Proceedings of the 1990 Symposium on Interactive 3D Graphics (I3D '90). Association for Computing Machinery. New York, NY, USA; 1990:235-241. Available from: https://doi.org/10.1145/91385.91451

[39] Morganbesser HB, Srinivasan MA. Force shading for shape perception in haptic virtual environments. In: Proceedings of the 5th Annual Symposium on Haptic Interfaces for Virtual Environment and Teleoperator Systems. Vol. 58. Atlanta, GA, DSC: ASME/IMECE; 1996

[40] Robles-De-La-Torre G, Hayward V. Virtual surfaces and haptic shape perception. In: Proceedings ASME IMECE Symposium on Haptic Interfaces for Virtual Environments and Teleoperator Systems. Vol. 69. 2000. p. 2

[41] Robles-De-La-Torre G, Hayward V. Force can overcome object geometry in the perception of shape through active touch. Nature. 2001;**412**(6845):445

[42] Hover R, Kósa G, Szekly G, Harders M. Data-driven haptic rendering—From viscous fluids to visco-elastic solids. IEEE Transactions on Haptics. 2009;**2**(1):15-27

[43] Yim S, Jeon S, Choi S. Data-driven haptic modeling and rendering of viscoelastic and frictional responses of deformable objects. IEEE Transactions on Haptics. 2016;**9**(4):548-559

[44] Romano JM, Yoshioka T, Kuchenbecker KJ. Automatic filter design for synthesis of haptic textures from recorded acceleration data. In: IEEE International Conference on Robotics and Automation (ICRA). 2010. pp. 1815-1821

[45] Choi S, Tan HZ. Toward realistic haptic rendering of surface textures. IEEE Computer Graphics and Applications. 2004;**24**(2):40-47

[46] Andrews S, Lang J. Haptic texturing based on real-world samples. In: IEEE International Workshop on Haptic, Audio and Visual Environments and Games (HAVE). 2007. pp. 142-147

[47] Campion G. On the synthesis of haptic textures. In: The Synthesis of Three Dimensional Haptic Textures: Geometry, Control, and Psychophysics. London: Springer; 2008. pp. 73-97

[48] Lang J, Andrews S. Measurement-based modeling of contact forces and textures for haptic rendering. IEEE Transactions on Visualization and Computer Graphics. 2011;**17**(3):380-391

[49] Culbertson H, Unwin J, Kuchenbecker KJ. Modeling and rendering realistic textures from unconstrained tool-surface interactions. IEEE Transactions on Haptics. 2014;**3**:1

[50] Shin S, Haghighi Osgouei R, Kim KD, Choi S. Data-driven modeling

of isotropic haptic textures using frequency-decomposed neural networks. In: World Haptics Conference (WHC). 2015. pp. 131-138

[51] Abdulali A, Jeon S. Data-driven rendering of anisotropic haptic textures. In: International Asia Haptics Conference. 2016. pp. 401-407

[52] Ilkhani G, Aziziaghdam M, Samur E. Data-driven texture rendering on an electrostatic tactile display. International Journal of Human–Computer Interaction. 2017;**33**(9):756-770

[53] Culbertson H, Lopez Delgado JJ, Kuchenbecker KJ. The Penn Haptic Texture Toolkit for modeling, rendering, and evaluating haptic virtual textures, Departmental Papers (MEAM). 299; 2014

[54] Haghighi Osgouei R, Shin S, Kim JR, Choi S. An inverse neural network model for data-driven texture rendering on electrovibration display. In: IEEE Haptics Symposium (HAPTICS). 2018. pp. 270-277

Chapter 5

Failure Mechanism and Property Modification of Insulation Material Used in Inverter-Fed Motors

Yan Yang and Guangning Wu

Abstract

The insulation system of inverter-fed motors is subjected to repetitive impulse voltages, which are generated by pulse width modulation (PWM) converters, and is expected to withstand partial discharge (PD) activity during service. PD-induced dielectric breakdown becomes one of the most important reasons for premature failures of motor insulation. Therefore, the dielectric properties of insulation material used in inverter-fed motors need to be improved to meet the requirement for rapid development of applications. Many approaches have been proposed to improve dielectric properties of insulation material used in inverter-fed motors. In this chapter, we will discuss the failure mechanism, modification approaches, and their effects on dielectric properties of insulation material of inverter-fed motors.

Keywords: inverter-fed motors, insulation material, failure mechanism, modification

1. Introduction

The insulation system of inverter-fed motors is subjected to repetitive impulse voltages that are generated by pulse width modulation (PWM) converters. For high-voltage motors with the rated voltage value greater than 700 V, which also be called as Type II motors, partial discharges (PD) may occur in all insulation systems during operation, but breakdown due to high-voltage endurance may not occur. The insulation materials are expected to withstand partial discharge (PD) activity during service. Under repetitive impulse voltages, accelerating of insulation deterioration would be caused [1]. PD-induced dielectric breakdown is one of the most important reasons for premature failures of motor insulation [2, 3]. In inverter-fed motors, most insulation failures are originated from the non-homogeneous distribution of voltage along the stator windings and are closely related with the short rise time and high frequency of the repetitive impulse voltages [4].

Polyimide (PI) is used as an insulating material for turn-to-turn insulation of inverter-fed traction motors because of its excellent electrical, thermal, and mechanical characteristics [1, 5]. Therefore, the dielectric properties of PI films still need to be improved to meet the requirement for rapid development of industry application [6, 7], especially for PD-dependent lifetime under repetitive impulse voltages. Many approaches have been proposed to improve PD-dependent lifetime of

IntechOpen

PI films. One effective approach is synthesizing PI/inorganic nanocomposites [8–11]. Polymer nanocomposites consisting of nanoparticles allow for a large tunability of the dielectric properties by varying the identity and content of nanoadditives. Moreover, lifetime under repetitive impulse voltages is much more closely related to the surface characteristics of PI films, for the degradation is always initiated from surface partial discharge, which is mainly caused by surface charge accumulation and local field enhancement, due to the mismatching of surface charge dissipation and voltage reversion [12–14]. Therefore, surface modification would be another effective approach to improve the lifetime of PI films under repetitive impulse voltages, by modifying the surface characteristics.

To provide a comprehensive understanding to this topic and to clarify the emerging problems, we review the recent progress in the failure mechanism and property modification of insulation material used in inverter-fed motors. Particular attention is paid on PD-dependent lifetime improvement under repetitive impulse voltages of PI films.

2. Failure mechanism of insulation material used in inverter-fed motors

2.1 Electrical stresses in inverter-fed motors

A typical schematic diagram and output waveform of inverter-fed traction motors used in high-speed railways are shown in **Figure 1** [5]. Different from traditional alternating current (AC) motors stressed by the sinusoidal (50 or 60 Hz) voltages, the insulation systems of inverter-fed motors are exposed to transient surge voltages with steep fronts generated by the switching of insulated gate bipolar transistor (IGBT) [6]. Besides, there are other factors that cause PDs, including overvoltage caused by impedance mismatch (between supply cable and motor) and uneven potential distribution could induce partial discharge (PD) in stator

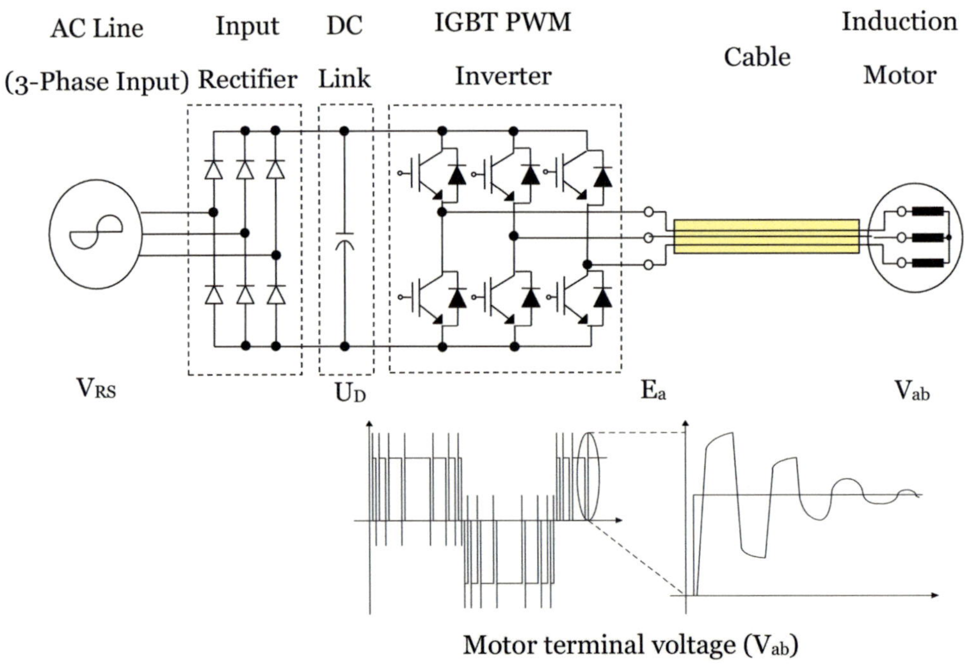

Figure 1.
Schematic diagram and output waveform of an inverter [5].

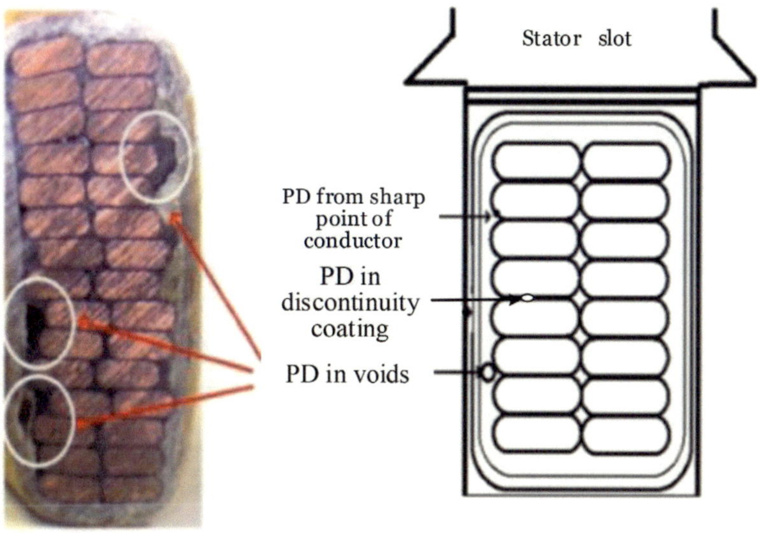

Figure 2.
Typical PD positions in the inner stator slot of an inter-fed motor [18].

insulation systems [15–17]. For Type II motors, formed windings are used in which the preformed coils are uniformly layered with rectangular section conductors. Form the typical cross section image of stators slot as shown in **Figure 2** [18], it can be inferred that PDs are easily occur from the sharp point of conductor, discontinuity coating, and voids formed during manufacture.

2.2 Failure mechanism under repetitive impulse voltage

Hereinafter, we will explain failure mechanism under repetitive impulse voltages though the investigation of the aging process and PD characteristics of turn-to-turn insulation of coils under repetitive impulse voltages [5]. In order to simulate the insulation structure of coils of an inverter-fed motor, the turn insulation specimens were manufactured according to the standard IEC 60034-18-42. In laboratory, bipolar repetitive square impulse voltages are used to take the place of pulse width modulated (PWM) voltage, with adjustable parameters such as amplitude, repetition frequency, and rising time. As bipolar repetitive square impulse voltages were adopted as the testing voltage, one cycle of test time is divided into 360°, and thus, we can analyze PD signals as they can be measured under AC power frequency voltage.

PD characteristics of turn insulation specimens were investigated, with the specimens be aged for 20, 60, and 100 h, respectively. Phase distribution of the discharge numbers and amplitude of PD pulses is shown in **Figure 3** [5]. Most of PD pulses appear next to the rising edge and the falling edge of square wave with a narrow distribution range from about 3 to 23° and 180 to 200°. PD pulses are rarely present during the flat region of repetitive square impulse voltages. This phase distribution characteristic is quite different from the PD characteristics under power frequency. The phase distribution appears widening tendency with the increase of aging time. Especially for the phase near 20°, the PD number and PD quantity increase obviously.

Storage effect of charges is obvious under repetitive impulse voltages. **Figure 4** [5] describes the change of space charge and electric field in the air gap at different time of bipolar continuous square impulse voltage. $E(t)$ is the applied field with electric field amplitude E_0. E_q is the electric field induced by space charge deposited

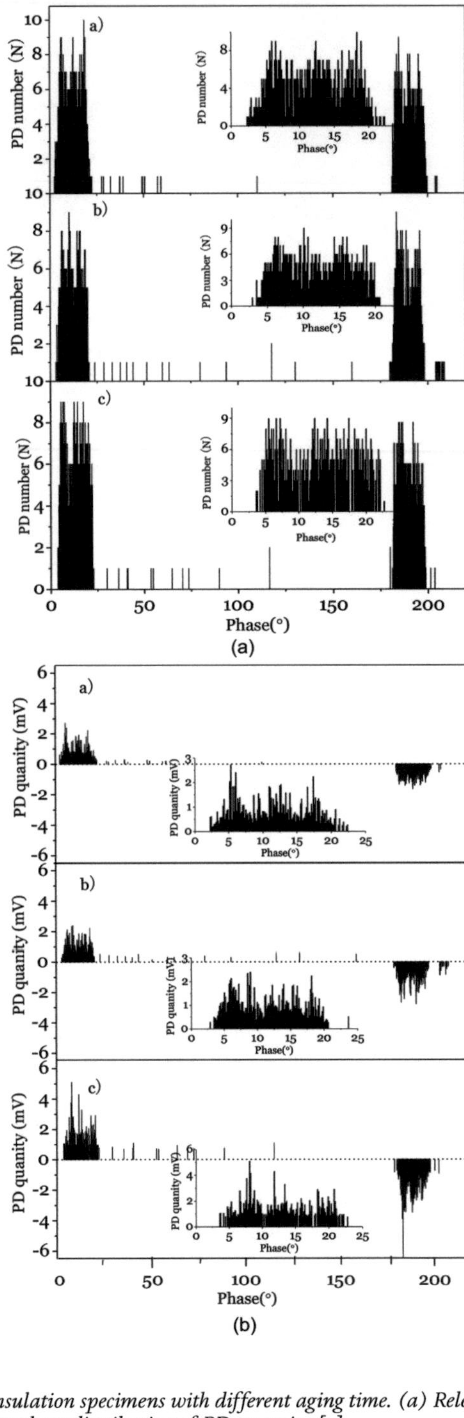

Figure 3.
PD characteristics of turn insulation specimens with different aging time. (a) Relevant phase distribution of the PD number. (b) Relevant phase distribution of PD quantity [5].

in the air gap. At the beginning, bipolar continuous square impulse voltage is applied on the sample, and the induced space charges gather in the air gap, as shown in **Figure 4a**. If the field E_i, difference between $E(t)$ and E_q, reaches PD inception field and the initiating electron is available, PD occurs, dropping the electric field E_i to a residual field value $Eres = E0 - Eq$, as shown in **Figure 4b**. When the polarity of impulse voltage reverses, the field value in the air gap changes to $Ei = -(E(t) + Eq)$, as shown in **Figure 4c**. Currently, space charge moves under the inverted electric

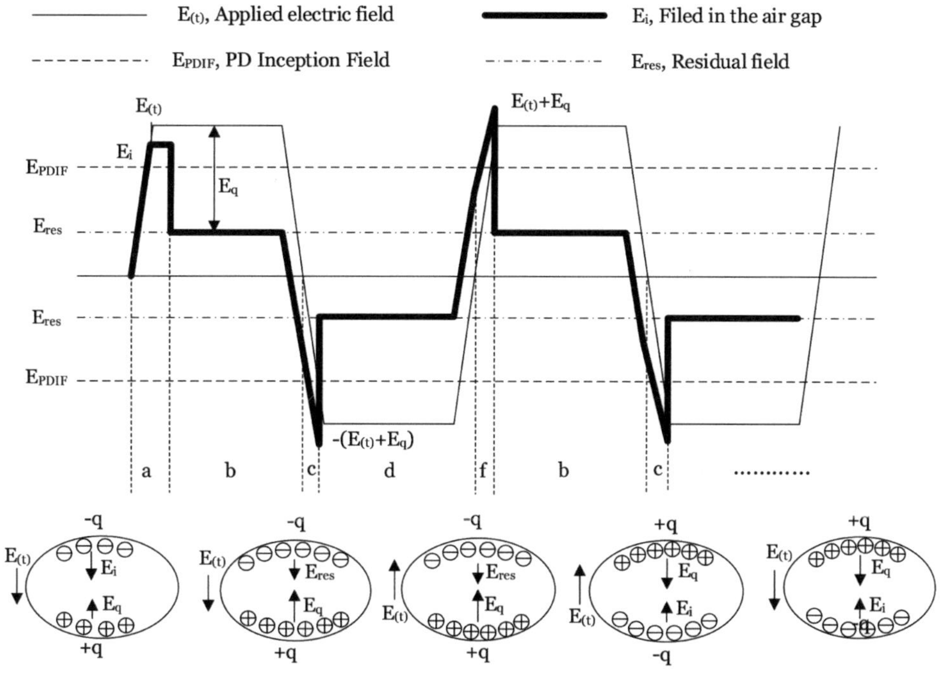

Figure 4.
Storage effect of charges under repetitive impulse voltages [5].

field. But this process is very weak and most of the charge cannot perform the invert motion because of the steep front of impulse voltage. The value of E_q decreases only a little, where Eq ≈ E0-Eres. So, at the instant of polarity reverses, Ei ≈ (E(t) + E0-Eres) would be larger than PD inception field, another PD could be promoted at the falling time of bipolar continuous square impulse voltage, as shown in **Figure 4c**. The charge in the air gap completes reverse movement during the process of PD and the electric field Ei drops again to the value -Eres, as shown in **Figure 4d**. When the applied field inverse again, which is shown in **Figure 4f**, the changes of charge and electric field Ei in the air are similar with that in **Figure 4c**. PD pulse would appear again. The behavior of charge and field could repeat in succession with the testing time, which is similar with the procedures from step b to step f.

As displayed in **Figure 4**, the electric field Ei in the air gap reaches to the maximum value at the rising stage and falling stage of bipolar continuous square impulse voltage. The main reason is that the movement of charge in the air gap could not follow the polarity reversion of impulse voltage; superposition effect between applied field E(t) and the field Eq in the air gap is obvious. Consequently, Most of PDs occur during the rising edge and falling edge of continuous square impulse voltage with a narrow distribution range, which can be observed in **Figure 3**. Residual field value Eres is not constant strictly because of the dissipation of charge. Individual PD appears during the flat region of bipolar square impulse voltage when the field Ei increases to PD inception filed accidentally because of the dissipation of charge in the air gap.

Partial discharge behavior is closely related to the degree of insulation aging. During PD activities, the organic polymer is eroded and degraded gradually, which could lead to two results: first, more new voids generate, which promote the increase of PD number, and second, the original voids were enlarged because of volatilization of organics, resulting in higher PDIV. When the polarity of bipolar continuous square impulse voltage inverses, applied voltage needs a longer delayed time to reach the higher PDIV, resulting in a higher energy of PD, which will be appear at the phase around 20° due to the time delay. This is consistent with what we discussed before that PD amplitudes and discharge numbers become higher near 20° in aged samples. Therefore, PD amplitudes and discharge numbers near 20° can be considered as critical characteristic parameters to evaluate the insulation condition under repetitive impulse voltages.

3. Property modification of polyimide films

3.1 Polyimide nanocomposites

Samples of PI/Al_2O_3 nanocomposites are usually synthetized by using in-situ polymerization. Corona-resistance lifetime under repetitive impulse voltages for nanocomposites with different content of Al_2O_3 nanofillers is investigated. And then, the effects of nanoparticles on reducing the degradation of polyimide nanocomposites will be analyzed in this section.

3.1.1 Lifetime under repetitive impulse voltages for polyimide nanocomposites

During lifetime measurement under repetitive impulse voltages, PDs occur in the air gap between the film sample and the upper rod electrode. PDs generate high-energy electron, ultraviolet rays, and high activity chemical groups, which would accelerate the insulation material aging [19, 20]. Reaching the lifetime, the failure of polymer nanocomposites would occur due to the fracture and degradation of the polymer chains [21].

As shown in **Figure 5** [22], the lifetime increases with the increasing nano-Al_2O_3 content. For the PI/Al_2O_3 nanocomposites with 25 wt.% content of nano-Al_2O_3, the lifetime is around 136 h, which is about 23 times longer than that of pure PI film, at

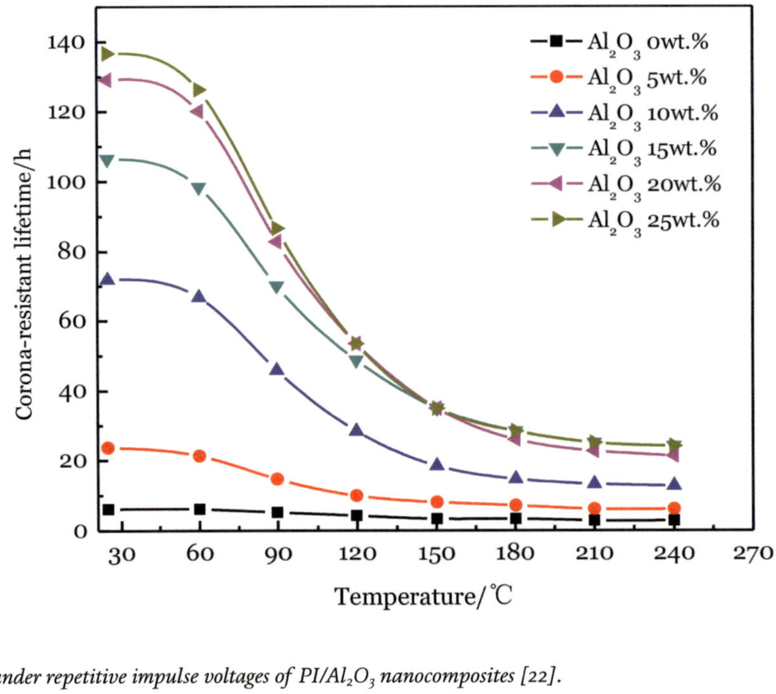

Figure 5.
Lifetime under repetitive impulse voltages of PI/Al_2O_3 nanocomposites [22].

the temperature of 25°C. Besides, the lifetime of PI/Al_2O_3 nanocomposites exhibits the obvious temperature dependence, which decreases gradually with increasing temperature. With increasing nano-Al_2O_3 content, the temperature dependence is much stronger.

3.1.2 Effects of nanoparticles on reducing the degradation of polyimide nanocomposites

The effects of nanoparticles on reducing the degradation of polyimide nanocomposites can be deriving from the following three aspects.

First, during the lifetime measurement under repetitive impulse voltages, PD-induced degradation of the polymer chains occurs. With higher melting point, inorganic particles would leave on the film surface, acting as obstacles in the erosion path of polyimide matrix. Then, the degradation of polymer matrix can be reduced, resulting in a prolonged lifetime. This statement can be proved through the element distribution analysis around the breakdown point, as shown in **Figure 6** and **Table 1** [23]. Next to the breakdown point, the mass percentage of Al is almost twice higher than the mean content (2.65%) in PI/Al_2O_3 nanocomposites with Al_2O_3 content of 5 wt.%. This is consistent with the statement that inorganic particles left on the surface of eroded regions.

Second, new chemistry bonds would be generated in the interfacial regions of polyimide nanocomposites, which is the lowest bonding energy. Under the effect of PDs, these weak bonds would be broken first, followed by the ether linkage, imide ring and aromatic ring of polyimide matrix, as shown in **Figure 7** [23].

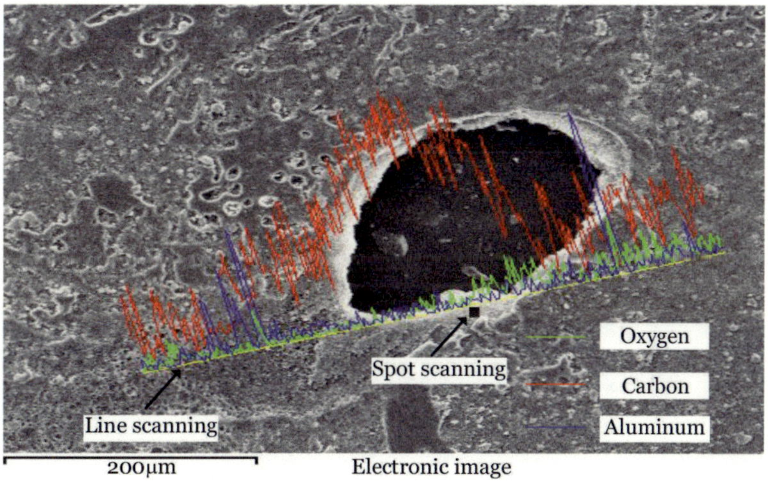

Figure 6.
EDS of spot and line scanning around the breakdown point of PI/Al_2O_3 nanocomposites with Al_2O_3 content of 5 wt.% [23].

Element	Spot scanning (Mass %)	Line scanning (Mass %)
Carbon	30.61	29.70
Oxygen	65.65	57.27
Aluminum	5.74	13.03

Table 1.
Mass percentage of elements in EDS result [23].

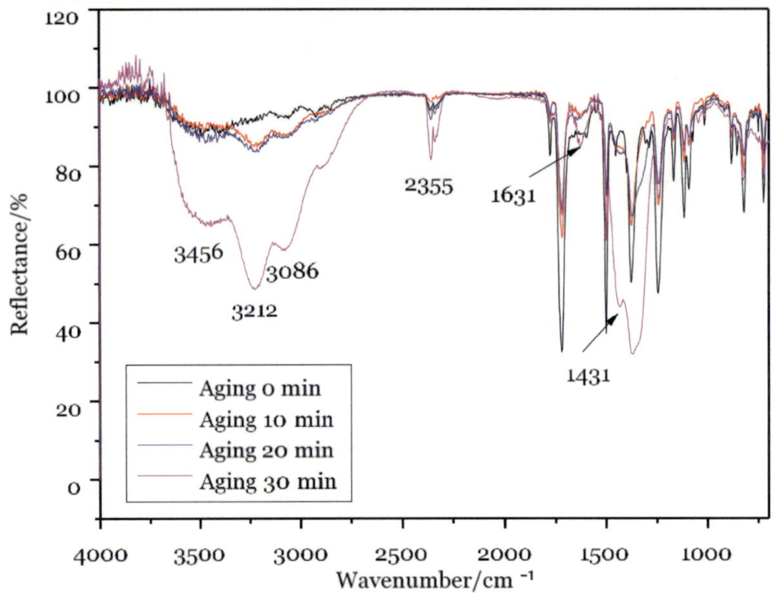

Figure 7.
FTIR spectrums of PI and PI/Al_2O_3 nanocomposites (5 wt.%) at different aging time [23].

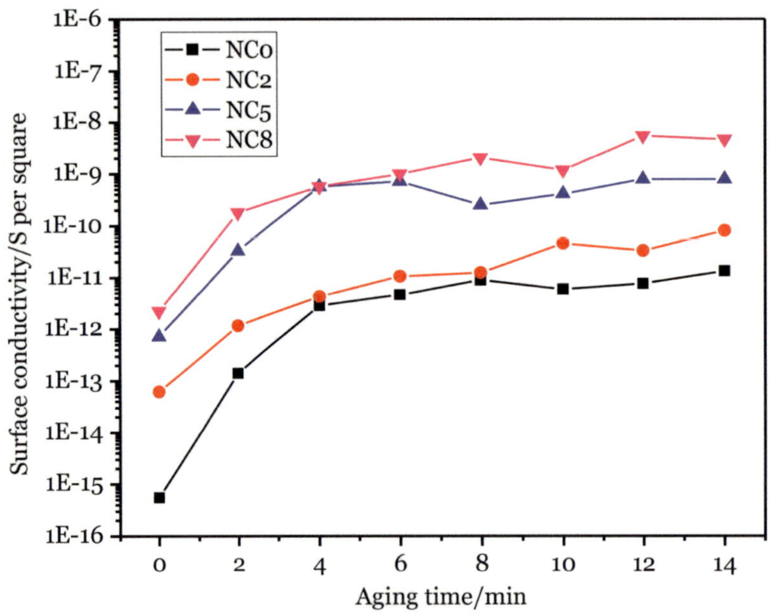

Figure 8.
Surface conductivity of PI and PI/Al_2O_3 nanocomposites [23].

Thus, weakest bonds in interfacial regions would be destroyed at the first stage of aging, which helped reduce the dissociations of other chemical bonds of polyimide molecules, by absorbing part of PD energy, resulting in a prolonged lifetime.

Third, surface conductivity of polyimide nanocomposites is higher than that of polyimide matrix, as shown in **Figure 8** [23], which facilitates surface charge dissipation, leading to lower PD intensity, and longer lifetime. The inorganic particles and the interfacial region near the film surface, with higher carrier

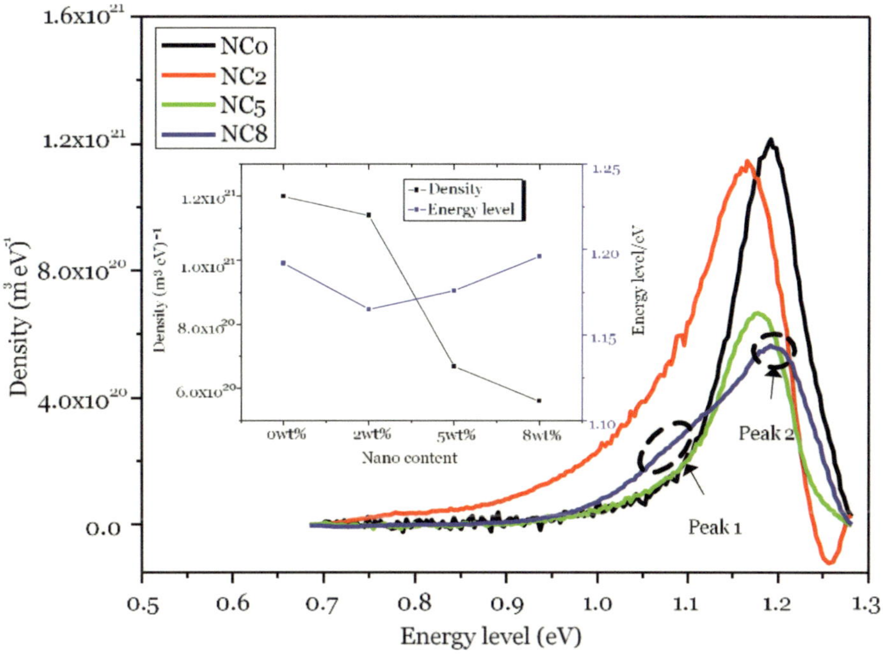

Figure 9.
Trap energy and density distributions of PI and PI/Al_2O_3 nanocomposites [23].

transfer ability, are mainly responsible for the increase in surface conductivity of polyimide nanocomposites. Improved surface conductivity can facilitate charge dissipation and mitigates the electric field distortion, resulting in higher PDIV and longer lifetime.

At last, dielectric characteristics of polyimide nanocomposites are greatly influenced by different trap density and energy distributions which depend on physical and chemical defects in dielectrics. As to deep traps, named as peak 2 in **Figure 9** [23], the trap energy nearly changes for polyimide nanocomposites, but trap density declines with increasing nanoparticle concentration, which means that the bonds between nanoparticles and polyimide molecules can restrict the movement of molecular and thus reduce the chemical disorder.

Besides, shallower traps are introduced in nanocomposites as nanocomposites content increases, as peak 1 appears on the curve of polyimide nanocomposites. With lower trap density, it is less possible for charge recombination in polyimide nanocomposites, thus lead to lower released energy, which is beneficial to reduce the degradation of polymer.

Many chemical defects are generated during PD aging due to the breakage of long molecular chain, which are reflected on the increase of both trap density and energy level. As the degree of electrical aging can be determined by the rate of increase in the density of stress-created traps [24], it can be concluded that the degree of PD aging is quite serious in PI after aging for 15 min, while that of PI/Al_2O_3 nanocomposites is much slighter. That is why no obvious increase trend can be observed in the trap density when the PD aging time is less than 15 min.

Besides, the trap-level distribution showed that both trap density and energy in PI/Al_2O_3 nanocomposites decreased as degradation, while the trend of PI was opposite. It indicated that the degradation of PI/Al_2O_3 nanocomposites was mitigated compared with PI, which could be attributed to the improved electrical characteristics of PI/Al_2O_3 nanocomposites.

3.2 Surface modification of polyimide films

Non-thermal plasma generated by dielectric barrier discharge (DBD) can be used to conduct surface modification of polyimide films [25]. This kind of plasma modification can efficiently cause the material surface cross-linking, etching, and the introduction of polar groups, without changing its bulk properties. Surface modification of polymer films using DBD are usually conducted to improve adhesion or hydrophilicity properties, as DBD exited high-energy electrons produce ions and reactive species that interact with film surface to modify the surface characteristics. Electrical properties such as surface flashover voltage and surface resistivity can also be improved by non-thermal plasma modification [26–28].

In this section, we will study lifetime improvement and its mechanism under repetitive impulse voltages, by considering the influence of trap energy distribution on carrier transportation and charge dissipation with voltage reversion. Thus, isothermal surface potential decay (ISPD) [29, 30] technique was conducted to study the charge transport characteristics of PI films before and after surface modification, revealing the energy distribution of both electron-type and hole-type traps [31].

3.2.1 Properties of plasma-modified polyimide films

Polyimide films were put between the air gap of DBD for surface modification, with different treating time set as 0, 10, 20, 30, 60, and 90 s. The DBD was excited by non-thermal plasma generator in atmospheric air, as shown in **Figure 10**.

The surface of the polyimide films was etched during plasma treatment. As shown in **Figure 11**, embossing, micro-voids, and obvious etched channels gradually appear on film surface, with increasing treating time. Based on the energy disperse spectroscopy (EDS) analysis results shown in **Table 2**, it is indicated that the oxygen content of plasma-treated films is higher than that of untreated film. This would be attributed to the introduction of rich-oxygen groups onto the film surface during plasma treatment.

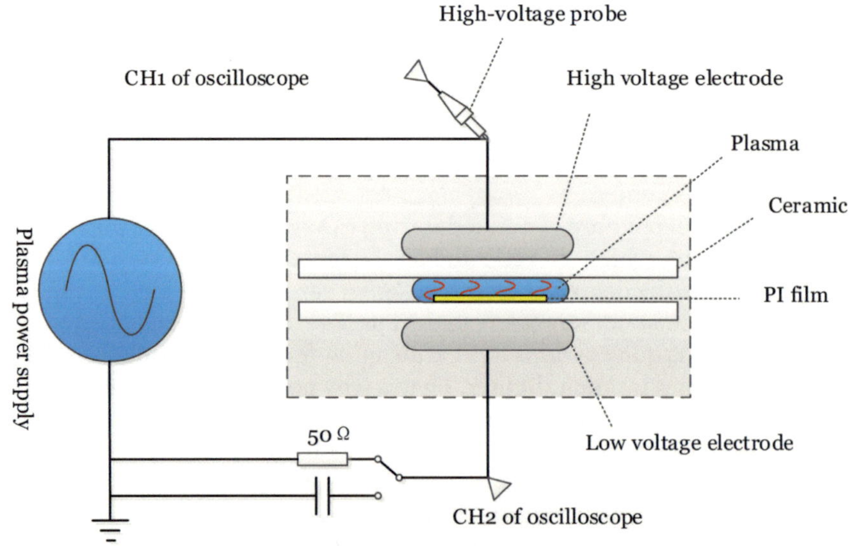

Figure 10.
Setup for dielectric barrier discharge (DBD) plasma treatment.

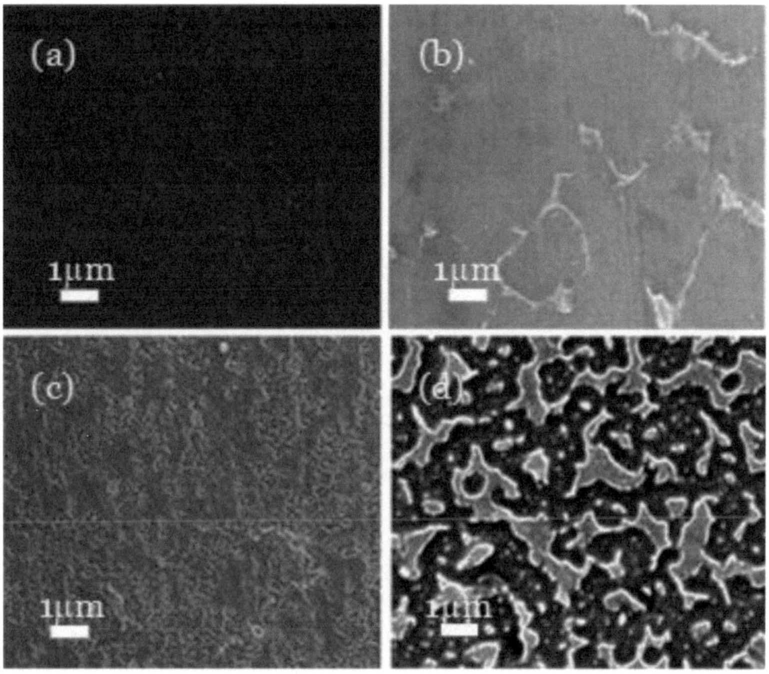

Figure 11.
SEM images of PI after different treating time. (a) PI. (b) PI-20. (c)PI-30. (d) PI-90.

Samples	Weight percentage of C (wt %)	Weight percentage of O (wt %)
PI	21.71	78.29
PI-90	18.70	81.30

Table 2.
Elemental analysis of PI films before and after plasma treatment.

Main molecular chains of PI have not been changed by plasma treatment [32, 33], while the intensities of these characteristic absorption peaks for plasma-treated films obviously decrease compared with untreated films, indicating some ring-opening reactions occurred during plasma treatment, as shown in **Figure 12**. Moreover, absorption bands between 3660 and 3200 cm^{-1}, which represent O—H stretching peaks in carboxyl acids [32, 33], can be observed in plasma-treated films. Peaks appear at 1410(—COOH) and 3220(—NH2) cm^{-1} also increase in plasma-treated films, indicating that reactive species, containing oxygen and nitrogen, have been introduced onto the film surface during plasma treatment. However, when the treating time is longer than 30 s, the intensities of characteristic absorption peaks representing —COOH and —NH2 obviously decrease. This would be attributed to further oxidizing reactions during plasma treatment, as working gas is air. Thus, appropriate treating time is an important factor to introduce reactive groups effectively.

Lifetime under repetitive impulse voltage for plasma-treated films can be improved when the plasma treating time is less than 20 s, but it decreases with increasing treating time, when the treating time is longer than 20 s, as shown in **Figure 13**. When the treating time is 20s, the lifetime reaches the maximum value of 23.5 min, which is 16.9% higher than that of untreated samples. When the treating time exceeds 30 s, the lifetime decreases sharply, by reaching the values even shorter than that of untreated samples.

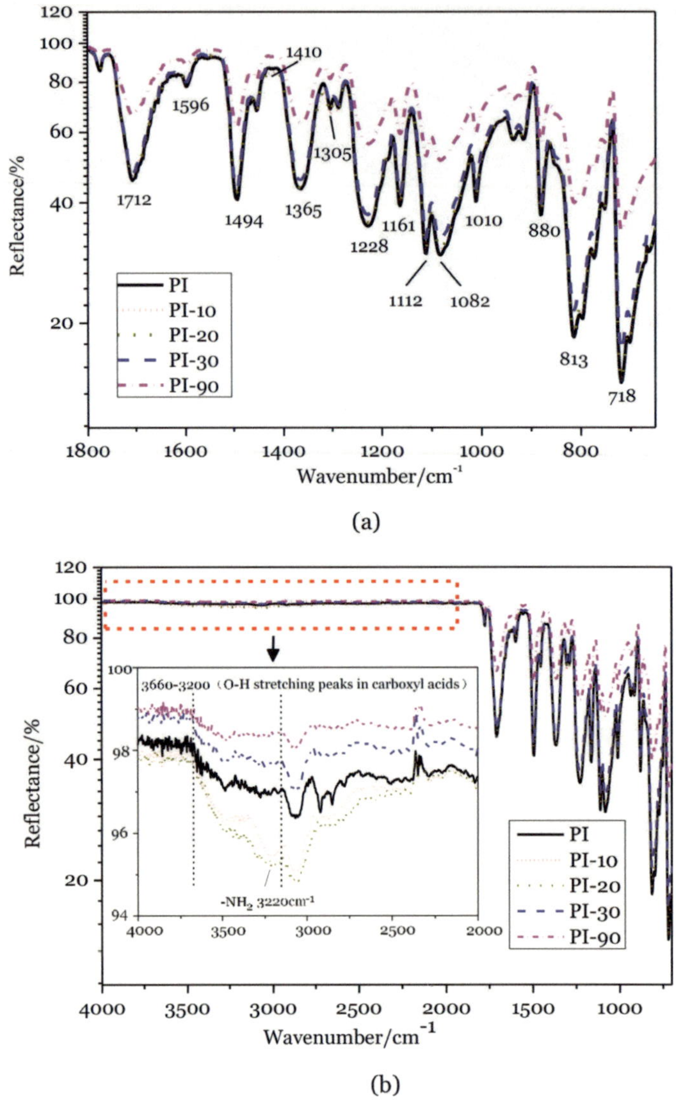

Figure 12.
FTIR spectra of PI films under different plasma treating time. (a) 1800–700 cm^{-1}*. (b) 4000–700 cm*$^{-1}$*.*

3.2.2 Modification mechanism

The energy levels of both electron-type and hole-type traps are lower in plasma-treated films than that in untreated ones, as shown in **Figure 14**. Apparent decrease of trap energies takes place when the treating time is shorter than 20 s. However, the trap energy levels increase when the treating time exceeds 20 s, but they are still lower than that of untreated samples. For each curve, two peaks can be observed, which represent the distribution of shallow and deep traps, respectively.

Shallow trap energy level for PI-20 is the lowest compared with the others, corresponding to the longest lifetime repetitive impulse voltages. Trap energy level dominates charge transfer characteristics rather than trap density. With lower shallow trap energy level, carriers are easier to transfer through the sample, causing less charge accumulation. For deeper traps, more energy is needed to detrap the carriers, which means more charges are trapped in samples. Only the near surface region with the depth about ~90 nm of the material has been modified by reactive

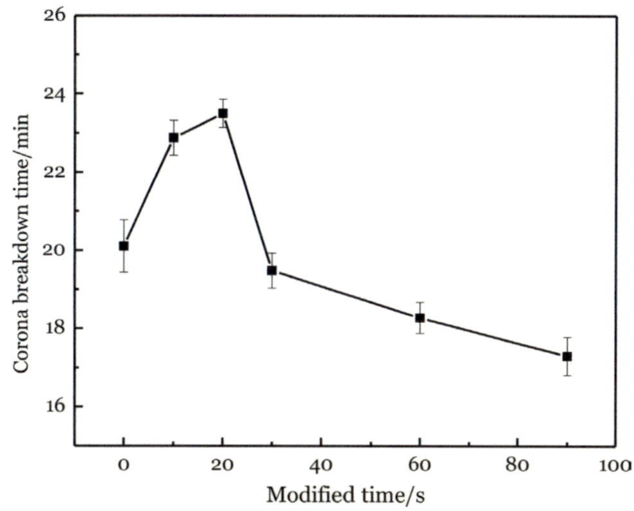

Figure 13.
Lifetime under repetitive impulse voltage with different plasma treating time.

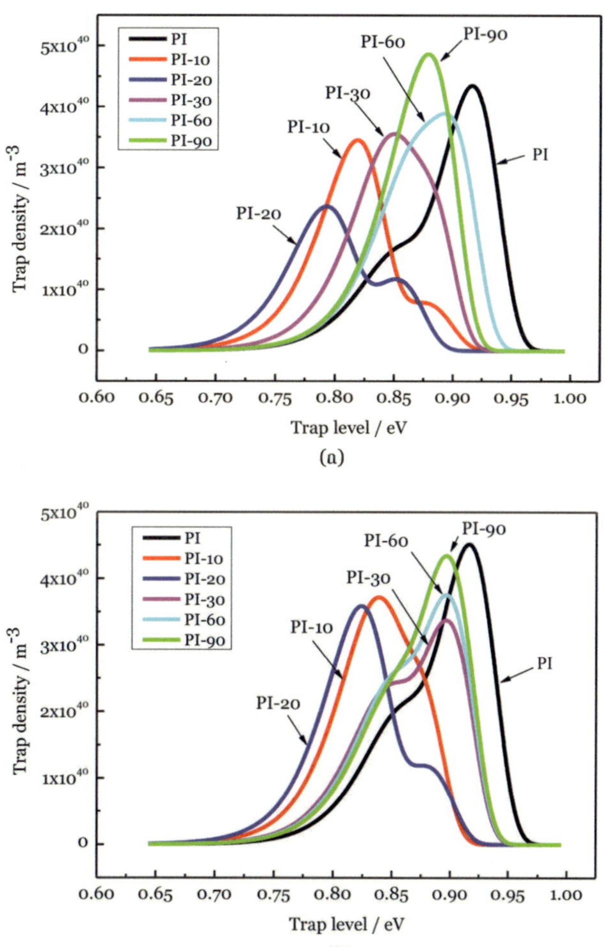

Figure 14.
Trap energy distribution of PI films under different treating time. (a) Energy distribution of hole-type traps. (b) Energy distribution of electron-type traps.

groups during plasma treatment [34]. Thus, the trap distribution modification mainly occurs in the near surface region, by introducing groups contained oxygen and nitrogen. Corresponding charge transportation and accumulation characteristics modification also takes place in this near surface region. When it comes to the repetitive impulse voltages, the trapped surface charges will be superposed to the applied voltage as polarity reverses, and local field would be further enhanced [23]. In this case, lifetime will be shortened due to intensive PD activity. For plasma-treated films, with lower energies of both shallow and deep traps, charge dissipation would be enhanced and the density of trapped charges would be reduced, leading to the local electric filed mitigation, resulting in suppressed PD intensity and corresponding longer lifetime.

However, when the treating time is too long, plasma-introduced oxygen and nitrogen rich groups would be broken by further oxidizing reactions, so less effective active groups have been introduced through plasma treatment. Furthermore, the degradation of PI films begins to occur, which means large molecular chains of PI began to break. It is more likely to form low-density region on the film surface. Although the charge dissipation is enhanced by the effect of shallower traps and corresponding PD activity might be weakened, this low-density region still cannot withstand the weakened PD intensity, and cleavages of the molecular chain will take place subsequently. Finally, the dielectric breakdown will occur in a short time.

4. Conclusion

The insulation system of inverter-fed motors is subjected to repetitive impulse voltages which are generated by pulse width modulation (PWM) converters. The insulation materials are expected to withstand partial discharge (PD) activity during service. Under repetitive impulse voltages, most of PD pulses appear next to the rising edge and the falling edge of applied voltage but rarely present during the flat region of repetitive square impulse voltages, which is quite different from the PD characteristics under power frequency. The phase distribution appears widening tendency with the increase of aging time. Especially for the phase near 20°, the PD number and PD quantity increase obviously.

As turn-to-turn insulation of inverter-fed traction motors, the dielectric properties of polyimide need to be improved to meet the requirement for rapid development of industry application, especially for PD-dependent lifetime under repetitive impulse voltages. Lifetime under repetitive impulse voltages can be improved by synthesizing polyimide/inorganic nanocomposites and plasma treatment.

For polyimide nanocomposites, during degradation process, inorganic particles would leave on the film surface of polyimide due to higher melting point, acting as obstacles in the erosion path of polyimide matrix. New chemistry bonds would be generated in the interfacial regions of polyimide nanocomposites, which is the lowest bonding energy. These weak bonds would be destroyed at the first stage of aging, reducing dissociations of other chemical bonds of polyimide molecules. With improved surface conductivity, charge dissipation of polyimide nanocomposites mitigates the electric field distortion, resulting in higher PDIV. As shallower traps are introduced in nanocomposites, it is less possible for charge recombination in polyimide nanocomposites, leading to lower released energy, which is beneficial to reduce the degradation of polymer.

For plasma-treated films, lower shallow trap energy level is beneficial to charge transfer; with lower deep trap energy level, less charge accumulation takes place. Trap energy level dominates the charge transportation and accumulation characteristics rather than trap density. Active groups contained oxygen and nitrogen, which are introduced during plasma treatment, are responsible for trap energy distribution modification of PI films. Under repetitive impulse voltage, faster surface potential decay rate and less charge accumulation can facilitate charge dissipation and local electric field mitigation, resulting in weakened PD intensity and a longer lifetime. Appropriate treating time is an important factor to introduce reactive groups and prolong the lifetime.

Acknowledgements

This work was financially supported by the National Natural Science Foundation of China (NSFC) under grant no. 51837009 and the State Key Laboratory of Electrical Insulation and Power Equipment (EIPE18212).

Author details

Yan Yang* and Guangning Wu
School of Electrical Engineering, Southwest JiaotongUniversity, China

*Address all correspondence to: yangyanyy@swjtu.edu.cn

IntechOpen

© 2019 The Author(s). Licensee IntechOpen. This chapter is distributed under the terms of the Creative Commons Attribution License (http://creativecommons.org/licenses/by/3.0), which permits unrestricted use, distribution, and reproduction in any medium, provided the original work is properly cited.

References

[1] Zhou L, Wu G, Gao B, et al. Study on charge transport mechanism and space charge characteristics of polyimide films. IEEE Transactions on Dielectrics and Electrical Insulation. 2009;**16**(4):1143-1149

[2] Hayakawa N. Partial discharge characteristics of inverter-fed motor coil samples under ac and surge voltage conditions. IEEE Electrical Insulation Magazine. 2005;**21**(1):5-10

[3] Kandev N, Castelan P, Lebey T, et al. Testing of low-voltage motor turn insulation intended for pulse-width modulated applications. IEEE Transactions on Dielectrics and Electrical Insulation. 2000;7(6):783-789

[4] Yin W. Failure mechanism of winding insulations in inverter-fed motors. IEEE Electrical Insulation Magazine. 1997;**13**(6):18-23

[5] Luo Y, Wu G, Liu J, et al. PD characteristics and microscopic analysis of polyimide film used as turn insulation in inverter-fed motor. IEEE Transactions on Dielectrics and Electrical Insulation. 2014;**21**(5):2237-2244

[6] Kaufhold M, Aninger H, Berth M, et al. Electrical stress and failure mechanism of the winding insulation in PWM-inverter-fed low-voltage induction motors. IEEE Transactions on Industrial Electronics. 2000;**47**(2):396-402

[7] Akram S, Wu G, Gao GQ, et al. Cavity and interface effect of PI-film on charge accumulation and PD activity under bipolar pulse voltage. Journal of Electrical Engineering and Technology. 2015;**10**(5):2089-2098

[8] Nehete K, Sharma RA, Chaudhari L, et al. Study of erosion resistance and mechanical properties of unsaturated polyester based nano-composites. IEEE Transactions on Dielectrics and Electrical Insulation. 2012;**19**(2):373-382

[9] Li Z, Okamoto K, Ohki Y, et al. Effects of nano-filler addition on partial discharge resistance and dielectric breakdown strength of micro-Al2O3/epoxy composite. IEEE Transactions on Dielectrics and Electrical Insulation. 2010;**17**(3):653-661

[10] Tanaka T, Ohki Y, Ochi M, et al. Enhanced partial discharge resistance of epoxy/clay nanocomposite prepared by newly developed organic modification and solubilization methods. IEEE Transactions on Dielectrics and Electrical Insulation. 2008;**15**(1):81-89

[11] Maity P, Basu S, Parameswaran V, et al. Degradation of polymer dielectrics with nanometric metal-oxide fillers due to surface discharges. IEEE Transactions on Dielectrics and Electrical Insulation. 2008;**15**(1):52-62

[12] Kwon OJ, Tang S, Myung SW, et al. Surface characteristics of polypropylene film treated by an atmospheric pressure plasma. Surface and Coatings Technology. 2005;**192**(1):1-10

[13] Park WJ, Yoon SG, Jung WS, et al. Effect of dielectric barrier discharge on surface modification characteristics of polyimide film. Surface & Coatings Technology. 2007;**201**(9):5017-5020

[14] Kostov KG, Nishime TMC, Hein LRO, et al. Study of polypropylene surface modification by air dielectric barrier discharge operated at two different frequencies. Surface and Coatings Technology. 2013;**234**:60-66

[15] Persson E. Transient effects in application of PWM inverters to induction motors. IEEE Transactions on Industry Applications. 2002;**28**(5):1095-1101

[16] Stone GC, Van Heeswijk RG, Bartnikas R. Investigation of the effect of repetitive voltage surges on epoxy insulation. IEEE Transactions on Energy Conversion. 1992;**7**(4):754-760

[17] Wheeler JCG. Effects of converter pulses on the electrical insulation in low and medium voltage motors. IEEE Electrical Insulation Magazine. 2005;**21**(2):22-29

[18] Wang P, Wu G, et al. Study of partial discharge characteristics at repetitive square voltages based on UHF method. Science China Technological Sciences. 2013;**56**(1):262-270

[19] Morshuis PHF. Degradation of solid dielectrics due to internal partial discharge: Some thoughts on Progress made and where to go now. IEEE Transactions on Dielectrics and Electrical Insulation. 2005;**12**(5):905-913

[20] Mayoux C. Degradation of insulating materials under electrical stress. IEEE Transactions on Dielectrics and Electrical Insulation. 2000;**7**(5):590-601

[21] Zhu Y, Otsubo M, Honda C. Degradation of polymeric materials exposed to corona discharges. Polymer Testing. 2006;**25**(3):313-317

[22] Luo Y, Wu G, Liu J, et al. Investigation of temperature effects on voltage endurance for polyimide/Al2O3 nanodielectrics. IEEE Transactions on Dielectrics and Electrical Insulation. 2014;**21**(4):1824-1834

[23] Zhong X, Wu G, Yang Y, et al. Effects of nanoparticles on reducing partial discharge induced degradation of polyimide/Al2O3 nanocomposites. IEEE Transactions on Dielectrics and Electrical Insulation. 2018;**25**(2):594-602

[24] Liufu D, Wang X, Tu D, et al. High-field induced electrical aging in polypropylene films. Journal of Applied Physics. 1998;**83**(4):2209-2214

[25] Shao T, Zhang C, Long K, et al. Surface modification of polyimide films using unipolar nanosecond-pulse DBD in atmospheric air. Applied Surface Science. 2010;**256**(12):3888-3894

[26] Shao T, Yang W, Zhang C, et al. Enhanced surface flashover strength in vacuum of polymethylmethacrylate by surface modification using atmospheric-pressure dielectric barrier discharge. Applied Physics Letters. 2014;**105**(7):082903-082264

[27] Shao T, Liu F, Hai B, et al. Surface modification of epoxy using an atmospheric pressure dielectric barrier discharge to accelerate surface charge dissipation. IEEE Transactions on Dielectrics and Electrical Insulation. 2017;**24**(3):1557-1565

[28] Zhao W, Xu R, Ren C, et al. Ion-implantation modification of surface flashover properties in vacuum of polytetrafluoroethylene. IEEE Transactions on Plasma Science. 2018:1-7

[29] Li J, Zhou F, Min D, et al. The energy distribution of trapped charges in polymers based on isothermal surface potential decay model. IEEE Transactions on Dielectrics and Electrical Insulation. 2015;**22**(3):1723-1732

[30] Min D, Cho M, Khan AR, et al. Surface and volume charge transport properties of polyimide revealed by surface potential decay with genetic algorithm. IEEE Transactions on Dielectrics and Electrical Insulation. 2012;**19**(2):600-608

[31] Han Y, Li S, Min D. Trap energy distribution in polymeric insulating materials through surface potential decay method. IEEE Transactions on Dielectrics and Electrical Insulation. 2018;**25**(2):639-648

[32] Park SJ, Lee HY. Effect of atmospheric-pressure plasma on adhesion characteristics of polyimide film. Journal of Colloid and Interface Science. 2005;**285**(1):267-272

[33] Yang Y, Yin D, Xiong R, et al. Ftir and dielectric studies of electrical aging in polyimide under AC voltage. IEEE Transactions on Dielectrics and Electrical Insulation. 2012;**19**(2):574-581

[34] Park B, Kim J, Cho M, et al. Ballistic-mode plasma-based ion implantation for surface-resistivity modification of polyimide film. IEEE Transactions on Plasma Science. 2012;**40**(6):1749-1752

Chapter 6

Dielectric Analysis Model for Measurement of Soil Moisture Water Content Using Electrical Capacitance Volume Tomography

Mukhlisin Muhammad and Saputra Almushfi

Abstract

Electromagnetic methods have been widely used in the measurement of the water content of the soil. These methods utilize the permittivity as electrical properties of the soil, to determine the moisture content of the soil. Since the measurements are carried out indirectly, a calibration between permittivity and the water content of the soil is needed. Generally, the calibration method is generated by using an empirical and mixing model. This study presents a proposed model of calibration by using a normalization approach to calibrate the value of the permittivity of the water content of the soil. Then the model was applied using electrical capacitance volume tomography (ECVT) to image soil water content during infiltration of water in a soil column. Granular and silty sand were used as soil material in the experiments. The result showed that the model for measuring moisture water content can be seen in each layer during soil water infiltration in the soil column.

Keywords: dielectric analysis model, soil moisture water content, ECVT

1. Introduction

Measurement of soil moisture water content has become an important part of the analysis of various fields of study, especially those involving irrigation in agriculture, forestry, hydrology, and land activity. For example, in agriculture study, it is required to determine water source and ensure the quality of crop [1]. Moreover, soil moisture water content is useful to analyze soil water contamination by observing changes in water content during the addition of substance [2]. Soil water content also plays an important role in slope stability analysis [3–6].

Various techniques in the measurement of contamination and water content of soil have been discussed in literatures (e.g., [7–12]). Based on previous studies, soil water content measurement techniques are widely used as an electromagnetic method [13] such as time domain reflectometry [14], ground-penetrating radar (GPR) [8], and electrical capacitance [15]. These all methods measure the value of relative permittivity of soil to find soil moisture water content.

Tomography is a promising technique for measurement of water content in soil, especially for capacitance-based tomography. It is because this technique is not only capable of measuring water content in the soil but also is capable of imaging the distribution of water in the soil. Tomography technique is also preferred because it is nondestructive and noninvasive. As a tomography technique, ECVT is a system used to view enclosed objects by measuring changes in capacitance and then compute relative permittivity distribution to create three-dimensional images in real time [16]. The shape of geometry sensor on ECVT is not confined to one form; it can be in the form of arbitrary shape of geometries [17]. This possibility gives an extra advantage of ECVT in measuring soil water content.

The previous study has successfully monitored the propagation of distribution of water in soil column [18]. In this study, the equation of normalized volumetric water content from ECVT method is proposed and compared with other equations. The proposed model is then used to analyze the volumetric water content during soil water infiltration in a vessel.

2. ECVT principle

The basic measurement of ECVT is derived from Poisson's equation:

$$\nabla \cdot (\varepsilon(x, y, z)\nabla\varphi(x, y, z) = -\rho(x, y, z)) \tag{1}$$

where ε is relative permittivity distribution, ϕ is electric potential, and ρ is charge distribution. From Eq. (1), capacitance value can be obtained by using the equation below:

$$C = \frac{1}{\Delta V}\oint_{\Gamma} \varepsilon(x, y, z)\nabla\varphi(x, y, z) \cdot \hat{n}\ dl \tag{2}$$

where ΔV is potential difference and C is capacitance. By using matrix expression, Eq. (2) can be written like the following equation:

$$\mathbf{C} = \mathbf{S}\ \mathbf{G} \tag{3}$$

where **C** is capacitance matrix, **G** is distribution of relative permittivity matrix, and **S** is sensitivity matrix. The sensitivity matrix is generated from the sensor and geometry design and number of sensors. In matrix operation, the value of **G** can be obtained by inversing matrix **S** and multiplying it with matrix **C**. For non-square matrix, matrix inversion is very difficult to solve, so the approximation could be attempted by using transpose matrix. The equation to calculate matrix **G** becomes

$$\mathbf{G} = \mathbf{S}^{\mathrm{T}}\ \mathbf{C} \tag{4}$$

Equations (3) and (4) are known as the forward and inverse problem, respectively. Inverse problem is used to reconstruct the capacitance measurements to become relative permittivity distribution. The simple method for reconstruction is using linear back projection (LBP) [19].

3. Models of soil moisture water content and relative permittivity relationship

The relationship between relative permittivity and volumetric water content has been used by previous researchers to determine the volumetric water content. Many numbers of functions have been proposed to describe the ε-θ relationship model across a range of soil water content. This model can be divided into two categories, which are model with one parameter and model with two or more parameters.

3.1 Model with one parameter

There are some models that proposed the ε-θ relationship. Topp et al. [20] introduced successfully the ε-θ relationship that is commonly used in geotechnical area. The relationship is as shown below:

$$\varepsilon = 3.03 + 9.3\theta + 146.0\theta^2 - 76.7\theta^3 \quad (5)$$

where ε is the relative permittivity or dielectric constant and θ is the volumetric water content of soil.

Equation (1) is derived empirically through experiments of various of mineral soil using a time-domain reflectometer (TDR) at a frequency between 1 MHz and 1 GHz, with an estimated error value of 0.013 [20]. In another form, Topp's equation can also be written as follows:

$$\theta = -5.3 \times 10^{-2} + 2.92 \times 10^{-2}\varepsilon - 5.5 \times 10^{-4}\varepsilon^2 + 4.3 \times 10^{-6}\varepsilon^3 \quad (6)$$

In addition, Topp et al. [20] also proposed equations of ε-θ relationship for organic soil and 450 μm glass beads.

$$\varepsilon = 1.74 - 0.34\theta + 135\theta^2 - 55.3\theta^3 \quad \text{Organic soil} \quad (7)$$

$$\varepsilon = 3.57 + 31.7\theta + 114\theta^2 - 68.2\theta^3 \quad 450 - \mu\text{m glass beads} \quad (8)$$

Then Roth et al. [21] proposed another empirical equation by an experiment using miniprobe TDR which has been used previously by [22]. The ε-θ relationship for mineral soil proposed by [21] is

$$\theta = -0.0728 + 0.0448\varepsilon - 0.00195\,\varepsilon^2 + 0.000036\,1\varepsilon^3 \quad (9)$$

while the ε-θ relationship for organic soil and material is

$$\theta = -0.0233 + 0.0285\varepsilon - 0.000431\,\varepsilon^2 + 0.00000304\varepsilon^3 \quad (10)$$

Calibration of these equations has a volumetric water content error value of 0.015 $cm^3\ cm^{-3}$ for mineral soil and 0.035 $cm^3\ cm^{-3}$ for organic soil [22].

Simple equation of ε-θ relationship was proposed by [23]. This equation resulted from the principal of dielectric mixing models and analyzed the TDR without coatings that are considered potential sources of error in measurement:

$$\theta = 0.1181\sqrt{\varepsilon} - 0.1841 \quad (11)$$

Schaap et al. [24] also introduce a simple equation by performing experiments of 505 measurements of organic forest floor sample by using TDR where the ε-θ relationship is

$$\theta = 0.136\sqrt{\varepsilon} - 0.119 \quad (12)$$

The next equation comes from [25], which is using coaxial transmission system and using soil samples with wide range of soil textures:

$$\theta = -0.0286 + 0.02435\varepsilon - 0.0003421\varepsilon^2 + 0.00000237\varepsilon^3 \quad (13)$$

Equations (5)–(8) are used for measurement at a frequency of 100 MHz.

[13] proposed an empirical model where permittivity measurement was measured based on capacitance. This experiment used a type of quartz sand with a range of particle sizes between 0.15 and 0.9 mm:

$$\varepsilon = A\left(\frac{1}{1 + (\alpha(1-\theta))^n}\right)^{1-\frac{1}{n}} + B \quad (14)$$

where A = 33, B = 2, α = 1.5, and n = 14.

3.2 Model with two or more parameters

Some relationship equations between permittivity and soil water content were also influenced by other parameters such as porosity and bulk density. By using the concept of mixing models and using data from others study [26–28], [29] proposed the following equations:

$$\varepsilon = \theta\left(\varepsilon_i + (\varepsilon_w - \varepsilon_i)\frac{\theta}{\theta_t}\gamma\right) + (\eta - \theta)\varepsilon_a + (1 - \eta)\varepsilon_r \quad (15)$$

Equation (11) is used for $\theta \le \theta_t$, while $\theta > \theta_t$ used the following equation:

$$\varepsilon = \theta_t(\varepsilon_i + (\varepsilon_w - \varepsilon_i)\gamma) + (\theta - \theta_t)\varepsilon_w + (\eta - \theta)\varepsilon_a + (1 - \eta)\varepsilon_r \quad (16)$$

where ε_i, ε_w, ε_a, and ε_r are the permittivity of ice, water, air and rock, respectively (i.e., ε_i = 3.2, ε_w = 80, and ε_a = 1), while θ_t is transition moisture (0.16–0.33), η is the porosity of soil (0.5), and γ is the fitting parameter (0.3–0.5) [29].

In [30], the equation based on dielectric mixing model, which has been described by [31], is proposed. Experiments carried out by measuring a wide range of soil types using TDR with the error value of soil water content is not more than 0.013 cm^3 cm^{-3} [30], with forms of the equation below:

$$\theta = \frac{\varepsilon^\gamma - (1-\eta)\varepsilon_s^\gamma - \eta\varepsilon_a^\gamma}{\varepsilon_w^\gamma - \varepsilon_a^\gamma} \quad ;\gamma = -1 \quad (17)$$

$$\theta = \frac{\varepsilon^\gamma - (1-\eta)\varepsilon_s^\gamma - \eta\varepsilon_a^\gamma}{\varepsilon_w^\gamma - \varepsilon_a^\gamma} \quad ;\gamma = 1 \quad (18)$$

where γ = −1 for three phases in series and γ = 1 for three phases in parallel.

Another model is proposed by [32]. They conducted the experiments by using TDR and 62 kinds of soil sample that consist of mineral soils, organic soil, standard pot soils, artificial peat loess and peat sand, sea and river sand, forest litter, etc. which differ in terms of texture and bulk density [32].

$$\theta = \frac{\sqrt{\varepsilon} - 3.47 + 6.22\eta - 3.82\eta^2}{7.01 + 6.89\eta - 7.83\eta^2} \quad (19)$$

Equation (13) gives the uncertainty of soil water content value of 0.03.

Gardner et al. [15] used capacitance measurement methods to obtain soil water content with the soil dry bulk density values ranging between 1.08 and 1.49, then using multiple linear regression analysis to best fit the measurement data, resulting in the following equation:

$$\theta = \frac{\sqrt{\varepsilon} + 1.208 - 2.454\rho}{9.93} \tag{20}$$

where ρ is dry bulk density.

Robinson et al. [14] give the equation used for coarse-textured, layered soils by using TDR and coarse-grained, glass bead, and quartz grains:

$$\theta = \eta\left(\frac{\sqrt{\varepsilon} - \sqrt{\varepsilon_{dry}}}{\sqrt{\varepsilon_{sat}} - \sqrt{\varepsilon_{dry}}}\right) \tag{21}$$

where ε_{dry} and ε_{sat} are the permittivity measured at oven dry soil and saturation soil (**Table 1**).

3.3 Comparison using existing data

Data from previous research (e.g., [20, 21, 25–27, 29, 31–35]) are used to compare the patterns of the equations which are discussed in this study. This data consists of various soil types with different properties.

Figure 1 shows several curves representing Eqs. (5)–(14). All equations look occupied by all the available data. However, each equation appears to have certain characteristics to the data. Eqs. (8) and (12) only cover the boundary area of the data, while the other equations lie mostly in the central part of the data. These would seem to depend on the properties of the soil types used. One of the characteristics analyzed in this study is the porosity of the soil.

Figure 2 shows the influence of porosity (η = 0.3 to η = 0.7), on the suitability of the equations (Eqs. (15), (16), (18) and (19)–(21)) with data, and also displayed some of the data with a value of porosity (0.33, 0.44, and 0.62) in order to see the effect of porosity on predictions of water content of the equation. From the image it can be seen that the different porosity values of the data will result in different patterns.

In **Figure 2a**, the equations already have the same pattern with the data, but a change of porosity in the equation does not give a significant effect on the pattern of the line, so it is only fit in certain small area of data though with different porosity.

Figure 2b shows that the equation is such as the linear equation that has not affected on the changes in porosity. These equations also appear not to follow the pattern of distribution of data. The same thing happened in **Figure 2c** with a shift in values on the x-axis which is more to the left.

An overestimated result is produced in **Figure 2e**, where the equation is not able to cover all areas of data; whereas in **Figure 2f**, it has been seen covering almost all areas except the data on water content values smaller than 0.2, but this equation has not been able to adjust to the data that have a value of porosity. The only equation that gives the better approach is **Figure 2d**. This equation produces a pattern in accordance with the existing data. The equation is also seen fit to data that has a value of porosity.

In this study the relative permittivity was analyzed by the ECVT system generated in the form of normalization. Normalized volumetric water content can be defined as

$$\Theta = \frac{\theta - \theta_r}{\theta_s - \theta_r} \tag{22}$$

Eq.	Source	Experimental method	Soil type	Properties of soil		
				Porosity (cm^3 cm^{-3})	Bulk density (g cm^{-3})	Particle density (g cm^{-3})
Model with one parameter						
(5) (6)	[20]	ε: using TDR Tektronix 7S12 model to perform 18 experiments with different treatments θ: using gravimetric technique	• Mineral soil • Organic soil • Vermiculite • Glass beads	—	• 1.04–1.44 • 0.422 • 1.08 • 1.49–1.61	—
(7)			Organic soil	—	0.422	—
(8)			450 μm glass beads	—	1.60–1.61	—
(9)	[21]	ε: TDR miniprobe 250 ps rise-time needle pulse θ: gravimetric technique	9 mineral soils	0.418–0.482	1.26–1.55	2.28–2.67
(10)			7 organic soils	0.527–0.785	0.2–0.77	0.70–1.63
(11)	[23]	Using model of inverse averaging for TDR method by analyzing the mixing model	—	—	—	—
(12)	[24]	ε: TDR Tektronix 1502B θ: gravimetric technique	25 samples of forest floors	—	0.086–0.263	1.3
(13)	[25]	Coaxial transmission/reflection apparatus controlled by a Hewlett-Packard 8510C Vector Network Analyzer system 45 MHz to 26.5 GHz	—	—	—	—
(14)	[13]	ε: based on capacitance measurement θ: gravimetric technique	Quartz sand	—	—	—
Model with two or more parameters						
(15) (16)	[29]	Modeling by using data from other studies [26–28]	22 different samples	0.4–0.6	1.1–1.7	2.6–2.75
(17) (18)	[30]	TDR	From 11 different field sites			
(19)	[32]	TDR CAMI	62 kinds of soil samples	0.33–0.95	0.13–1.66	1.06–2.7
(20)	[15]	Capacitance probe 80–150 MHz	• Brown earths • Silica materials	—	• 1.08–1.49 • 1.24–1.63	—
(21)	[14]	TDR Tektronix 1502B	Coarse-grained, quartz grain, sandy soil	—	—	—

Table 1.
Summary of all equations of ε-θ relationship.

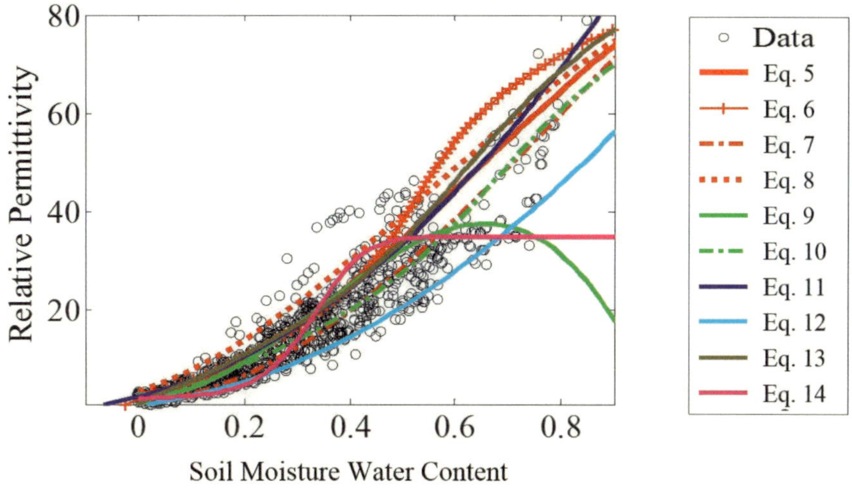

Figure 1.
Comparisons using all data for Eqs. (5)–(14).

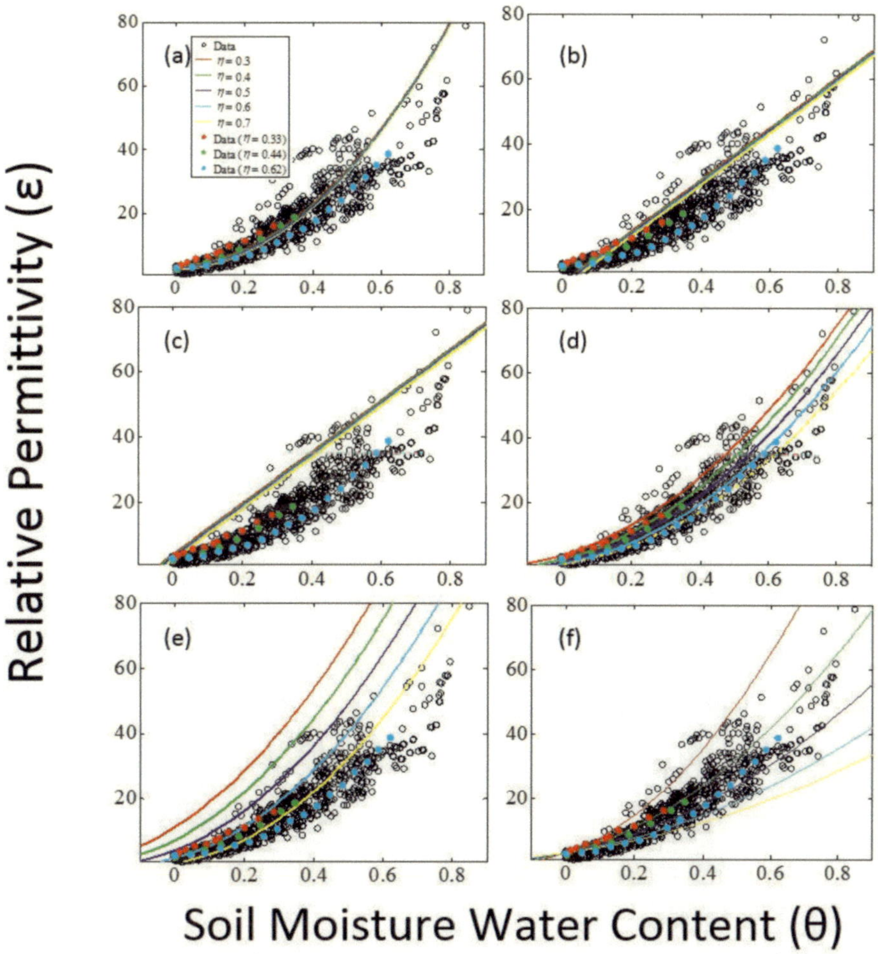

Figure 2.
Comparison all equations with different porosity, by using: a) Eq. 15; b) Eq. 16; c) Eq. 18; d) Eq. 19; e) Eq. 20; f) Eq. 21.

Material	Relative permittivity	Elements
Air	1	N_2, O_2
Water	80	H_2O
Dry soil	3–10	N, P, K, Ca, Mg, S, Cu, Zn, Fe, Mn, B, Cl, Na, H

Table 2.
Material properties.

where θ is the volumetric water content, θ_r is the residual volumetric water content, and θ_s is the saturated volumetric water content.

The normalization of relative permittivity gives privilege to define normalized volumetric water content, in this study it is assumed in three models:

$$\begin{aligned} \Theta &= {\varepsilon_N}^{0.5} && \text{(a)} \\ \Theta &= \varepsilon_N && \text{(b)} \\ \Theta &= {\varepsilon_N}^{2} && \text{(c)} \end{aligned} \tag{23}$$

where ε_N is normalized permittivity which can be calculated as

$$\varepsilon_N = \frac{\varepsilon - \varepsilon_{air}}{\varepsilon_{water} - \varepsilon_{air}} \tag{24}$$

where ε, ε_{air}, and ε_{water} are actual relative permittivity measurement, relative permittivity of air, and relative permittivity of water, respectively (**Table 2**).

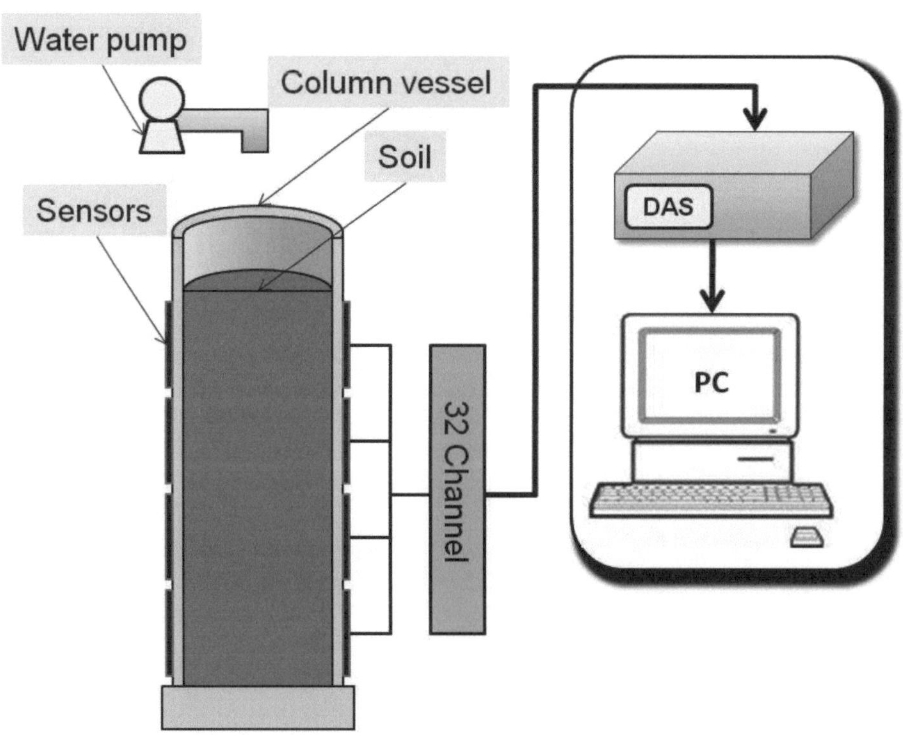

Figure 3.
Experimental setup [15].

4. Experimental setup

ECVT system consists of three parts: (i) sensors, (ii) data acquisition, and (iii) system reconstruction and visualization as shown in **Figure 3**. In this experiment we used 32 channel hexagonal sensors for the first experiment and 24 channel hexagonal sensors for the second experiment, as a soil column with diameter and height 11.5 and 27 cm, respectively. This column is divided into 32 layers with the 1st layer as the top layer and 32nd layer as the bottom layer.

In the first experiment, 3 l by volume of soil in the column was supplied. The soil material that was used in this study was sand collected from the Cisadane River in Tangerang, Indonesia. The soil contained 17% fine sand and 83% medium sand with porosity of soil 41.79%. The specific gravity and soil density were 2.663 and 1.55 g cm^{-3}, respectively. In this experiment, the soil in the vessel was supplied with water flow with a discharge of 7.2 ml/s until ponded condition, and the discharge was stopped when the pond of water level was at 2 cm above the surface soil. During ponded condition, the data capacitances were measured iteratively and sent to the computer. The data acquisition frequency was set to one frame per second.

In the second experiment, 3 kg of silty sand was supplied into the column. After that, 1.4 liter of water was filled into the soil column using constant head method, in which height of the water is maintained constant by a distance of 4 cm from the surface of soil.

5. Result and discussion

Normalized volumetric water content and relative permittivity relationship of several equations is shown in **Figure 4**. The first model of the proposed model seems to have a similar pattern with Topp et al. [20] and Malicki et al. [32] models. By adding some constants value to the first model, this model will be quite fit with

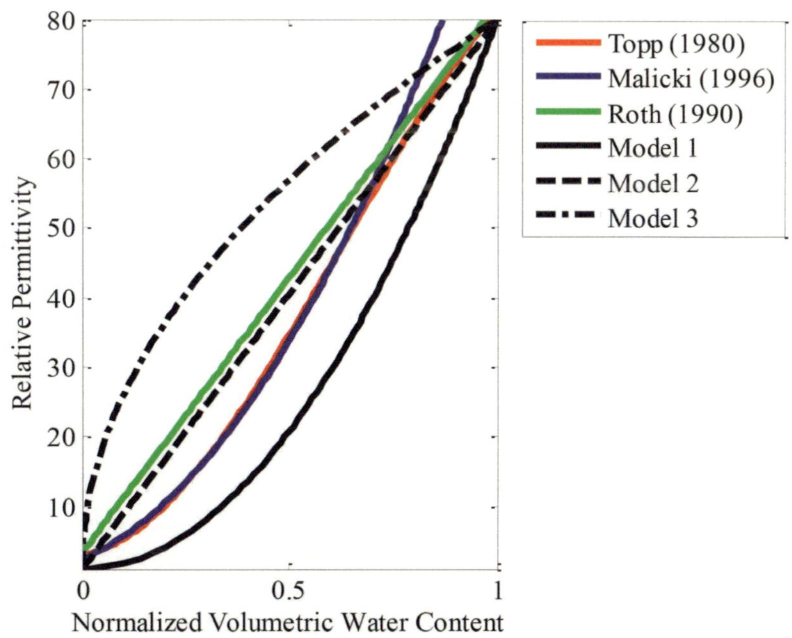

Figure 4.
Normalized volumetric water content and relative permittivity relationship of the three proposed models compared with established models.

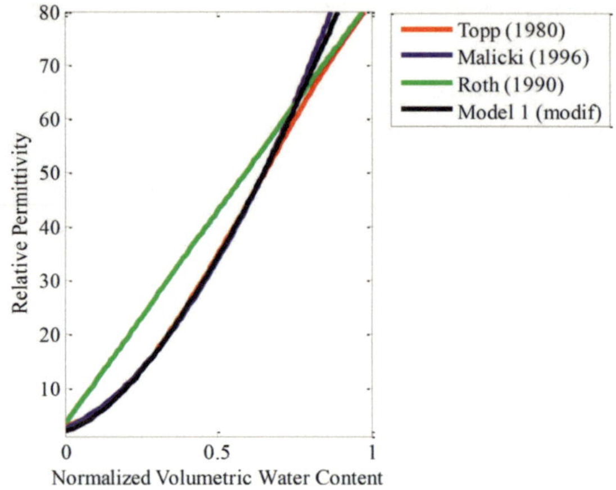

Figure 5.
Normalized volumetric water content and relative permittivity relationship of the first model modification compared with established models.

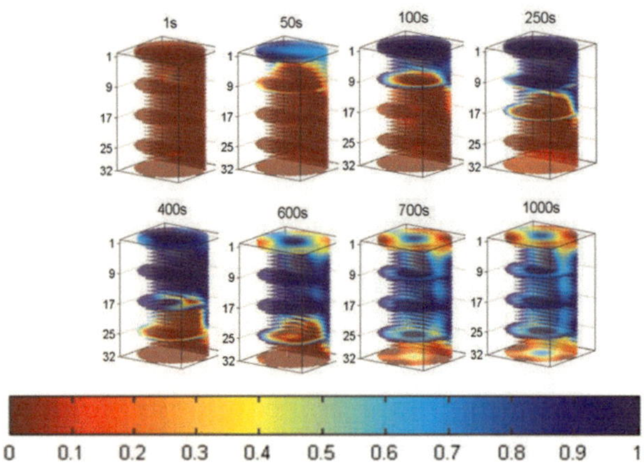

(a)

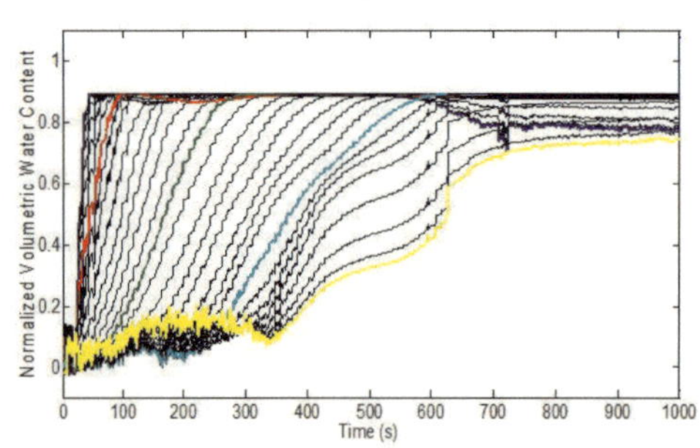

(b)

Figure 6.
Result from experiment 1, soil water infiltration (a) and normalized volumetric water content of soil (blue line, 1st layer; red line, 9th layer; green line, 17th layer; cyan line, 25th layer; and yellow line, 32nd layer) (b).

Topp and Malicki models. However, the second model looks similar to Roth et al. [21] model, which gives a linear relationship between normalized volumetric water content and relative permittivity. In contrast, the third model is out of fit with other models.

Modification to the first model (Eq. (23a)) to fit with previous models can be done by adding the constants by trial and error. The modification is shown below:

$$\Theta = 0.9 \times (\varepsilon_N\text{-}0.015)^{0.65} \tag{25}$$

Figure 5 shows the first model with modification (Eq. (25)) quite fit with the Topp and Malicki models.

Based on this result, Eq. (25) was used to analyze the normalized volumetric water content during water infiltration in soil. The results of water infiltration in soil can be seen in **Figure 6**. **Figure 6a** shows the images of normalized volumetric water content from red (i.e., dry condition, ε_N = 0) to blue (i.e., saturated condition, ε_N = 1) colors. The scale of the color means normalization value of relative permittivity distribution in image.

Figure 6a shows the image sequencing of water infiltration methods from 1, 50, 100, 250, 400, 600, 700, and 1000 s, respectively. In this figure, the position and movement of water per second can be seen clearly. **Figure 6b** shows normalized

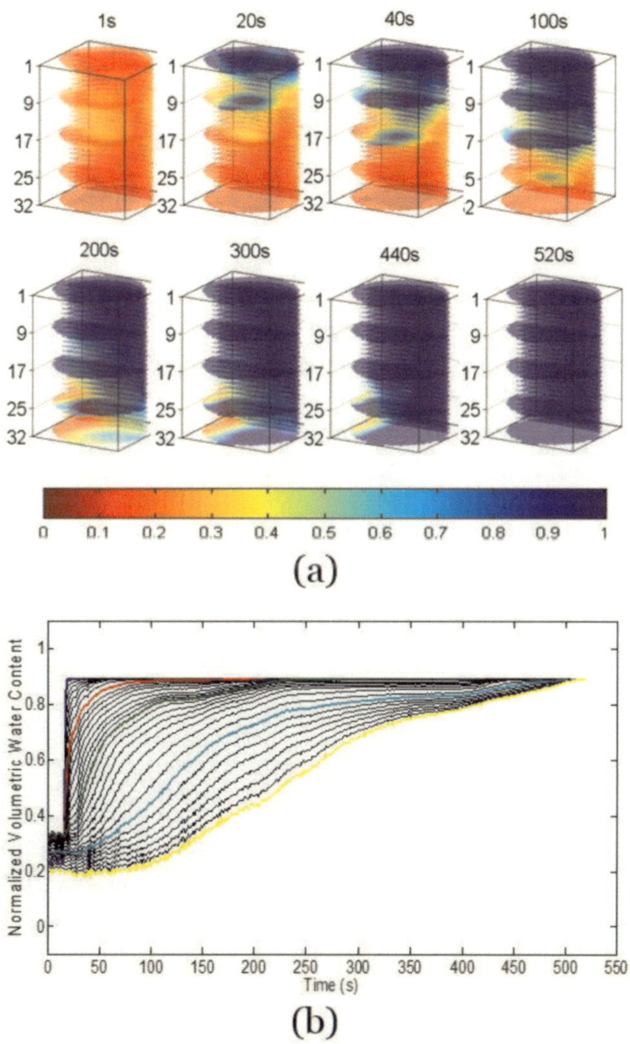

Figure 7.
Result from experiment 2, soil water infiltration (a) and normalized volumetric water content of soil (blue line, 1st layer; red line, 9th layer; green line, 17th layer; cyan line, 25th layer; and yellow line, 32nd layer) (b).

volumetric water content of soil for 32 layers during water infiltration in the soil. Blue, green, red, cyan, and yellow lines indicated 1, 9, 17, 25, and 32 soil layers, respectively. In the first layer, the normalized volumetric water content increases very fast and reaches a saturated condition at around 100–300 s. The degree of saturation for 1st layer starts to decrease after supplying of water was stopped at 300 s and relative constant was at around 0.3 from 700 to 1000 s. For 9th and 17th layers, the degree of saturation increases sequencing and reaches a saturated condition at a similar time (i.e., at around 450 s) and starts to decrease at 500 s, while for the 25th layer, the degree of saturation increases at around 300 s and reaches a stable condition at 700 s with a degree of saturation at around 0.85. Moreover at the 32nd layer, the degree of saturation increases at 630 s, and the degree of saturation reaches 0.2 at the end of the experiment (i.e., 1000 s). This experiment showed clearly the availability of air trapped at the bottom of soil in the vessel (see **Figure 6a**).

Figure 7 also shows the image sequencing of water infiltration (**Figure 7a**) and normalized volumetric water content of each layer of the soil column (**Figure 7b**). In **Figure 7a**, it can be seen that the silty sand has a normalized relative permittivity value around 0.3 before water infiltrate to the soil column. It can be caused by two possibilities, (i) because of the moisture content stored in the soil and (ii) because the soil particles are very small and compact, so the porosity of soil is also small which causes no air cavity in the soil. From **Figure 7b** we can see how the water infiltrates into the each layer clearly. The first layer has increased drastically around the first 30 s. It is easily understood that the top layers will reach the maximum value of normalized volumetric water content first, because these layers get the first supply of water.

6. Conclusion

Some equations between the relative permittivity and volumetric water content have been described in this study. From this study, there is an equation that has demonstrated efficacy in conformity with the data of experimental results, that is, equation proposed by [32]. This equation uses the porosity factor as a parameter in the relationship between volumetric water content and relative permittivity.

Normalized volumetric water content of soil has been analyzed in this study using ECVT system. Normalized volumetric water content can be shown and analyzed layer per layer of soil column for every second. We found that the ECVT system has advantages in measuring soil water content which are nondestructive and noninvasive to the sample object, 3D image, and real-time monitoring for water infiltration.

Acknowledgements

The authors wish to thank the Ministry of Research, Technology and Higher Education of Republic of Indonesia, for providing the research grand.

Conflict of interest

The authors declare that there is no conflict of interest.

Author details

Mukhlisin Muhammad[1*] and Saputra Almushfi[2]

1 Politeknik Negeri Semarang, Semarang, Indonesia

2 CTECH Labs Edwar Technology, Tangerang, Indonesia

*Address all correspondence to: mmukhlis2@gmail.com

IntechOpen

© 2019 The Author(s). Licensee IntechOpen. This chapter is distributed under the terms of the Creative Commons Attribution License (http://creativecommons.org/licenses/by/3.0), which permits unrestricted use, distribution, and reproduction in any medium, provided the original work is properly cited.

References

[1] Moroizumi T, Hamada H, Sukchan S, Ikemoto M. Soil water content and water balance in rainfed fields in Northeast Thailand. Agricultural Water Management. 2009;**96**(1):160-166

[2] Kottler BD, White JC, Kelsey JW. Influence of soil moisture on the sequestration of organic compounds in soil. Chemosphere. 2001;**42**(8):893-898

[3] Osman N, Barakbah SS. Parameters to predict slope stability—Soil water and root profiles. Ecological Engineering. 2006;**28**(1):90-95

[4] Mukhlisin M, Taha MR, Kosugi K. Numerical analysis of effective soil porosity and soil thickness effects on slope stability at a hillslope of weathered granitic soil formation. Geosciences Journal. 2008;**12**(4):401-410

[5] Mukhlisin M, Taha MR. Slope stability analysis of a weathered granitic hillslope as effects of soil thickness. European Journal of Scientific Research. 2009;**30**(1):36-44

[6] Mukhlisin M, Baidillah MR, Taha MR, El-Shafie A. Effect of soil water retention model on slope stability analysis. International Journal of Physical Sciences. 2011;**6**(19):4629-4635

[7] Topp GC. State of the art of measuring soil water content. Hydrological Processes. 2003;**17**(14): 2993-2996

[8] Huisman JA, Hubbard SS, Redman JD, Annan AP. Measuring soil water content with ground penetrating radar: A review. Vadose Zone Journal. 2003; **2**(4):476-491

[9] Basinger JM, Kluitenberg GJ, Ham JM, Frank JM, Barnes PL, Kirkham MB. Laboratory evaluation of the dual-probe heat-pulse method for measuring soil water content. Vadose Zone Journal. 2003;**2**(3):389

[10] Gaskin GJ, Miller JD. Measurement of soil water content using a simplified impedance measuring technique. Journal of Agricultural Engineering Research. 1996;**63**(2):153-159

[11] Ramírez V, Sánchez JA, Hernández G, Solís S, Torres J, Antaño R, et al. A promising electrochemical test for evaluating the hydrocarbon-type pollutants contained in industrial waste soils. International Journal of Electrochemical Science. 2011;**6**(5): 1415-1437

[12] Ruíz C, Anaya J, Ramírez V, Alba GI, García MG, Carrillo-Chávez A, et al. Soil Arsenic Removal by a Permeable Reactive Barrier of Iron Coupled to an Electrochemical Process. 2011

[13] Wu SY, Zhou QY, Wang G, Yang L, Ling CP. The relationship between electrical capacitance-based dielectric constant and soil water content. Environment and Earth Science. 2011; **62**(5):999-1011

[14] Robinson DA, Jones SB, Wraith JM, Or D, Friedman SP. A review of advances in dielectric and electrical conductivity measurement in soils using time domain reflectometry. Vadose Zone Journal. 2003;**2**(4):444-475

[15] Gardner CMK, Dean TJ, Cooper JD. Soil water content measurement with a high-frequency capacitance sensor. Journal of Agricultural Engineering Research. 1998;**71**(4):395-403

[16] Warsito W, Marashdeh Q, Fan L-S. Electrical capacitance volume tomography. IEEE Sensors Journal. 2007;7(4):525-535

[17] Warsito W, Marashdeh Q, Fan LS. Real time volumetric imaging of multiphase flows using electrical capacitance volume-tomography

(ECVT). In: 5th World Congress in Industrial Process Tomography; 2007. pp. 755-760

[18] Mukhlisin M, Baidillah MR, El-Shafie A, Taha MR. Real time monitoring of soil water infiltration using electrical capacitance volume-tomography (ECVT). In: 4th International Conference on Water Resources and Arid Environments (ICWRAE 4); 2010. pp. 749-756

[19] Yang WQ, Spink DM, York TA, McCann H. An image-reconstruction algorithm based on Landweber's iteration method for electrical-capacitance tomography. Measurement Science and Technology. 1999;**10**(11): 1065-1069

[20] Topp GC, Davis JL, Annan AP. Electromagnetic determination of soil water content: Measurements in coaxial transmission lines. Water Resources Research. 1980;**16**(3):574-582

[21] Roth CH, Malicki MA, Plagge R. Empirical evaluation of the relationship between soil dielectric constant and volumetric water content as the basis for calibrating soil moisture measurements by TDR. Journal of Soil Science. 1992; **43**(1):1-13

[22] Malicki MA, Plagge R, Renger M, Walczak RT. Application of time-domain reflectometry (TDR) soil moisture miniprobe for the determination of unsaturated soil water characteristics from undisturbed soil cores. Irrigation Science. 1992;**13**(2): 65-72

[23] Ferré PA, Rudolph DL, Kachanoski RG. Spatial averaging of water content by time domain reflectometry: Implications for twin rod probes with and without dielectric coatings. Water Resources Research. 1996;**32**(2):271-279

[24] Schaap MG, de Lange L, Heimovaara TJ. TDR calibration of organic forest floor media. Soil Technology. 1997;**11**(2):205-217

[25] Curtis JO. Moisture effects on the dielectric properties of soils. IEEE Transactions on Geoscience and Remote Sensing. 2001;**39**(1):125-128

[26] Lundien JR. Terrain Analysis by Electromagnetic Means. Vicksburg, MS: U.S. Army Engineer Waterways Experiment Station; 1971

[27] Newton RW. Microwave Remote Sensing and its Application to Soil Moisture Detection. College Station, TX: Texas A&M University, College Station; 1977

[28] Wang J, Schmugge T, Williams D. Dielectric constants of soils at microwave frequencies—II. National Aeronautics and Space Administration Technical Paper. 1978

[29] Wang JR, Schmugge TJ. An empirical model for the complex dielectric permittivity of soils as a function of water content. IEEE Transactions on Geoscience and Remote Sensing. 1980;**GE-18**(4):288-295

[30] Roth K, Schulin R, Fluhler H, Attinger W. Calibration of time domain reflectometry for water content measurement using a composite dielectric approach. Water Resources Research. 1990;**26**:2267-2273

[31] Dobson MC, Ulaby FT, Hallikainen MT, El-Rayes MA. Microwave dielectric behavior of wet soil—Part II: Dielectric mixing models. IEEE Transactions on Geoscience and Remote Sensing. 1985; **GE-23**:35-45

[32] Malicki MA, Plagge R, Roth CH. Improving the calibration of dielectric TDR soil moisture determination taking into account the solid soil. European Journal of Soil Science. 1996;**47**:357-366

[33] Friedman SP. A saturation degree-dependent composite spheres model for

describing the effective dielectric constant of unsaturated porous media. Water Resources Research. 1998;**34**: 2949-2961

[34] Hilhorst MA, Dirksen C, Kampers FWH, Feddes RA. New dielectric mixture equation for porous materials based on depolarization factors. Soil Science Society of America Journal. 2000;**64**:1581-1587

[35] Robinson DA, Jones SB, Blonquist JM, Friedman SP. A physically derived water content/permittivity calibration model for coarse-textured, layered soils. Soil Science Society of America Journal. 2005;**69**:1372-1378

Chapter 7

Techniques for Calculating Two Interesting Types of Dielectric Materials in Straight Rectangular Waveguides and Their Applications

Zion Menachem

Abstract

This chapter presents two interesting types of dielectric materials in the straight rectangular waveguides. Five examples of the different discontinuous cross sections and complementary shapes will demonstrate. We will introduce in all case the effective technique to calculate the dielectric profile in the cross section. The first type will demonstrate where the dielectric material is located in the center of the cross section. The second type will demonstrate where the hollow core is located in the center of the cross section in the case of the hollow waveguide. The two different types are complementary shapes for two different applications. The proposed techniques relate to the method based on Laplace and Fourier transforms and the inverse Laplace and Fourier transforms. The method is based also on Fourier transform, thus we need use with the image method to calculate the dielectric profile in the cross section. The image method and periodic replication are needed for fulfilling the boundary condition of the metallic waveguide. The applications are useful for straight waveguides in millimeter regimes, in the cases where the dielectric profile is located in the center of the cross section, for cases where the hollow rectangle is located in the center of the cross section, and also for complicated and discontinuous profiles in the cross section.

Keywords: wave propagation, dielectric profiles, rectangular waveguides, dielectric materials

1. Introduction

Review of numerical and approximate methods for the modal analysis of general optical dielectric waveguides with emphasis on recent developments has been published [1]. In this review, interesting methods are given such as, the finite-difference and the finite-element methods. Review of numerical methods for the analysis of the homogeneous and inhomogeneous, isotropic and anisotropic, microwave and optical dielectric waveguides with arbitrarily-shaped cross sections has been published [2]. The main approaches as the integral equations, finite difference, and finite element have been discussed.

IntechOpen

A fundamental and accurate approach to compute the attenuation of electromagnetic waves propagating has been proposed [3]. The propagation constant was found by substituting the values of transverse wave numbers into the dispersion relation. A Green's dyadic for describing the propagating electromagnetic waves in a rectangular dielectric waveguide has been developed [4]. The use of Green's dyadic developed in order to calculate the effect of small perturbations upon the system.

An analytical method for solution of one-dimensional optical systems, based on the differential transfer matrices has been presented [5]. An approach to solve the problem of the propagation of electromagnetic waves in unidimensional media with an arbitrary variation of their dielectric permittivity has been proposed [6]. This method was deduced from the Maxwell equations with a minimum of approximations and allows a full vectorial description of both the electric and magnetic fields through the direct calculation of their cartesian coordinates. The equations permit the simulation of materials with a continuous variation of their dielectric permittivity without approximating them by discontinuous layered media, reducing so the computational effort of the models.

An analytical method for solution of non-homogeneous anisotropic optical systems, based on the extension of transfer matrices into differential form has been presented [7]. This approach can be used for exact calculation of various functions including reflection and transmission coefficients, band structures and bound states. A full-wave analysis of lossy dielectric waveguides using a hybrid vector finite element method has been presented [8]. The direct matrix solution technique with minimum degree of reordering has been combined with the modified Lanczos algorithm to solve for the resultant sparse generalized eigenmatrix equation efficiently.

Three-dimensional finite-element method with edge elements for electromagnetic waveguide discontinuities has been proposed [9]. This paper shows that the finite-element method using edge elements succeeds in suppressing spurious solutions and moreover that this method succeeds in the analysis of three-dimensional electromagnetic waveguide problems with metal wedges. An analytical approach based on scalar wave approximation to estimate the modal dispersion characteristics and cutoff condition of an optical waveguide has been presented [10]. This approach has an arbitrary and uniform core cross-section. The structure represents the core of a circular waveguide which is compressed at both the ends of a diameter.

Propagation characteristics of modes in some rectangular waveguides using the finite-difference time-domain method have been studied [11]. The method in this paper was used to determine the modal characterization of rectangular waveguide structures by means of a least-square non-linear fitting to a theoretical modal expansion. Wave propagation along a rectangular waveguide with slowly varying width has been investigated [12] with the help of field theory and approximate circuit theory. Many properties of the modulated periodic structure, e.g., the frequency dependence of the propagation constant, group and phase velocities, and the electric field axial variation for the fundamental space harmonic and its filter-like property have been investigated.

A method for measuring samples using a partially-filled waveguide has been presented [13]. A mode-matching technique was used to determine the fields in the three regions. An advantageous finite element method for the rectangular waveguide problem has been developed [14] by which complex propagation characteristics may be obtained for arbitrarily shaped waveguide.

A method has been introduced for the frequency domain analysis of arbitrary longitudinally inhomogeneous waveguides [15]. The integral equations of the longitudinally inhomogeneous waveguides, converted from their differential equations, were solved using the method of moments. A method has been introduced for the frequency domain analysis of arbitrarily loaded lossy and dispersive nonuniform transmission lines [16]. In this method, all distributed primary parameters of the line and also the voltage and current distribution along the line were considered as a Taylor's series.

A technique based on the two-dimensional Fourier transform has been presented [17] and applied to the study of nonlinear wave propagation phenomena in one-dimensional, finite, nonlinear transmission lines.

A method for calculating the modes of arbitrarily shaped dielectric waveguides has been presented [18]. It consists of expanding the field in a two-dimensional Fourier series. The expansion has been used to convert the scalar wave equation into a matrix eigenvalue equation. A transfer matrix function for the analysis of electromagnetic wave propagation along the straight dielectric waveguide with arbitrary profiles has been proposed [19]. This method is based on the Laplace and Fourier transforms and the inverse Laplace and Fourier transforms.

2. Complicated and discontinuous profiles in the cross section

In this chapter, we present some examples of dielectric structures as shown in **Figure 1a–e**. The method is based on Fourier transform, thus we need use with the image method and periodic replication for fulfilling the boundary conditions of the metallic waveguide. We relate also to the complementary shapes for different applications. The periodicity and the symmetry properties are chosen to force the boundary conditions at the location of the walls in real problem, by extending the

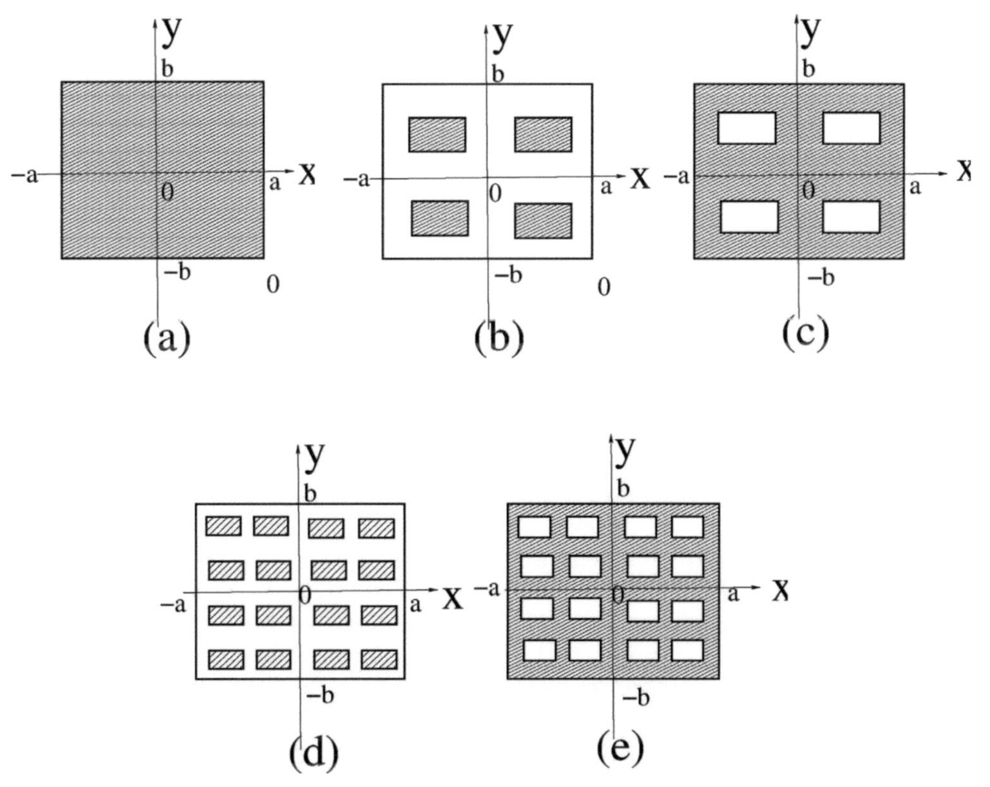

Figure 1.
The image method and periodic replication for five examples. (a). The cross section entirely filled with the dielectric material, (b). The dielectric material is located in the center of the cross section, (c). The hollow waveguide where the hollow rectangle is located in the center of the cross section, (d). The cross section consists with four dielectric profiles, and (e). The hollow waveguide with four hollow rectangles.

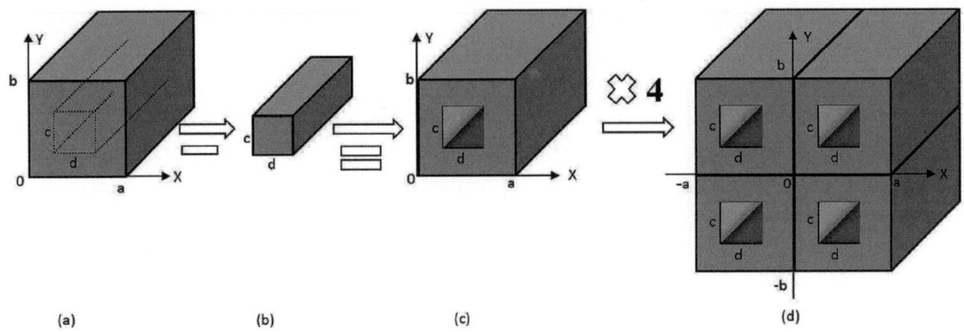

Figure 2.
Example of receiving a hollow waveguide where the hallow rectangle is located in the center (c) by substracting the waveguide with dielectric material located at the center of the cross section (b) from the waveguide entirely filled with the dielectric material (a), and by using the image method (d).

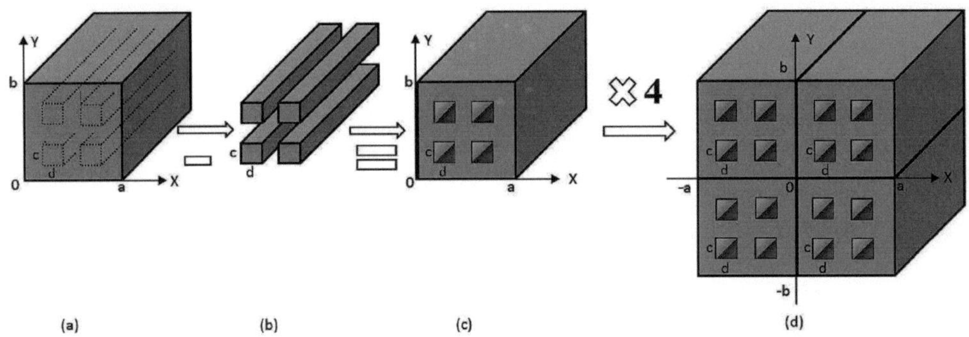

Figure 3.
Example of receiving a hollow waveguide with four hollow rectangles (c) by substracting the four dielectric rectangles (b) from the waveguide entirely filled with the dielectric material (a), and by using the image method (d).

waveguide region ($0 \le x \le a$, and $0 \le y \le b$) to a fourfold larger regions ($-a \le x \le a$, and $-b \le y \le b$), as larger regions, as described in [19]. **Figure 1a** shows the case where the cross section entirely filled with the dielectric material. The dielectric material in **Figure 1b** is located in the center of the cross section. **Figure 1c** represents the hollow waveguide where the hollow rectangle is located in the center of the cross section. The cross section of **Figure 1d** consists with four dielectric profiles in the cross section. **Figure 1e** represents the hollow waveguide with four hollow rectangles.

Note that the geometries from **Figure 1a–e** become more complex. Calculating of the dielectric profile for **Figure 1a** is the simplest and for **Figure 1e** is the most complicated in relation to the examples described in these examples. The integrals of the dielectric profiles of the geometries described in **Figure 1a** and **b** can be calculated by analytical form and without solving numerical form. The calculation of the dielectric profile for the geometry of **Figure 1c** for a hollow rectangle in the center is interesting in the case of hollow waveguide, but it is more complicated than the case of **Figure 1b**. Therefore, in this case we should calculate the dielectric profile by substracting the dielectric profile of **Figure 1b** from the dielectric profile of **Figure 1a**. The calculation of the integrals of the dielectric profiles for the geometries depicted in **Figure 1d** and **e** are already more complicated, and in such cases the proposed techniques require that the integrals be solved numerically. We will explain the proposed technique for calculating the dielectric profile for each case.

Figure 2c represents an example of receiving a hollow waveguide where the hollow rectangle is located in the center by substracting the waveguide with dielectric material located at the center of the cross section (**Figure 2b**) from the waveguide entirely filled with the dielectric material (**Figure 2a**), and by using the image method (**Figure 2d**). **Figure 3c** represents an example of receiving a hollow waveguide with four hollow rectangles by substracting the four dielectric rectangles (**Figure 3b**) from the waveguide entirely filled with the dielectric material (**Figure 3a**), and by using with the image method (**Figure 3d**).

3. The analytical technique to calculate the dielectric profiles

This section presents a technique to calculate the dielectric profile for the two inhomogeneous geometries of the cross section, as shown in **Figure 4a** and **b**.

Figure 4a and **b** show the cross section of **Figure 1b** and **c**, respectively, with the relevant parameters. These figures represent two examples of the complementary shapes in the cross section. The periodic replication is shown in **Figure 4c** by using with the image method. The dielectric material in **Figure 4a** is located in the center of the cross section. The hollow rectangle in **Figure 4b** is located in the center of the cross section in the case of the hollow waveguide.

The dielectric profile $g(x, y)$ is given according to $\epsilon(x, y) = \epsilon_0(1 + g(x, y))$. If $g(x, y) = g_0$ and $\epsilon_r(x, y) = \epsilon_r$, then we obtain that g0 = $\epsilon_r - 1$. According to **Figure 4a** and **c** for $g(x, y) = g_0$, we obtain

$$\begin{aligned}
g(n, m) =& \frac{g_0}{4ab}\int_{-a}^{a} dx\int_{-b}^{b} \exp\left[-j(k_x x + k_y y)\right]dy \\
=& \frac{g_0}{4ab}\left\{\int_{x_{11}}^{x_{12}} dx\int_{y_{11}}^{y_{12}} \exp\left[-j(k_x x + k_y y)\right]dy + \int_{-x_{12}}^{-x_{11}} dx\int_{y_{11}}^{y_{12}} \exp\left[-j(k_x x + k_y y)\right]dy\right. \\
&\left. + \int_{-x_{12}}^{-x_{11}} dx\int_{-y_{12}}^{-y_{11}} \exp\left[-j(k_x x + k_y y)\right]dy + \int_{x_{11}}^{x_{12}} dx\int_{-y_{12}}^{-y_{11}} \exp\left[-j(k_x x + k_y y)\right]dy\right\} \\
=& \frac{g_0}{4ab}\left\{\int_{x_{11}}^{x_{12}} dx\int_{y_{11}}^{y_{12}} \exp\left[-j(k_x x + k_y y)\right]dy + \int_{x_{12}}^{x_{11}} -dx\int_{y_{11}}^{y_{12}} \exp\left[-j(-k_x x + k_y y)\right]dy\right. \\
&\left. + \int_{x_{12}}^{x_{11}} -dx\int_{y_{12}}^{y_{11}} - \exp\left[-j(-k_x x - k_y y)\right]dy + \int_{x_{11}}^{x_{12}} dx\int_{y_{12}}^{y_{11}} - \exp\left[-j(k_x x - k_y y)\right]dy\right\} \\
=& \frac{g_0}{4ab}\left\{\int_{x_{11}}^{x_{12}} \exp\left[j(k_x x)\right]dx\int_{y_{11}}^{y_{12}} \exp\left[j(k_y y)\right]dy + \int_{x_{11}}^{x_{12}} \exp\left[j(k_x x)\right]dx\int_{y_{11}}^{y_{12}} \exp\left[-j(k_y y)\right]dy\right. \\
&\left. + \int_{x_{11}}^{x_{12}} \exp\left[-j(k_x x)\right]dx\int_{y_{11}}^{y_{12}} \exp\left[j(k_y y)\right]dy + \int_{x_{11}}^{x_{12}} \exp\left[-j(k_x x)\right]dx\int_{y_{11}}^{y_{12}} \exp\left[-j(k_y y)\right]dy\right\} \\
=& \frac{g_0}{4ab}\left\{\int_{x_{11}}^{x_{12}} dx\int_{y_{11}}^{y_{12}} \exp\left[-j(k_x x + k_y y)\right]dy + \int_{x_{11}}^{x_{12}} dx\int_{y_{11}}^{y_{12}} \exp\left[j(k_x x - k_y y)\right]dy\right. \\
&\left. + \int_{x_{11}}^{x_{12}} dx\int_{y_{11}}^{y_{12}} \exp\left[j(k_x x + k_y y)\right]dy + \int_{x_{11}}^{x_{12}} dx\int_{y_{11}}^{y_{12}} \exp\left[-j(k_x x - k_y y)\right]dy\right\} \\
=& \frac{g_0}{2ab}\left\{\int_{x_{11}}^{x_{12}} \left(\exp\left(jk_x x\right) + \exp\left(-jk_x x\right)\right)dx\int_{y_{11}}^{y_{12}} \cos\left(k_y y\right)dy\right\},
\end{aligned} \tag{1}$$

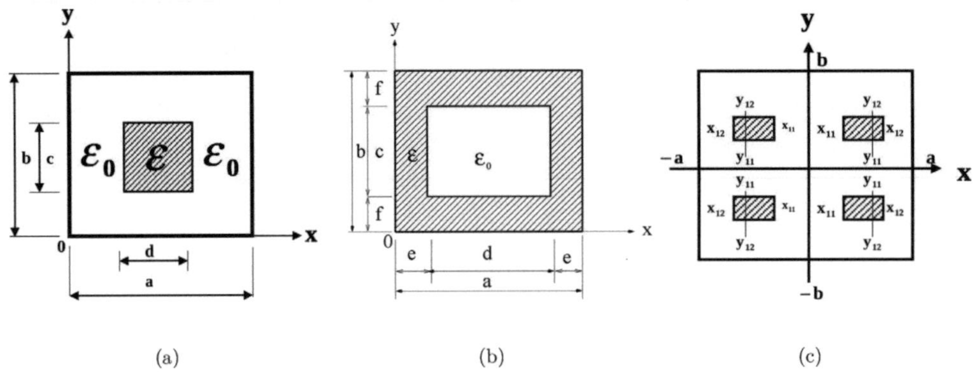

Figure 4.
Two examples of the complementary shapes of profiles in the cross section and the periodic replication, (a). The dielectric material is located in the center of the cross section with the relevant parameters of ***Figure 1b****, (b). The hollow rectangle is located in the center of the cross section in the case of the hollow waveguide with the relevant parameters of* ***Figure 1c****, (c). The periodic replication according to the image method.*

where $k_x = (n\pi x)/a$, and $k_y = (m\pi y)/b$. The derivatives of the dielectric profile are calculated according to $g_x(x, y) = (1/\epsilon(x, y))(d\epsilon(x, y)/dx)$, and $g_y(x, y) = (1/\epsilon(x, y))(d\epsilon(x, y)/dy)$, where $\epsilon(x, y) = \epsilon_0(1 + g(x, y))$.

For the cross section as shown in **Figure 4a** and according to **Figure 4c**, the center of the dielectric rectangle is located at (a/2, b/2). According to **Figure 4a** and **c**, the Fourier components of the dielectric profile are given by

$$g_1(n, m) = \begin{cases} \frac{g_0}{4ab}(4cd) & n = 0, m = 0 \\ \frac{g_0}{4ab}\left(\frac{8d}{k_{0y}m}\sin\left(\frac{k_{0y}mc}{2}\right)\cos\left(\frac{k_{0y}mb}{2}\right)\right) & n = 0, m \neq 0 \\ \frac{g_0}{4ab}\left(\frac{8c}{k_{0x}n}\sin\left(\frac{k_{0x}nd}{2}\right)\cos\left(\frac{k_{0x}na}{2}\right)\right) & n \neq 0, m = 0 \\ \frac{g_0}{4ab}\left(\frac{16}{k_{0x}k_{0y}nm}\sin\left(\frac{k_{0x}nd}{2}\right)\cos\left(\frac{k_{0x}na}{2}\right)\sin\left(\frac{k_{0y}mc}{2}\right)\cos\left(\frac{k_{0y}mb}{2}\right)\right) & n \neq 0, m \neq 0 \end{cases} \quad (2)$$

Similarly, for the cross section entirely filled with the dielectric material as shown in **Figure 1a**, the Fourier components of the dielectric profile are given by

$$g_2(n, m) = \begin{cases} g_0 & n = 0, m = 0 \\ \frac{g_0}{4ab}\left(\frac{8a}{k_{0y}m}\sin\left(\frac{k_{0y}mb}{2}\right)\cos\left(\frac{k_{0y}mb}{2}\right)\right) & n = 0, m \neq 0 \\ \frac{g_0}{4ab}\left(\frac{8b}{k_{0x}n}\sin\left(\frac{k_{0x}na}{2}\right)\cos\left(\frac{k_{0x}na}{2}\right)\right) & n \neq 0, m = 0 \\ \frac{g_0}{4ab}\left(\frac{16}{k_{0x}k_{0y}nm}\sin\left(\frac{k_{0x}na}{2}\right)\cos\left(\frac{k_{0x}na}{2}\right)\sin\left(\frac{k_{0y}mb}{2}\right)\cos\left(\frac{k_{0y}mb}{2}\right)\right) & n \neq 0, m \neq 0 \end{cases} \quad (3)$$

The Fourier components of the dielectric profile ($g(n, m)$) of the hollow rectangular waveguide with the hollow rectangle in the center of the cross section

(**Figure 4b**) are calculated by subtracting the dielectric profile of the waveguide with the dielectric material in the core ($g_1(n, m)$) as shown in **Figure 4a** from the dielectric profile of the waveguide entirely with the dielectric material ($g_2(n, m)$) as shown in **Figure 1a**. Namely, the Fourier components of the dielectric profile ($g(n, m)$) are given by $g_2(n, m) - g_1(n, m)$. The two different types of the cross sections are complementary shapes for two different applications.

4. The numerical technique to calculate the dielectric profiles

4.1 Calculation of the dielectric profile according to the ω_ε function

Figure 5a and **b** show the cross section of **Figure 1d** and **e**, respectively, with the relevant parameters. These figures represent two examples of the complementary shapes in the cross section. The cross section consists with four dielectric profiles as shown in **Figure 5a**. The hollow waveguide with four hollow rectangles is shown in **Figure 5b**. The centers of the first, the second, the third and the firth dielectric rectangles in **Figure 5a** and hollow rectangles in **Figure 5b** are located at the points (a/4, b/4), (3a/4, b/4), (a/4, 3b/4), and (3a/4, 3b/4), respectively. The calculation of the elements of the matrix in **Figure 5b** is already more complicated. Thus, the Fourier components of the dielectric profile ($g(n, m)$) of the hollow waveguide with four hollow rectangles in **Figure 5b** are calculated by subtracting the dielectric profile of the cross section with four dielectric profiles in **Figure 5a** from the dielectric profile of the waveguide entirely with the dielectric material ($g_2(n, m)$) as shown in **Figure 1a**, and according to Eq. (3).

In order to calculate the elements of the dielectric profiles of the inhomogeneous geometry of the cross section in **Figure 5a**, we can use with the ω_ε function [20], as shown in **Figure 6**. The ω_ε function is defined as $\omega_\varepsilon(r) = C_\varepsilon \exp\left[-\varepsilon^2/(\varepsilon^2 - |r|^2)\right]$ for $|r|>\varepsilon$, where C_ε is a constant, and $\int \omega_\varepsilon(r)dr = 1$. In the limit $\varepsilon \to 0$, the ω_ε function is shown in **Figure 6**.

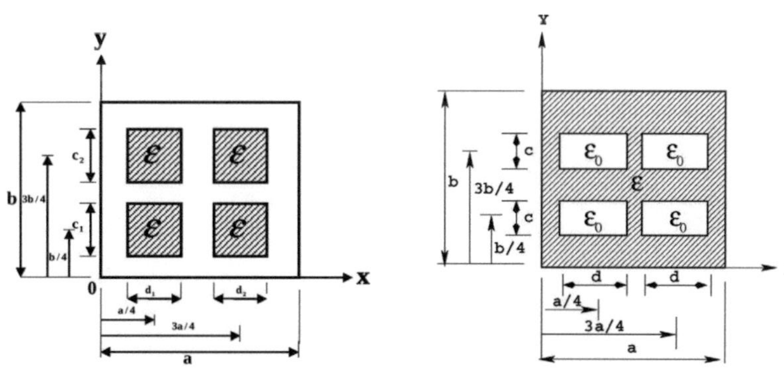

(a) (b)

Figure 5.
*Two examples of the complementary shapes of profiles in the cross section and their relevant parameters. The centers of the first, the second, the third and the firth rectangles are located at the points (a/4, b/4), (3a/4, b/4), (a/4, 3b/4), and (3a/4, 3b/4), respectively: (a) the cross section consists with four dielectric profiles with the relevant parameters of **Figure 1d**; (b) the hollow waveguide with four hollow rectangles with the relevant parameters of **Figure 1e**.*

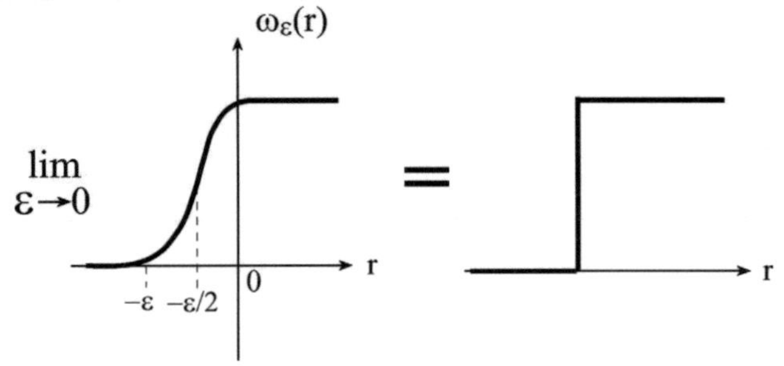

Figure 6.
The ω_ε function to calculate the elements of the dielectric profiles of the inhomogeneous geometry.

4.2 The technique based on ω_ε function for the inhomogeneous cross section

It is good idea to calculate the dielectric profile of **Figure 4a** again, by using the ω_ε function, and before considering to the more complicated geometry as shown in **Figure 5a**. The second way to calculate the dielectric profile of **Figure 4a** is given according to the ω_ε function, where the dielectric profile is located at (a/2, b/2) and is given by

$$g(x) = \begin{cases} g_0 \exp\left(1 - g_1(x)\right) & (a-d-\varepsilon)/2 \le x < (a-d+\varepsilon)/2 \\ g_0 & (a-d+\varepsilon)/2 < x < (a+d-\varepsilon)/2 \\ g_0 \exp\left(1 - g_2(x)\right) & (a+d-\varepsilon)/2 \le x < (a+d+\varepsilon)/2 \\ 0 & else \end{cases}, \quad (4)$$

and

$$g(y) = \begin{cases} g_0 \exp\left(1 - g_3(y)\right) & (b-c-\varepsilon)/2 \le y < (b-c+\varepsilon)/2 \\ g_0 & (b-c+\varepsilon)/2 < y < (b+c-\varepsilon)/2 \\ g_0 \exp\left(1 - g_4(y)\right) & (b+c-\varepsilon)/2 \le y < (b+c+\varepsilon)/2 \\ 0 & else \end{cases}, \quad (5)$$

where

$$g_1(x) = \frac{\varepsilon^2}{\varepsilon^2 - [x - (a-d+\varepsilon)/2]^2}, \quad g_2(x) = \frac{\varepsilon^2}{\varepsilon^2 - [x - (a+d-\varepsilon)/2]^2},$$

$$g_3(y) = \frac{\varepsilon^2}{\varepsilon^2 - [y - (b-c+\varepsilon)/2]^2}, \quad g_4(y) = \frac{\varepsilon^2}{\varepsilon^2 - [y - (b+c-\varepsilon)/2]^2}.$$

Similarly, we can calculate the rectangular dielectric profile according to the location of the profile in the cross section of the waveguide.

The elements of the matrix in Fourier space of the inhomogeneous geometry described in **Figure 4a** are given in the case of $b \neq c$ by

$$g(n,m) = \frac{g_0}{ab}\left\{\int_{(a-d-\varepsilon)/2}^{(a-d+\varepsilon)/2} \exp\left(1-g_1(x)\right)\cos\left(\frac{n\pi x}{a}\right)dx + \int_{(a-d+\varepsilon)/2}^{(a+d-\varepsilon)/2}\cos\left(\frac{n\pi x}{a}\right)dx + \int_{(a+d-\varepsilon)/2}^{(a+d+\varepsilon)/2}\exp\left(1-g_2(x)\right)\cos\left(\frac{n\pi x}{a}\right)dx\right\}$$
$$\left\{\int_{(b-c-\varepsilon)/2}^{(b-c+\varepsilon)/2}\exp\left(1-g_3(y)\right)\cos\left(\frac{m\pi y}{b}\right)dy + \int_{(b-c+\varepsilon)/2}^{(b+c-\varepsilon)/2}\cos\left(\frac{m\pi y}{b}\right)dy + \int_{(b+c-\varepsilon)/2}^{(b+c+\varepsilon)/2}\exp\left(1-g_4(y)\right)\cos\left(\frac{m\pi y}{b}\right)dy\right\}. \quad (6)$$

4.3 The technique based on ω_ε function for the inhomogeneous cross section

Figure 5a shows the cross section where the centers of the first, the second, the third and the firth rectangles are located at the points (a/4, b/4), (3a/4, b/4), (a/4, 3b/4), and (3a/4, 3b/4), respectively, and for ϵ_r = 1.1, 1.2, 1.3 and 1.4, respectively. We assume for simplicity that $d_1 = d_2 = d$ and $c_1 = c_2 = c$, according to **Figure 5a**.

The dielectric profile of **Figure 5a** is given according to the ω_ε function by

$$g(x) = \begin{cases} g_0 \exp\left(1-q_1(x)\right) & ((a/2)-d_1-\varepsilon)/2 \le x < ((a/2)-d_1+\varepsilon)/2 \\ g_0 & ((a/2)-d_1+\varepsilon)/2 < x < ((a/2)+d_1-\varepsilon)/2 \\ g_0 \exp\left(1-q_2(x)\right) & ((a/2)+d_1-\varepsilon)/2 \le x < ((a/2)+d_1+\varepsilon)/2 \\ g_0 \exp\left(1-q_3(x)\right) & ((3a/2)-d_2-\varepsilon)/2 \le x < ((3a/2)-d_2+\varepsilon)/2, \\ g_0 & ((3a/2)-d_2+\varepsilon)/2 < x < ((3a/2)+d_2-\varepsilon)/2 \\ g_0 \exp\left(1-q_4(x)\right) & ((3a/2)+d_2-\varepsilon)/2 \le x < ((3a/2)+d_2+\varepsilon)/2 \\ 0 & else \end{cases} \quad (7)$$

and

$$g(y) = \begin{cases} g_0 \exp\left(1-q_5(y)\right) & ((b/2)-c_1-\varepsilon)/2 \le y < ((b/2)-c_1+\varepsilon)/2 \\ g_0 & ((b/2)-c_1+\varepsilon)/2 < y < ((b/2)+c_1-\varepsilon)/2 \\ g_0 \exp\left(1-q_6(y)\right) & ((b/2)+c_1-\varepsilon)/2 \le y < ((b/2)+c_1+\varepsilon)/2 \\ g_0 \exp\left(1-q_7(y)\right) & ((3b/2)-c_2-\varepsilon)/2 \le y < ((3b/2)-c_2+\varepsilon)/2, \\ g_0 & ((3b/2)-c_2+\varepsilon)/2 < y < ((3b/2)+c_2-\varepsilon)/2 \\ g_0 \exp\left(1-q_8(y)\right) & ((3b/2)+c_2-\varepsilon)/2 \le y < ((3b/2)+c_2+\varepsilon)/2 \\ 0 & else \end{cases} \quad (8)$$

where

$$q_1(x) = \frac{\varepsilon^2}{\varepsilon^2 - [x-((a/2)-d_1+\varepsilon)/2]^2}, \quad q_2(x) = \frac{\varepsilon^2}{\varepsilon^2 - [x-((a/2)+d_1-\varepsilon)/2]^2},$$
$$q_3(x) = \frac{\varepsilon^2}{\varepsilon^2 - [x-((3a/2)-d_2+\varepsilon)/2]^2}, \quad q_4(x) = \frac{\varepsilon^2}{\varepsilon^2 - [x-((3a/2)+d_2-\varepsilon)/2]^2}$$
$$q_5(y) = \frac{\varepsilon^2}{\varepsilon^2 - [y-((b/2)-c_1+\varepsilon)/2]^2}, \quad q_6(y) = \frac{\varepsilon^2}{\varepsilon^2 - [y-((b/2)+c_1-\varepsilon)/2]^2},$$
$$q_7(y) = \frac{\varepsilon^2}{\varepsilon^2 - [y-((3b/2)-c_2+\varepsilon)/2]^2}, \quad q_8(y) = \frac{\varepsilon^2}{\varepsilon^2 - [y-((3b/2)+c_2-\varepsilon)/2]^2}.$$

The elements of the matrix for the inhomogeneous geometry of the cross section of **Figure 5a** are calculated in Fourier space by

$$
\begin{aligned}
g(n,m) = \frac{g_0}{ab}\Bigg\{ &\int_{((a/2)-d_1-\varepsilon)/2}^{((a/2)-d_1+\varepsilon)/2} \exp\left(1-q_1(x)\right)\cos\left(\frac{n\pi x}{a}\right)dx \\
&+\int_{((a/2)-d_1+\varepsilon)/2}^{((a/2)+d_1-\varepsilon)/2} \cos\left(\frac{n\pi x}{a}\right)dx + \int_{((a/2)+d_1-\varepsilon)/2}^{((a/2)+d_1+\varepsilon)/2} \exp\left(1-q_2(x)\right)\cos\left(\frac{n\pi x}{a}\right)dx \\
&+\int_{((3a/2)-d_2-\varepsilon)/2}^{((3a/2)-d_2+\varepsilon)/2} \exp\left(1-q_3(x)\right)\cos\left(\frac{n\pi x}{a}\right)dx \\
&+\int_{((3a/2)-d_2+\varepsilon)/2}^{((3a/2)+d_2-\varepsilon)/2} \cos\left(\frac{n\pi x}{a}\right)dx + \int_{((3a/2)+d_2-\varepsilon)/2}^{((3a/2)+d_2+\varepsilon)/2} \exp\left(1-q_4(x)\right)\cos\left(\frac{m\pi y}{b}\right)dx \Bigg\} \\
\Bigg\{ &\int_{((b/2)-c_1-\varepsilon)/2}^{((b/2)-c_1+\varepsilon)/2} \exp\left(1-q_5(y)\right)\cos\left(\frac{m\pi y}{b}\right)dy \\
&+\int_{((b/2)-c_1+\varepsilon)/2}^{((b/2)+c_1-\varepsilon)/2} \cos\left(\frac{m\pi y}{b}\right)dy + \int_{((b/2)+c_1-\varepsilon)/2}^{((b/2)+c_1+\varepsilon)/2} \exp\left(1-q_6(y)\right)\cos\left(\frac{m\pi y}{b}\right)dy \\
&+\int_{((3b/2)-c_2-\varepsilon)/2}^{((3b/2)-c_2+\varepsilon)/2} \exp\left(1-q_7(y)\right)\cos\left(\frac{m\pi y}{b}\right)dy \\
&+\int_{((3b/2)-c_2+\varepsilon)/2}^{((3b/2)+c_2-\varepsilon)/2} \cos\left(\frac{m\pi y}{b}\right)dy + \int_{((3b/2)+c_2-\varepsilon)/2}^{((3b/2)+c_2+\varepsilon)/2} \exp\left(1-q_8(y)\right)\cos\left(\frac{m\pi y}{b}\right)dy \Bigg\}.
\end{aligned}
\tag{9}
$$

According to Ref. [19]. The cyclic matrix G is given by the form

$$
G = \begin{bmatrix}
g_{00} & g_{-10} & g_{-20} & \cdots & g_{-nm} & \cdots & g_{-NM} \\
g_{10} & g_{00} & g_{-10} & \cdots & g_{-(n-1)m} & \cdots & g_{-(N-1)M} \\
g_{20} & g_{10} & \ddots & \ddots & \ddots & & \\
\vdots & g_{20} & \ddots & \ddots & \ddots & & \\
g_{nm} & \ddots & \ddots & \ddots & g_{00} & \vdots & \\
\vdots & & & & & & \\
g_{NM} & \cdots & \cdots & \cdots & \cdots & \cdots & g_{00}
\end{bmatrix}
\tag{10}
$$

Similarly, the $\mathbf{G_x}$ and $\mathbf{G_y}$ matrices are obtained by the derivatives of the Fourier components of the dielectric profile. These matrices relate to the method that based on the Laplace and Fourier transforms, the inverse Laplace and Fourier transforms [19].

The proposed technique to calculate the elements of the matrix **G** relates to the method [19]. This method becomes an improved method by using the proposed technique also for discontinuous problems of the hollow rectangular waveguide.

5. Numerical results

This section presents several examples for the different geometries in the cross section. Five examples of the different discontinuous cross sections and complementary shapes are demonstrated in **Figure 1a–e**. All the next graphical results are demonstrated as a response to a half-sine (TE_{10}) input-wave profile and the inhomogeneous geometries of the cross section.

Figure 7a–e shows the results of the output field as a response to a half-sine (TE_{10}) input-wave profile for the rectangular dielectric profile (**Figure 4a**) in the rectangular cross section for ϵ_r = 3, 5, 7, and 10, respectively, where a = b = 20 mm, and c = d = 5 mm. The center of the dielectric rectangle is located in the center of the cross section at the point (a/2, b/2), where ϵ_r = 3, 5, 7, and 10, respectively. The output field in the same cross section of **Figure 7a–d** is shown in **Figure 7e** for the x-axis where y = b/2 = 10 mm, where ϵ_r = 3, 5, 7, and 10, respectively. The other parameters are z = 15 cm, k_0 = 167 $1/m$, λ = 3.75 cm, and β = 58 $1/m$.

The results of **Figure 7a–e** are demonstrated for ϵ_r = 3, 5, 7, and 10, respectively. The results are strongly affected by the half-sine (TE_{10}) input-wave profile, and the location of the center of the dielectric material.

The output field for the hollow rectangular waveguide where the dielectric material is located between the hollow rectangle and the metal is shown in

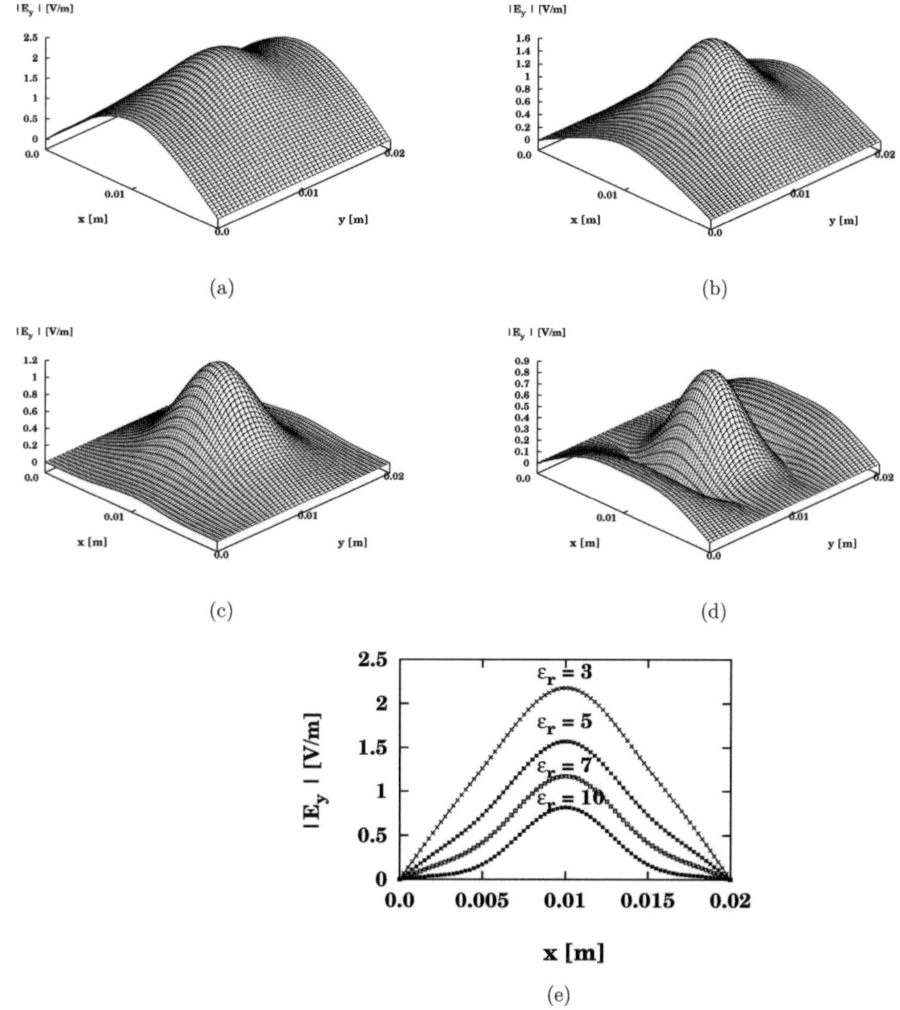

Figure 7.
The output field as a response to a half-sine (TE_{10}) input-wave profile for the rectangular dielectric profile ***Figure 4a*** *in the rectangular cross section, where a = b = 20 mm, and c = d = 5 mm. The center of the rectangle dielectric profile is located in the center of the cross section, for (a) ϵ_r = 3, for (b) ϵ_r = 5, for (c) ϵ_r = 7, and for (d) ϵ_r = 10. (e). The output field in the same cross section of the results (a–d) for x-axis and where y = b/2 = 10 mm, for the values of ϵ_r = 3, 5, 7, and 10, respectively.*

Figure 8a–d, where a = b = 20 mm, and c = d = 14 mm. In this case the thickness of the dielectric layer is equal to 3 mm (e = f = 3 mm). The results are demonstrated for $\epsilon_r = 2.5$, $\epsilon_r = 3$, $\epsilon_r = 3.5$, and $\epsilon_r = 4$, respectively. The output field in the same cross section of **Figure 8a–d** is shown in **Figure 8e**, for $\varepsilon_r = 2.5$, 3, 3.5, and 4, respectively, for x-axis where y = b/2 = 10 mm. In this case, z = 15 cm, $k_0 = 167\ 1/m$, $\lambda = 3.75$ cm, and $\beta = 58\ 1/m$. These results are shown for $\epsilon_r = 2.5$, 3, 3.5, and 4, respectively, and these results are strongly affected by the half-sine (TE_{10}) input-wave profile, the location of the center of the dielectric material, and the thickness of the dielectric material between the hollow rectangle and the metal.

We can find the relevant parameters to obtain the Gaussian behavior of the output field. We can show that the Gaussian behavior is obtained when the thickness of the dielectric layer is equal to 3 mm, as shown in the results of **Figure 8a–d**. The output results are strongly affected by the thickness of the dielectric layer.

Figure 9a–d shows the results of the output field as a response to a half-sine (TE_{10}) input-wave profile for the cross section with four dielectric profiles, as

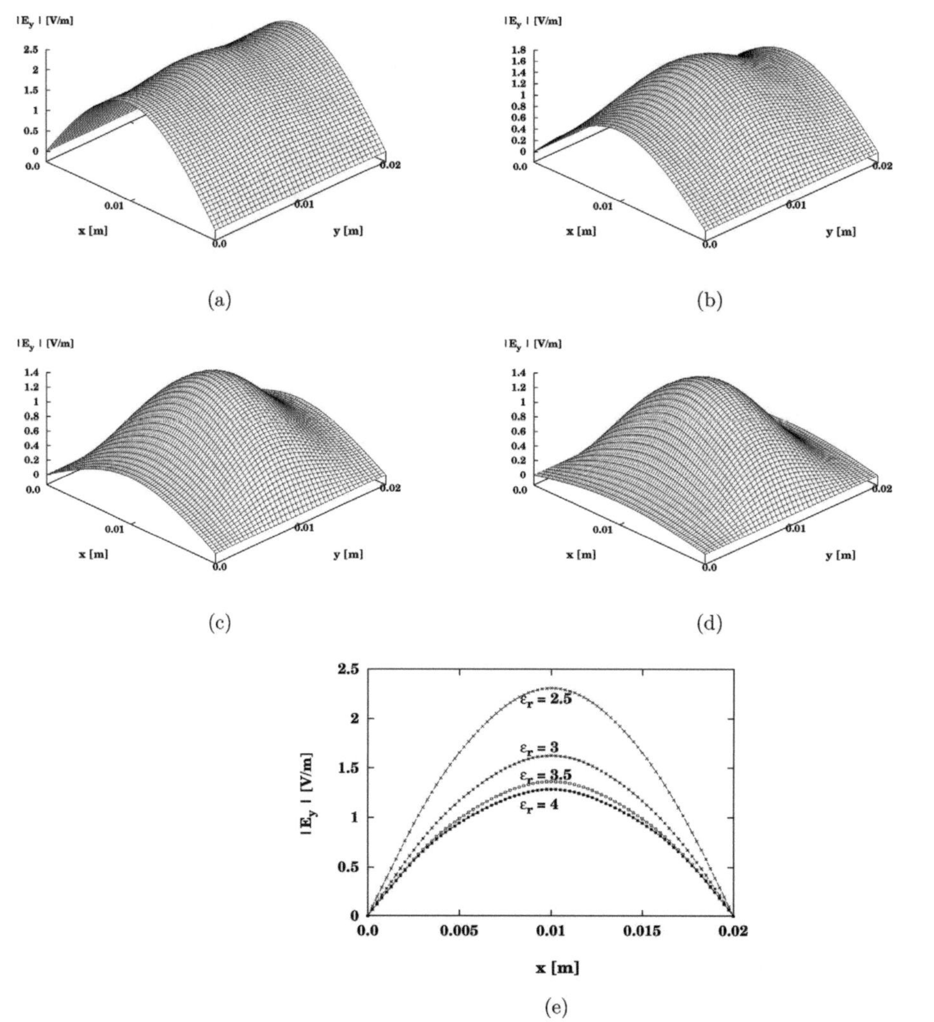

Figure 8.
*The output field as a response to a half-sine (TE_{10}) input-wave profile for the hollow rectangular waveguide where the hollow rectangle is located in the center of the cross section **Figure 4b**, where a = b = 20 mm, c = d = 14 mm. The thickness of the dielectric layer is equal to 3 mm (e = f = 3 mm). The results are shown for (a) $\epsilon_r = 2.5$, for, (b) $\epsilon_r = 3$, for, (c) $\epsilon_r = 3.5$, and for, (d) $\epsilon_r = 4$, (e) the output field in the same cross section of the results (a–d) is shown for x-axis and where y = b/2 = 10 mm, for $\epsilon_r = 2.5$, 3, 3.5, and 4, respectively.*

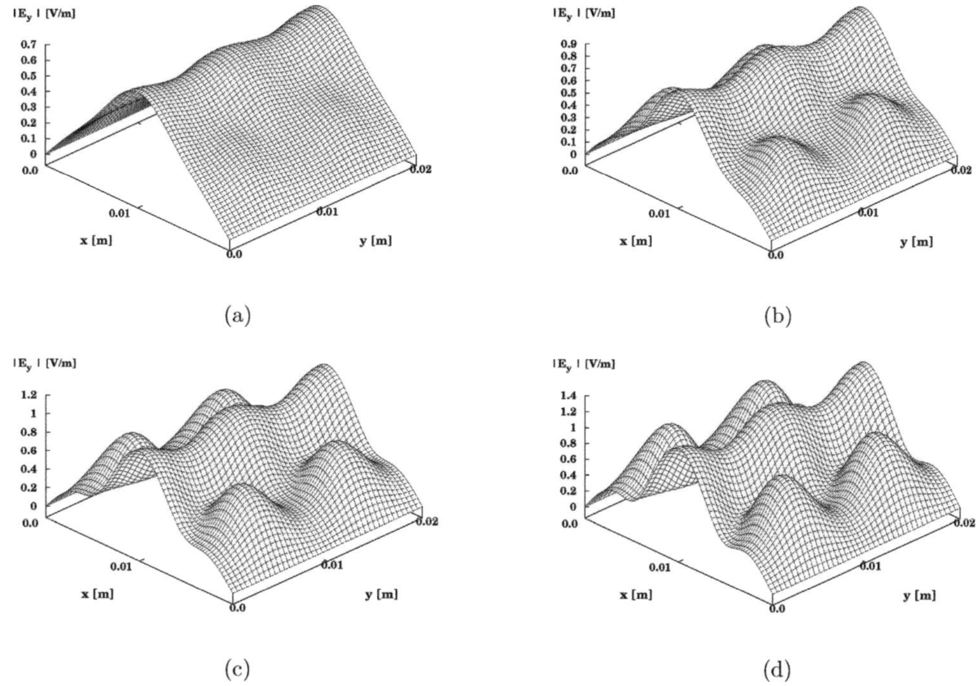

Figure 9.
*The output field as a response to a half-sine (TE_{10}) input-wave profile for the cross section with four dielectric profiles, as shown in **Figure 5a**. The centers of the first, the second, the third and the firth dielectric rectangles are located at the points (a/4, b/4), (3a/4, b/4), (a/4, 3b/4), and (3a/4, 3b/4), respectively. The parameters are a = b = 20 mm, c = d = 0.4 mm, and z = 15 cm, where (a). ϵ_r = 1.1; (b). ϵ_r = 1.2; (c). ϵ_r = 1.3; (d). ϵ_r = 1.4.*

shown in **Figure 5a**. In this example we suppose that a = b = 20 mm, c = d = 0.4 mm, z = 15 cm, the centers of the first, the second, the third, and the firth dielectric rectangles are located at the points (a/4, b/4), (3a/4, b/4), (a/4, 3b/4), and (3a/4, 3b/4), respectively, and for ε_r = 1.1, 1.2, 1.3 and 1.4, respectively. The other parameters are k_0 = 167 $1/m$, λ = 3.75 cm and β = 58 $1/m$. By increasing the parameter ϵ_r from 1.1 to 1.4, the output dielectric profile increased, the output profile of the half-sine (TE_{10}) profile decreased, and the output amplitude increased. These results are strongly affected by the half-sine (TE_{10}) input-wave profile, and the locations of the rectangular profiles along x-axis and y-axis.

Figure 10a–d shows the results of the output field as a response to a half-sine (TE_{10}) input-wave for the hollow waveguide with four hollow rectangles, where a = b = 20 mm, c = d = 3.3 mm, and z = 15 cm, ϵ_r = 1.2, 1.3, 1.4, and 1.5, respectively. The centers of the first, the second, the third and the firth hollow rectangles are located at the points (a/4, b/4), (3a/4, b/4), (a/4, 3b/4), and (3a/4, 3b/4), respectively. The other parameters are k_0 = 167 $1/m$, λ = 3.75 cm and β = 58 $1/m$. These results are strongly affected by the half-sine (TE_{10}) input-wave profile, and the locations of the rectangular profiles along x-axis and y-axis.

By increasing only the parameter ϵ_r, the output dielectric profile increased, the output profile of the half-sine (TE_{10}) profile decreased, and the output amplitude increased. These results are strongly affected by the half-sine (TE_{10}) input-wave profile, and the locations of the rectangular profiles along x-axis and y-axis.

The applications are useful for straight waveguides in millimeter regimes, in the cases where the dielectric profile is located in the center of the cross section, for the cases where the hollow rectangle is located in the center of the cross section, and also for complicated and discontinuous profiles in the cross section.

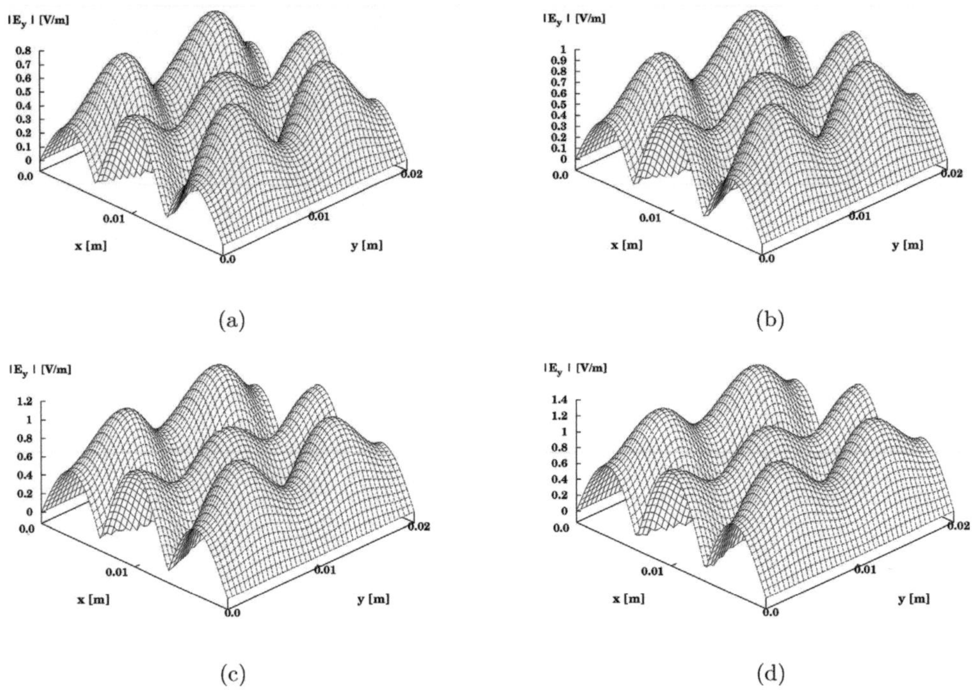

Figure 10.
*The output field as a response to a half-sine (TE_{10}) input-wave profile for the hollow waveguide with four hollow rectangles, as shown in **Figure 5b**. The centers of the first, the second, the third and the firth hollow rectangles are located at the points (a/4, b/4), (3a/4, b/4), (a/4, 3b/4), and (3a/4, 3b/4), respectively. The parameters are a = b = 20 mm, c = d = 3.3 mm, and z = 15 cm, where hollow waveguide with four hollow rectangles, as shown in **Figure 1e**, (a). ϵ_r = 1.5; (b). ϵ_r = 2; (c). ϵ_r = 2.5; (d). ϵ_r = 3.*

6. Conclusions

This chapter presents two interesting types of dielectric materials in the straight rectangular waveguides and their applications. Five examples of the different discontinuous cross section were demonstrated. The effective technique was proposed for all case of **Figure 1a–e**, in order to calculate the dielectric material in the specific cross section.

The proposed techniques are very important to understand the influence of the dielectric materials and the hollow rectangles in all case of discontinuous geometry in the cross section. All the graphical results are demonstrated as a response to a half-sine (TE_{10}) input-wave profile. The proposed techniques relate to the method based on Laplace and Fourier transforms and the inverse Laplace and Fourier transforms.

The method is based on Fourier transform, thus we need use with the image method to calculate the dielectric profile in the cross section. The image method and periodic replication are needed for fulfilling the boundary condition of the metallic waveguide.

Figure 8a–e relates to the hollow rectangular waveguide where the hollow rectangle is located in the center of the cross section as shown in **Figure 4b**. We can find the relevant parameters to obtain the Gaussian behavior of the output field. From the results of **Figure 8a–e**, the Gaussian behavior is obtained when the thickness of the dielectric layer is equal to 3 mm. The output results are strongly affected by the thickness of the dielectric layer.

By increasing only the parameter ϵ_r, the output dielectric profile increased, the output profile of the half-sine (TE_{10}) profile decreased, and the output amplitude increased. The results are strongly affected by the half-sine (TE_{10}) input-wave profile, the location of the center of the dielectric material, and the thickness of the dielectric material between the hollow rectangle and the metal.

The applications are useful for straight waveguides in millimeter regimes, in the cases where the dielectric profile is located in the center of the cross section, for cases where the hollow rectangle is located in the center of the cross section, and also for complicated and discontinuous profiles in the cross section.

Author details

Zion Menachem
Department of Electrical Engineering, Sami Shamoon College of Engineering, Beer Sheva, Israel

*Address all correspondence to: zionme@sce.ac.il; zionmm@gmail.com

IntechOpen

© 2019 The Author(s). Licensee IntechOpen. This chapter is distributed under the terms of the Creative Commons Attribution License (http://creativecommons.org/licenses/by/3.0), which permits unrestricted use, distribution, and reproduction in any medium, provided the original work is properly cited.

References

[1] Chiang KS. Review of numerical and approximate methods for the modal analysis of general optical dielectric waveguides. Optical and Quantum Electronics. 1994;**26**:S113-S134

[2] Saad M. Review of numerical methods for the analysis of arbitrarily-shaped microwave and optical dielectric waveguides. IEEE Transactions on Microwave Theory and Techniques. 1985;**MTT-33**:894-899

[3] Yeap KH, Teh KH, Yeong KC, Lai KC, Loh MC. Propagation in dielectric rectangular waveguides. Optica Applicata. 2016;**XLVI**:317-330

[4] Bartling JQ. Propagation of an electromagnetic wave in an infinite rectangular dielectric wave guide. Journal of the Franklin Institute. 1969; **287**:389-407

[5] Khorasani S. Differential transfer-matrix method for solution of one-dimensional linear nonhomogeneous optical structures. Optical Society of America. 2003;**20**:91-96

[6] Rojas JAM, Alpuente J, Pineiro J, Sanchez R. Rigorous full vectorial analysis of electromagnetic wave propagation in 1D inhomogeneous media. Progress in Electromagnetics Research. 2006;**63**:89-105

[7] Mehrany K, Khorasani S. Analytical solution of non-homogeneous anisotropic wave equations based on differential transfer matrices. Journal of Optics A. 2002;**4**:624-635

[8] Lee JF. Finite element analysis of lossy dielectric waveguides. IEEE Transactions on Microwave Theory and Techniques. 1994;**42**:1025-1031

[9] Ise K, Inoue K, Koshiba M. Three-dimensional finite-element method with edge elements for electromagnetic waveguide discontinuities. IEEE Transactions on Microwave Theory and Techniques. 1991;**39**:1289-1295

[10] Singh V, Joshi M, Prasad B, Ojha P. Modal dispersion characteristics and waveguide dispersion of an optical waveguide having a new unconventional core cross-section. Journal of Electromagnetic Waves and Applications. 2004;**18**:455-468

[11] Hernandez-Lopez MA, Quintillan M. Propagation characteristics of modes in some rectangular waveguides using the finite-difference time-domain method. Journal of Electromagnetic Waves and Applications. 2000;**14**:1707-1722

[12] Mallick AK, Sanyal GS. Electromagnetic wave propagation in a rectangular waveguide with sinusoidally varying width. IEEE Transactions on Microwave Theory and Techniques. 1978;**MTT-26**:243-249

[13] Bogle A, Havrilla M, Nyquis D, Kempel L, Rothwell E. Electromagnetic material characterization using a partially-filled rectangular waveguide. Journal of Electromagnetic Waves and Applications. 2005;**19**:1291-1306

[14] Vaish A, Parthasarathy H. Analysis of a rectangular waveguide using finite element method. Progress in Electromagnetics Research C. 2008;**2**: 117-125

[15] Khalaj-Amirhosseini M. Analysis of longitudinally inhomogeneous waveguides using the method of moments. Progress In Electromagnetics Research. 2007;**74**:57-67

[16] Khalaj-Amirhosseini M. Analysis of nonuniform transmission lines using Taylor's series expansion. International

Journal of RF and Microwave Computer-Aided Engineering. 2006;**10**: 536-544

[17] Molina-Fernandez I, Camacho-Penalosa C, Ramos JI. Application of the two-dimensional Fourier transform to nonlinear wave propagation phenomena. IEEE Transactions on Microwave Theory and Techniques. 1994;**42**:1079-1085

[18] Henry CH, Verbeek BH. Solution of the scalar wave equation for arbitrarily shaped dielectric waveguides by two-dimensional Fourier analysis. Journal of Lightwave Technology. 1989;7:308-313

[19] Menachem Z, Jerby E. Transfer matrix function (TMF) for propagation in dielectric waveguides with arbitrary transverse profiles. IEEE Transactions on Microwave Theory and Techniques. 1998;**46**:975-982

[20] Vladimirov V. Equations of Mathematical Physics. New York, NY: Marcel Dekker; 1971

Chapter 8

Research Progress on Synergistic Effect between Insulation Gas Mixtures

Su Zhao and Dengming Xiao

Abstract

Synergetic effect is a special gas discharge phenomenon among insulating gas mixtures, which has important reference value for gas selection of future gas-insulated power equipment. The research progress and investigation methods of synergistic effect and insulation characteristics of different gas mixtures at home and abroad are reviewed in this chapter. The synergistic effect between different kinds of gas mixtures including SF_6 gas mixtures and some new insulation gases such as c-C_4F_8, CF_3I, and C_4F_7N is presented. Combined with the results of multiple studies, it can be seen that the synergistic effect of the gas mixture has a certain relationship with the electronic transport parameters and discharge patterns. Besides, the synergistic effect of the same gas mixture may change with the change of external conditions such as gas pressure, voltage type, and electrode distance.

Keywords: gas discharge, synergistic effect, gas mixture, insulation, environmental friendly gas

1. Introduction

With the continuous development of power industry, gas-insulated power equipment has been used more and more widely in electrical transmission system. Gas-insulated power equipment mainly includes gas-insulated transformer (GIT), gas-insulated transmission line (GIL), and gas-insulated switchgear (GIS). These gas-insulated equipment are noncombustible, nonexplosive, safe, and stable and have long maintenance period [1]. At present, SF_6 gas is the most widely used gas in power equipment, which has excellent insulation performance and arc extinguishing ability. As early as the 1940s, SF_6 gas has been used as insulation gas in power equipment [2]. Since the 1970s, researchers have studied the use of gas mixture including SF_6 and some buffer gases such as N_2, CO_2, Ar, He, air, etc. to replace pure SF_6 as the insulation medium [3–5]. With further study, researchers found that the insulation characteristics of gas mixtures containing SF_6 gas did not increase linearly according to the mixing ratio of SF_6, and the electrical strength of the mixture was higher than the weighted average of the electrical strength of the two gas components. In 1980, Wootton and Chantry first used synergism to describe this phenomenon [6]. In the study of synergistic effect, scholars have found that there are also super synergistic effect and negative synergistic effect, which are shown in **Figure 1**.

Suppose that the composition of gas mixture is gas 1 and gas 2, the breakdown voltage of gas 1 is U_1, the breakdown voltage of gas 2 is U_2, and $U_1 > U_2$, the

IntechOpen

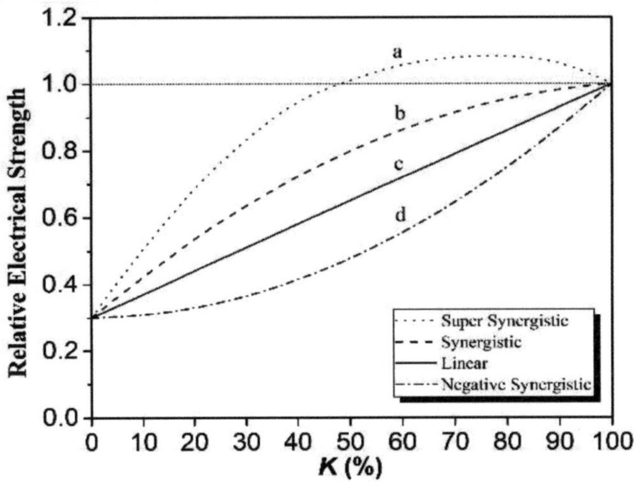

Figure 1.
Types of synergistic effect.

breakdown voltage of gas mixture is U_m, and K is the mixing ratio of gas 1 in the gas mixture. For super synergistic effect, $U_m > U_1$ and $U_m > U_2$ exist in some mixing ratios, as shown in curve a; for synergistic effect, $U_m > U_1 + U_2$, as shown in curve b; for linear relationship, $U_m = U_1 + U_2$, as shown in curve c; for negative synergistic effect, $U_m < U_1 + U_2$, as shown in curve d in **Figure 1**.

At first, researchers studied the synergistic effect of SF_6 gas mixture in order to solve the problems of high liquefaction temperature, sensitivity to electric field, and the high price of SF_6 gas [7]. However, with the deepening understanding of SF_6, scholars have found that SF_6 is a strong greenhouse gas. It is estimated that the global annual production of SF_6 gas is more than 20,000 tons and 80% of SF_6 gas produced globally each year is used in the power industry. Although SF_6 has many advantages, its greenhouse effect on the earth cannot be ignored. The global warming potential (GWP) value of SF_6 gas is 23,900; it means that the emission of 1 kg SF_6 is equivalent to the emission of 23,900 kg of CO_2. What is more serious is that SF_6 has a very stable chemical property, which is difficult to decompose after it spreads to the outside environment, and can exist for up to 3200 years. The environmental impact and greenhouse effect generated by SF_6 will continue to accumulate [8–10].

Affected by climate change, more and more international cooperation has been carried out to reduce greenhouse gas emissions, so as to curb global climate change and maintain the sustainable development of the environment. In the Kyoto protocol of the United Nations framework convention on climate change signed in Kyoto, Japan, in 1997, SF_6 has been clearly regulated as one of the six greenhouse gases and requires developed countries to freeze and reduce the total greenhouse gas emissions [11]. It means that the use of SF_6 in the industrial field will be increasingly restricted and pressured. Therefore, it is an urgent task to study a new gas insulation scheme to replace SF_6.

Although at present there have been many studies about alternative SF_6 gas, no gas can thoroughly replace SF_6 gas in the form of a single gas; they all have to be mixed with buffer gas for industrial application. The reasonable use of synergistic effect can effectively improve dielectric strength of gas mixtures and reduce the use of insulation gas, so in this paper, the research progress and methods of synergistic effect with gas mixtures are introduced; the prospect and the difficulties in the field were also discussed. This paper is expected to provide help and reference for future research on synergistic effect.

2. Synergistic effect in SF_6 gas mixtures

In view of the synergistic effect and insulation strength of SF_6 mixture, scholars studied and analyzed it through theoretical calculation and experimental research. **Figure 2a** is a simple and intuitive calculation method proposed by Wieland et al. for the insulation strength of SF_6 gas mixture, and **Figure 2b** is the comparison of the calculation results with the actual values and weighted values [12].

Christophorou et al. thought that the preferred gas mixture should include effective electron-attaching gas and/or electron-slowing down gas [13]. The attachment cross section of electron-attaching components should be as wide as possible, or the attachment cross section of different gases is in different energy interval, so the attachment cross section of gas mixtures is also wide. The effect of electron-slowing down gas is to slow down the free electrons, making them easier to attach and reducing secondary ionization. Based on this theory, they believed that when an electronegative gas and a gas with a large dipole moment are mixed, the synergistic effect and insulation strength of the gas mixture will be better. **Figure 3** shows the experimental results of SF_6 gas mixtures with CF_4, CHF_3, and 1,1,1-CH_3CF_3 gas, and the electric dipole moments of these three gases are 0, 1.65D, and 2.32D, respectively.

From the figure, it can be seen that the breakdown voltage curve of SF_6-CF_4 gas mixtures shows almost a linear trend, and the electric dipole moments of CF_4 is 0. When it comes to SF_6-CHF_3 gas mixture, as the content of SF_6 gas increases, the breakdown voltage of the gas mixtures does not increase in a straight line, and there is synergistic effect that occurs. The electric dipole moments of 1,1,1-CH_3CF_3 gas is 2.32D, which is the one with the largest electric dipole moment among the three gases; from **Figure 3c** it can be seen that the synergistic effect of this gas mixture is the most pronounced.

Okubo et al. investigated the partial discharge (PD) and breakdown characteristics of SF_6-N_2 gas mixtures in order to analyze the relationship between electronegativity, additive gases, and the insulation strength [14]. They believed that the synergistic effect of gas mixture is related to the change of discharge form. **Figure 4** shows the

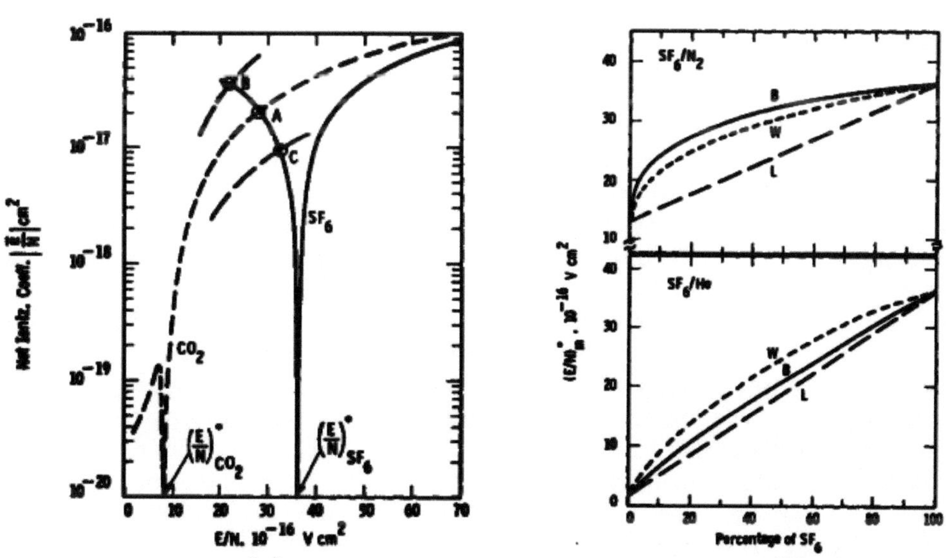

(a) (b)

Figure 2.
Experimental results of SF_6 gas mixtures. (a) Wieland calculation method, (b) Results comparison of different methods.

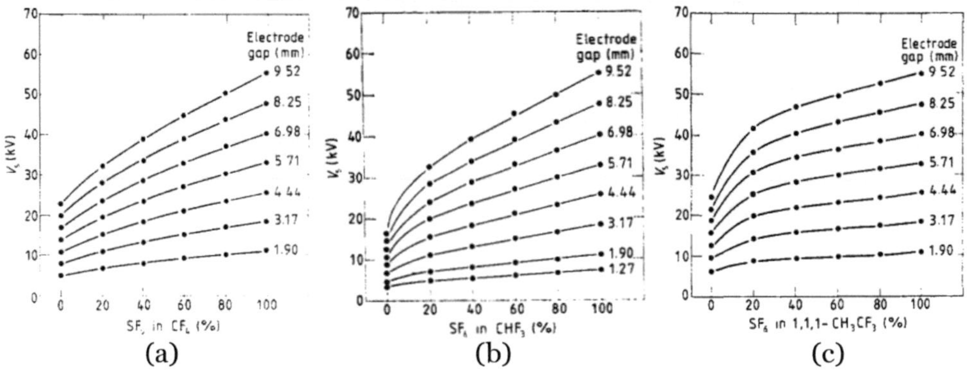

Figure 3.
Experimental results of SF_6 gas mixtures. (a) SF_6-CF_4, (b) SF_6-CHF_3, (c) SF_6-1,1,1-CH_3CF_3.

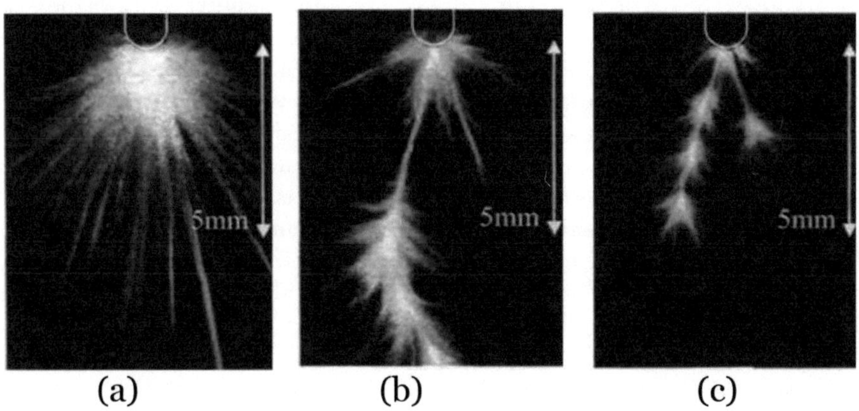

Figure 4.
Impulse PD in SF_6-N_2 gas mixture. (a) 0.1MPa, (b) 0.2MPa, (c) 0.3MPa.

impulse PD of SF_6-N_2 gas mixture at different pressure. The needle-plate electrode was used in the experiment, and the content of SF_6 was 10% and N_2 was 90%.

It can be seen that with the change of gas pressure, the development of impulse PD has changed. When the gas pressure is 0.1 MPa, as shown in **Figure 5a**, the brush-like partial discharge occurred around the needle electrode, and it can be thought of as streamer discharge. When the pressure increases to 0.2 MPa, it can be seen from **Figure 5b** that the development of discharge process becomes longer; streamer discharge turns into leader discharge. As the pressure continues to increase, when the pressure is 0.3 MPa, the PD type of the gas mixture is still the leader discharge, and the path of discharge development becomes shorter with the increase of the pressure.

Yamada et al. studied the insulation properties of a kind of gas mixture containing ultra-dilute SF_6 gas [15]. It has been found that trace SF_6 has a significant effect on the streamer discharge of the gas mixture. As the SF_6 content increases, the number of the discharge channels decreased significantly, and the number of channels that can reach the plane electrode also reduced, as shown in **Figure 4**. Except for the effect on discharge characteristics, the results show that the breakdown voltage and the PD voltage of SF_6/N_2 gas mixture have a significant synergistic effect. Yamada T thought that the addition of trace SF_6 inhibits the development of streamer discharge process, which leads to synergistic effect.

Osmokrovic et al. conducted an in-depth study on the synergistic effect of SF_6/N_2 under impulse voltage. The experimental results show that the synergistic

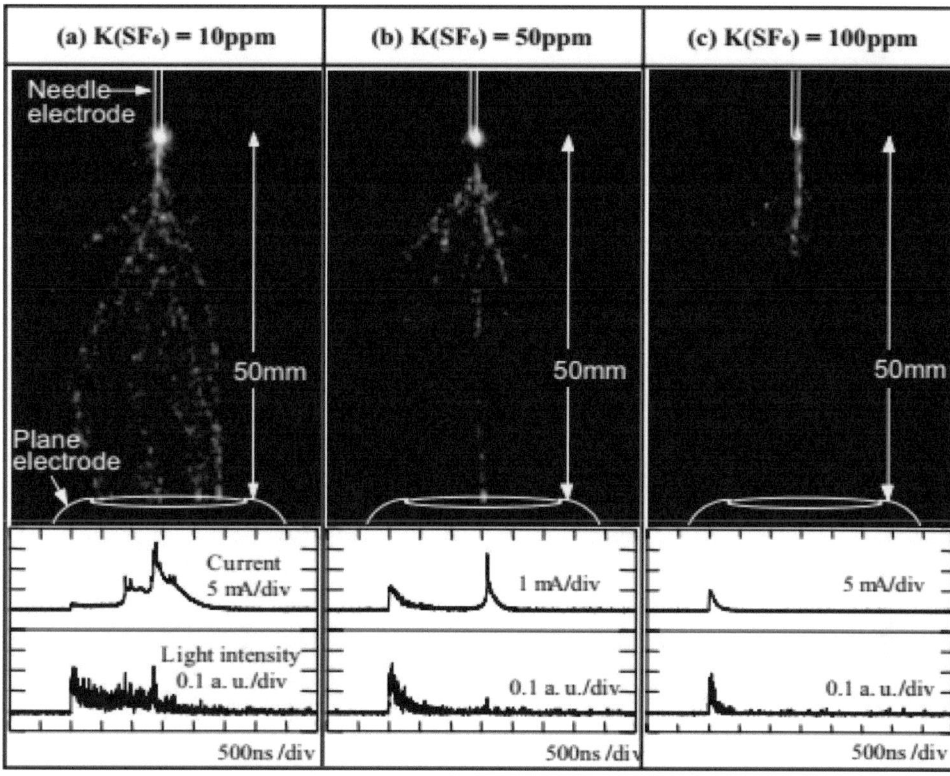

Figure 5.
PD characteristics in SF_6-N_2 gas mixture.

effect of SF_6/N_2 gas mixture related to the rising rate of the impulse voltage. As the impulse voltage rise rate increases, the synergistic effect is gradually weakened. The synergistic effect of SF_6 gas mixture is very weak and almost completely disappears under some impulse voltage with very high rise rate [16]. The insulation characteristics of SF_6 and SF_6/N_2 gas mixture under impulse voltage with different rise rate are shown in **Figure 5**. The rise rates of shock voltage in **Figure 6a–c** are 1, 50 and 800 kv/ms, respectively.

Based on the above phenomenon, Osmokrovic proposed that after adding N_2 to the SF_6 gas, electrons can make N_2 vibration and rotation dynamics excited or dissociated, this process will make electrons lose energy, and the effective temperature decreases, to realize the modulation of electron energy spectrum and increase the probability that the electrons are captured. The rise rate of impulse voltage has influence on the modulation of the electron energy spectrum, which in turn affects the synergistic effect of SF_6 gas mixture.

Hayakawa et al. studied the generation and development of PD characteristics of SF_6/N_2 gas mixture under positive lightning impulse [17]. PD and breakdown characteristics with different SF_6 content are shown in **Figure 7**. Hayakawa proved through the film of the streak camera and ICCD that the discharge process in SF_6/N_2 gas mixture did change with the increase of gas pressure. Streamer discharge and leader discharge are the two types of PD process in SF_6/N_2 gas mixture, which all have relationship with gas pressure and SF_6 content. With the increase of gas pressure and the content of SF_6 gas, the leader discharge process gradually takes the leading position, and the streamer discharge process of gradually weakens.

Chen studied the discharge characteristics of SF_6/N_2 gas mixture under DC voltage and lightning impulse in extremely uneven electric field. The experimental

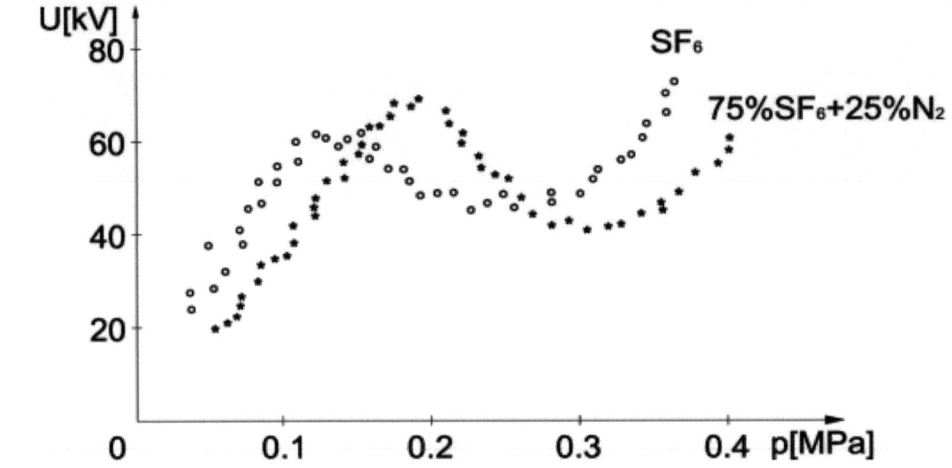

(a)

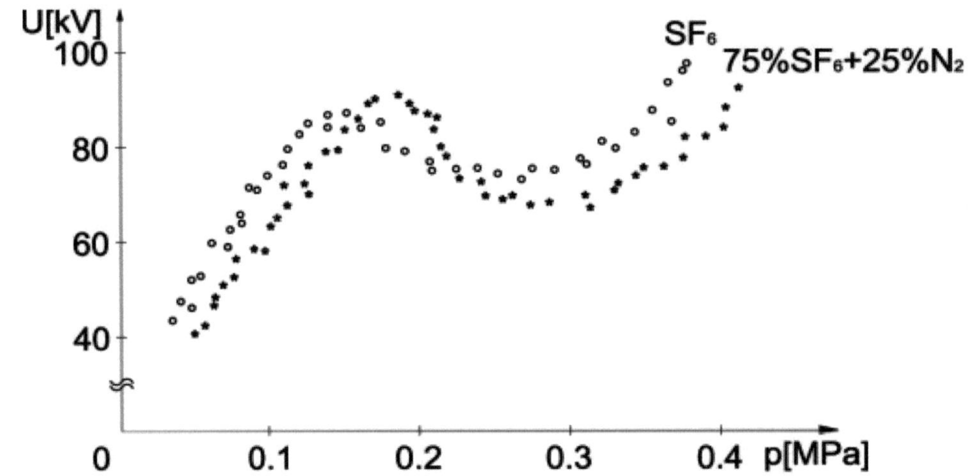

(b)

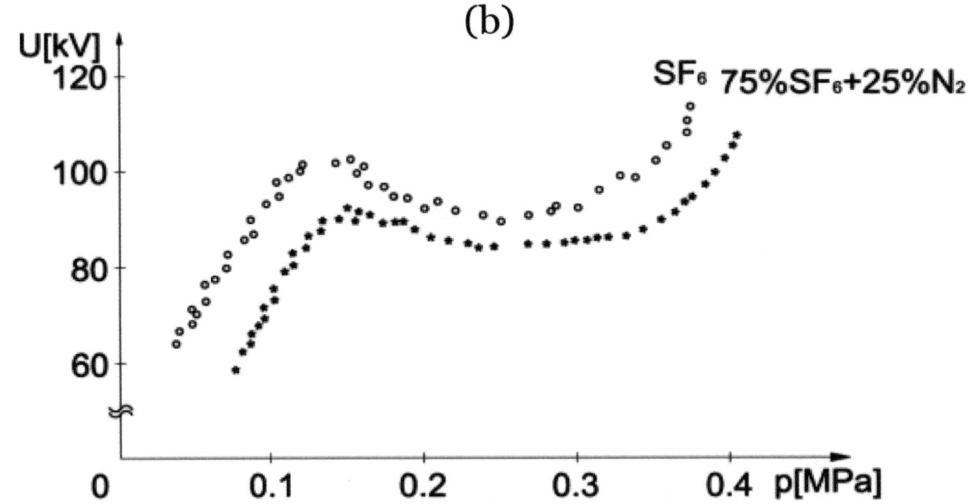

(C)

Figure 6.
Insulation characteristics of SF_6 and SF_6/N_2 gas mixture under different rising rate impulse voltages. (a) 1kV/ms, (b) 50kv/ms, (C) 800kV/ms.

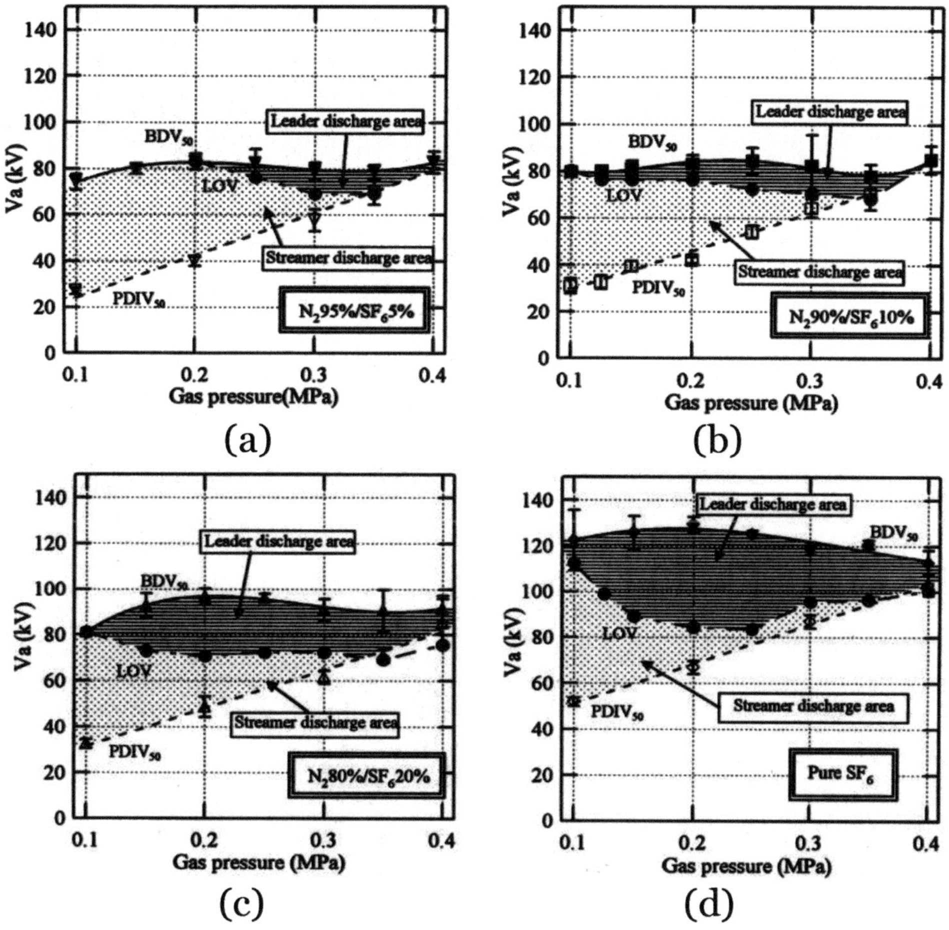

Figure 7.
PD and breakdown characteristics with different SF_6 content. (BDV_{50}, breakdown voltage; $PDIV_{50}$, 50% probability PD inception voltage; LOV, leader discharge onset voltage). (a) $SF_6$5–$N_2$95%, (b) $SF_6$10–$N_2$90%, (c) $SF_6$20–$N_2$80%, (d) pure SF_6.

results show that there was a significant decrease of breakdown voltage with the increase of gas pressure under DC voltage and the abnormal discharge characteristics pressure range of SF_6/N_2 gas mixture is larger than SF_6 gas. This phenomenon leads to the breakdown voltage of SF_6/N_2 gas mixture is higher than that of SF_6 gas. So under this abnormal range, positive synergistic effect occurs. However, under lighting impulse voltage, no abnormal discharge phenomenon was found, but positive synergistic effect still existed, as shown in **Figure 8** [18].

Tagashira et al. believe that synergistic effects can be divided into three categories: a-synergistic effect (SF_6 + SiH_4), η-synergistic effect (SF_6 + c-C_4F_8), and γ-synergistic effect (N_2 + CH_4). The research found that the curves of SF_6/SiH_4, SF_6/c-C_4F_8, SF_6/C_3F_6 gas mixtures with SF_6 gas content all decreased first and then increased, that is, the curve had a minimum point. For the curve of SF_6/SiH_4 gas mixture, the falling part of the curve is due to the decrease of ionization coefficient α, that is, α-synergistic effect, and for the curve of SF_6/c-C_4F_8 gas mixture, the rising part is caused by the increase of attachment coefficient η and that is η-synergistic effect. In addition to these two synergistic effects, they also proposed a synergistic effect of γ on the secondary ionization coefficient of N_2 + CH_4 gas mixture [19].

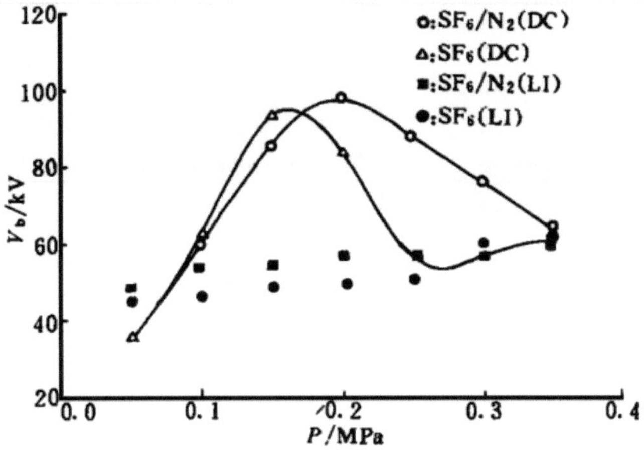

Figure 8.
Abnormal discharge phenomenon in SF_6/N_2 and SF_6.

Takuma et al. studied the synergistic effect of gas mixtures such as SF_6/N_2, CCl_2F_2/N_2, etc. They assumed that the effective ionization coefficient of the gas mixture is equal to the sum of the coefficients of the two component gases multiplied by their respective partial pressure ratios, suggesting an empirical formula for breakdown voltage of SF_6/N_2 gas mixture under slight uneven electric field [20]:

$$U_m = U_2 + \frac{k}{k + C(1-k)}(U_1 - U_2) \tag{1}$$

where U_1 and U_2 are the breakdown voltage of component gas 1 and gas 2, U_m is the breakdown voltage of gas mixture ($U_1 > U_2$), k is the partial pressure ratio of component gas 1, and C is the synergistic effect coefficient, which is independent of the partial pressure ratio k. When Constant $C = 0.08$, the calculated and experimental values of the SF_6/N_2 gas mixture by this formula are shown in **Figure 5**. The breakdown voltage values are basically the same, which can reflect the synergistic effect of SF_6 gas mixture.

According to formula (1), the formula for calculating the synergy coefficient proposed by Takuma is

$$C = \frac{k(U_1 - U_M)}{(1-k)(U_M - U_2)} \tag{2}$$

As can be seen from Eq. (2), when $C = 1$, the breakdown voltage of the gas mixture is equal to the weighting value of breakdown voltage of the two components according to the mixing ratio. That is, the breakdown voltage of the gas mixture exhibits a linear relationship that increases as the mixing ratio of the component gas increases. When $0 < C < 1$, the breakdown voltage of the gas mixture reflects synergistic effect phenomenon, and the smaller the value of C, the more significant the nonlinear increase of the breakdown voltage of the gas mixture, which means synergistic effect becomes more significant. When $C = 0$, the breakdown voltage of the gas mixture equals to the breakdown voltage of gas 1, which means $U_M = U_1$.

If all the situations of synergistic effect, i.e., positive synergistic effect, synergistic effect, and negative synergistic, are considered at the same time, then the above formula is no longer applicable. Assuming a positive synergistic effect of the gas mixture, it can be seen from the calculation that the coefficient $C < 0$ under positive synergistic. When the breakdown voltage of the gas mixture is lower than all the breakdown voltage of the component gas, that is, the phenomenon of "super negative synergistic effect" appears, then value of the coefficient C is still less than 0, so the various synergistic effects of the gas mixture cannot be clearly distinguished by formula (2).

Guo et al. introduced a normalization coefficient h to investigate the synergistic effect of SF_6/N_2 gas mixture under lightning impulse. The definition of coefficient h is as follows:

$$\begin{aligned} h &= \frac{(U_m - U_2) - k(U_1 - U_2)}{0.5(U_1 + U_2)} & 0 < k < 1 \\ h &= 0 & k = 0 \text{ or } 1 \end{aligned} \quad (3)$$

where U_1, U_2, U_{m}, and k in Eq. 3 have the same meaning as those in Eq. 2. When $h > 0$, it means that gas mixture has synergistic effect, and when $h < 0$, it represents that the gas mixture has negative synergistic effect. The relationship between the

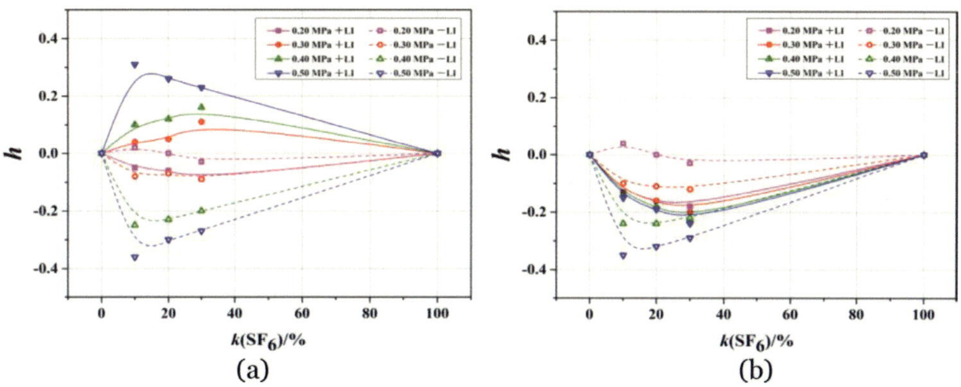

Figure 9.
Coefficient h *for SF_6/N_2 gas mixture under lightning impulse with different needle-plane electrode* (R *is radius curvatures of needle electrodes). (a) R = 2 mm, (b) R = 1 mm.*

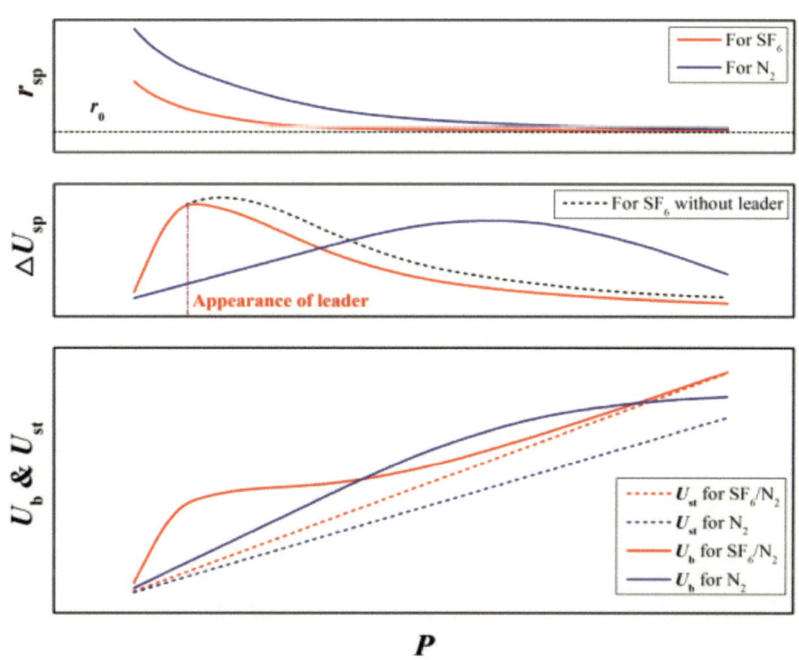

Figure 10.
Breakdown voltage differences between N_2 and SF_6/N_2.

coefficient h of SF_6/N_2 and k under lightning impulse with different needle-plane electrodes is shown in **Figure 9** [21].

The analysis results show that under the action of negative lightning impulse voltage, the negative synergistic effect increases with the increase of gas pressure. The synergistic effect under the positive impact voltage decreases with the decrease of gas pressure. With the increase of the electric field inhomogeneity coefficient, the synergistic effect has a negative synergistic effect. The analysis of the development process of the flow discharge shows that there are three reasons for the negative synergistic effect: similar flow corona starting voltage, different space charge effects, and different N_2 and SF_6/N_2 mixed gas discharge forms. The difference of breakdown voltage between N_2 and SF_6/N_2 gas mixture is shown in **Figure 10**. r_{sp} represents the range of space charge, ΔU_{SP} is the influence of space charge on breakdown voltage, U_{st} is the streamer corona onset voltage, and $U_b = U_{st} + \Delta U_{SP}$.

3. Synergistic effect in SF_6 alternative gas mixtures

3.1 c-C_4F_8

c-C_4F_8 is colorless, odorless, nonflammable, and nonexplosive; the GWP of c-C_4F_8 is about 8700, and the molecular structure of c-C_4F_8 is shown in **Figure 11**. The insulation performance of pure c-C_4F_8 gas is better than SF_6, and the gas mixture of c-C_4F_8 gas has similar insulation characteristics to SF_6, which can meet the requirements of practical application. The liquefaction temperature of c-C_4F_8 is about −6°C, higher than that of SF_6, which is −63.8°C [22]. Therefore, c-C_4F_8 can only be mixed with the gas in a certain proportion to reduce the overall liquefaction temperature of the gas mixture for application.

While studying the synergistic effect of SF_6 gas mixture, Christophorou et al. also conducted a comparative study on the synergistic effect of c-C_4F_8 gas mixtures.

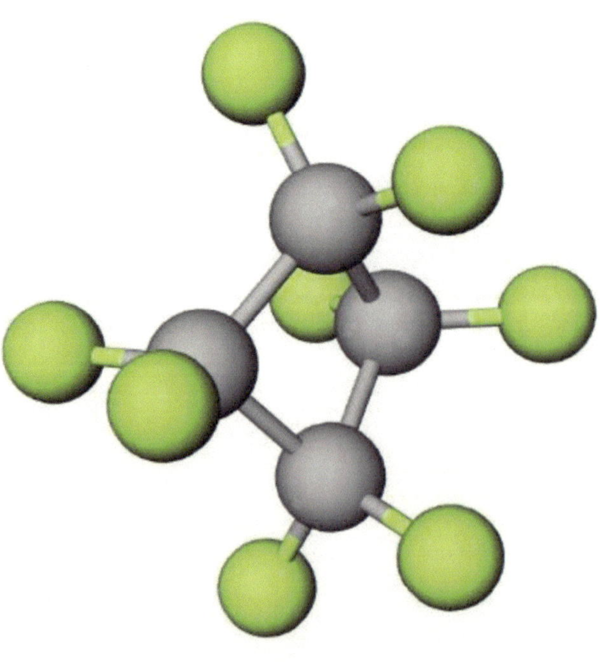

Figure 11.
Molecular structure of c-C_4F_8.

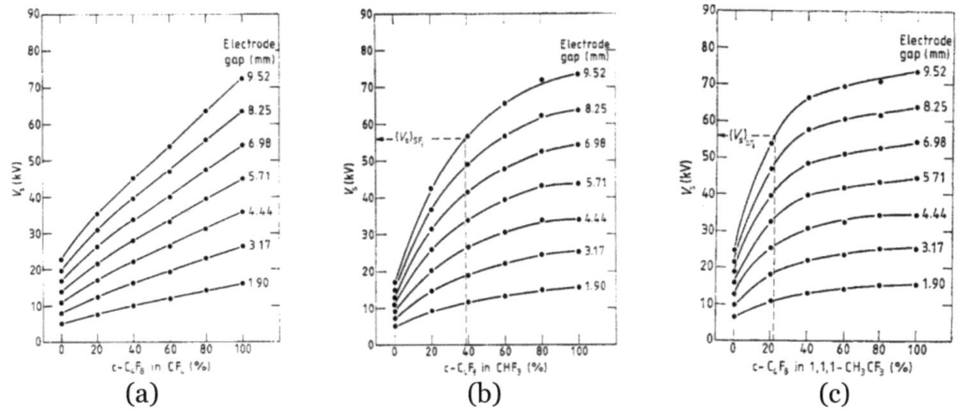

Figure 12.
Breakdown voltage of c-C_4F_8 gas mixtures for various electrode gaps. (a) c-C_4F_8/CF_4, (b) c-C_4F_8/CHF_3, (c) c-C_4F_8/1,1,1-CH_3CF_3.

The insulation characteristics and synergistic effect of c-C_4F_8/CF_4 and c-C_4F_8/CHF_3 gas mixture are shown in **Figure 12** [13].

It can be seen from the figure that the addition of a small amount of c-C_4F_8 can greatly improve the insulation strength of gas mixture. Mixing 20% c-C_4F_8 in CHF_3 can double the breakdown voltage. The insulation strength of c-C_4F_8/1,1,1-CH_3CF_3 gas with the same mixing ratio is the same as that of pure SF_6 gas. By comparison, it can be seen that the synergistic effect phenomenon will also occur when c-C_4F_8 gas and the gas with large dipole moment are mixed.

Xing et al. discussed the feasibility of replacing SF_6 gas with c-C_4F_8/N_2 mixture for gas insulation equipment from the perspective of PD performance. The initial PD voltage of c-C_4F_8/N_2 gas mixture was measured under different pressure, different mixing ratio, and different electrode distance. The influence of these three factors on the PD performance of the gas mixture was obtained and compared with the initial PD voltage of pure SF_6 gas. The results show that the initial PD voltage of pure c-C_4F_8 gas is 1.3 times that of pure SF_6 gas [23]. c-C_4F_8 and N_2 have synergistic effect, and the synergistic coefficient calculated by Takuma's equation under different electrode distance and gas pressure is shown in **Table 1**.

Yamamoto et al. examined the insulation characteristics of c-C_4F_8 gas mixtures such as c-C_4F_8/N_2, c-C_4F_8/air, and c-C_4F_8/CO_2 under different electric field [24]. The experimental results show that c-C_4F_8/CO_2 gas mixture has better synergistic effect than the other two mixtures, and improve gas pressure or gap distance can greatly

Electrode distance/mm	Gas pressure/ Mpa	C			
		5% c-C_4F_8	10% c-C_4F_8	15% c-C_4F_8	20% c-C_4F_8
10	0.20	0.48	0.50	0.54	0.53
	0.25	0.20	0.40	0.44	0.63
20	0.20	0.47	0.86	0.45	0.52
	0.25	0.34	0.50	0.49	0.54
30	0.20	0.15	0.25	0.30	0.36
	0.25	0.24	0.32	0.34	0.41

Table 1.
Synergistic coefficient for c-C_4F_8/N_2.

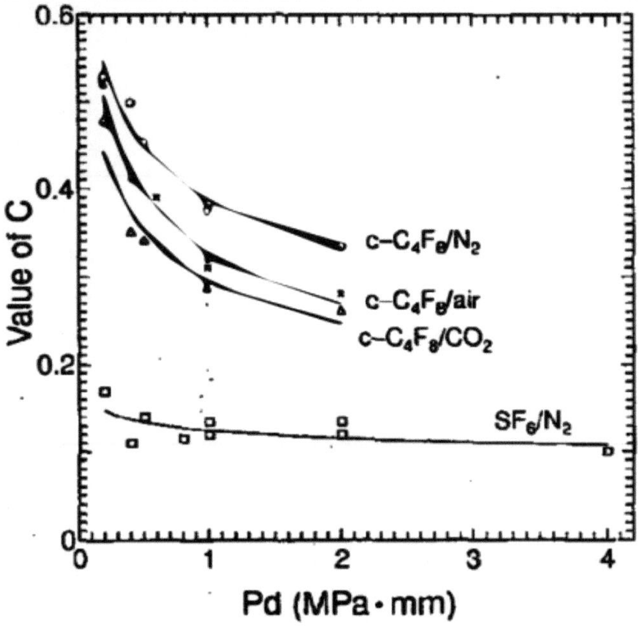

Figure 13.
Relationship of synergistic effect coefficient C *with Pd.*

enhance the synergistic effect of c-C_4F_8 gas mixture. The relationship of synergistic effect coefficient C with Pd is shown in **Figure 13**.

3.2 CF_3I

CF_3I gas is odorless, nonflammable, chemically stable, and material compatible. The molecular structure of CF_3I is shown in **Figure 14**. CF_3I is considered as one of the ideal alternatives to conventional Freon refrigerant. In terms of environmental characteristics, CF_3I is an extremely environmental friendly gas, and its GWP value is about 1–5, much lower than SF_6 gas. At the same time, the C-I chemical bond in the molecular structure of CF_3I is easy to photolysis under solar radiation, resulting in the very short existence time of CF_3I in the atmosphere, and the ozone destruction potential of the gas can also be ignored. The boiling point of CF_3I at normal pressure is −22.5°C, which indicates that CF_3I gas will transform from gaseous to liquid when the temperature is lower than −22.5°C [25]. Therefore, CF_3I gas must be mixed with buffer gas before it can be used in power equipment.

Urquijo et al. experimentally measured the electron transport parameters of CF_3I and CF_3I/N_2 gas mixture. They analyzed and studied several parameters including electron drift velocity, diffusion coefficient, electron ionization coefficient, and attachment coefficient by means of pulsed Townsend method. Experimental results show that the critical electric field intensity $(E/N)_{\lim}$ of CF_3I was 437Td, and the insulation strength of the gas is about 1.2 times that of SF_6 gas. When the content of CF_3I is 70%, the dielectric strength of CF_3I/N_2 gas mixture is basically the same with pure SF_6 gas [26]. The insulation characteristic comparison of CF_3I/N_2 and SF_6/N_2 gas mixtures IS shown in **Figure 15**. From the figure it can be seen that CF_3I/N_2 and SF_6/N_2 gas mixture both have synergistic effect phenomenon, but the phenomenon of CF_3I/N_2 gas mixture is weak when compared with SF_6/N_2.

In addition to CF_3I/N_2 mixture, there is also a synergistic effect of CF_3I/CO_2 mixture. Jiao et al. studied the gas mixture of CF_3I/CO_2 with low content of CF_3I,

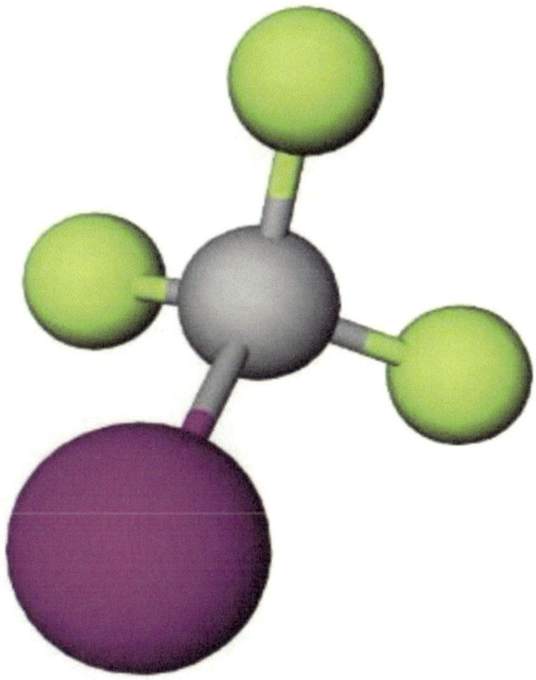

Figure 14.
Molecular structure of CF_3I.

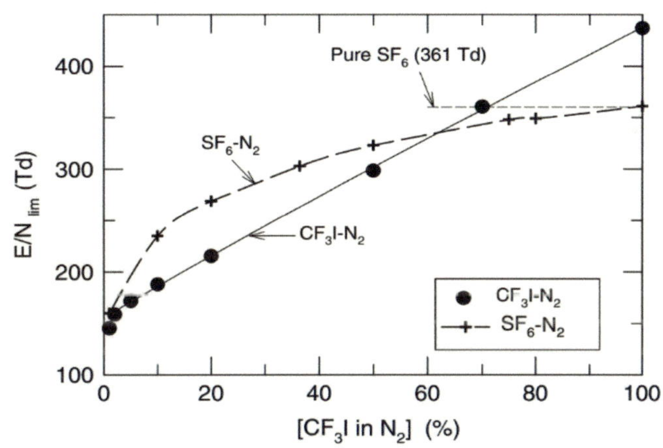

Figure 15.
Comparison of insulation strength between CF_3I/N_2 and SF_6/N_2.

and conducted breakdown test on CF_3I/CO_2 gas mixture with different mixing ratio, different gas pressure, and different discharge gap distance under extremely uneven electric field [27]. The experimental results show that the mixing of trace CF_3I can significantly increase the breakdown voltage, and the breakdown voltage tends to be stable when the content of CF_3I gas is over 6%. Although there is still a gap of the breakdown voltage between CF_3I/CO_2 and SF_6, the trace amount of CF_3I has a good synergistic relationship with CO_2 mixture. The relationship between breakdown voltage and CF_3I content under different electrode distances is shown in **Figure 16**, and the synergistic coefficient of CF_3I/CO_2 gas mixture calculated by Takuma's equation is shown in **Table 2**. The author believes that the negative synergetic effect coefficient is caused by the saturation of the breakdown voltage and

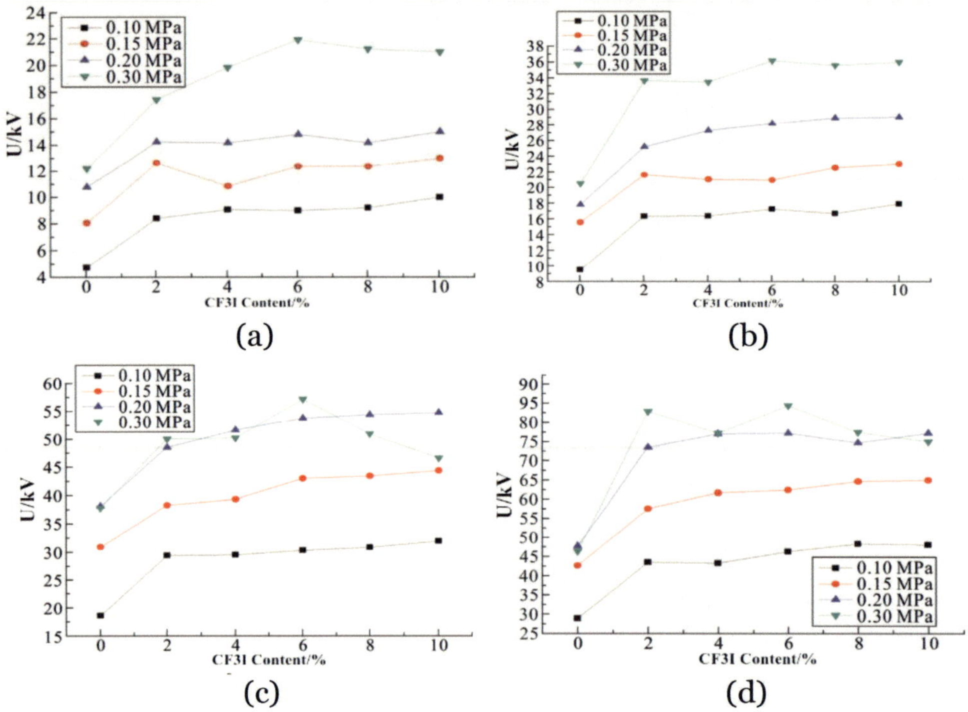

Figure 16.
Relationship between breakdown voltage and CF_3I content under different electrode distance. (a) 5 mm, (b) 10 mm, (c) 20 mm, (d) 30 mm.

Gap distance /mm	Pressure /MPa	Synergetic coefficient			
		2%CF_3I	4%CF_3I	6%CF_3I	8%CF_3I
5	0.10	0.11	0.14	0.35	0.72
	0.15	0.02	0.52	0.22	0.60
	0.20	0.06	0.16	0.08	0.99
	0.30	0.17	0.10	−0.14	−0.09
10	0.10	0.06	0.15	0.13	0.68
	0.15	0.06	0.24	0.58	0.27
	0.20	0.13	0.12	0.12	0.04
	0.30	0.05	0.15	0.02	0.20
20	0.10	0.06	0.15	0.22	0.36
	0.15	0.21	0.40	0.18	0.31
	0.20	0.15	0.16	0.10	0.04
	0.30	−0.07	−0.20	−0.82	−1.32
30	0.10	0.08	0.22	0.15	−0.05
	0.15	0.12	0.11	0.19	0.06
	0.20	0.04	0.01	0.01	0.38
	0.30	−0.05	−0.05	−0.37	−0.32

Table 2.
Synergetic coefficients of CF_3I/CO_2 gas mixtures using Takuma's equation.

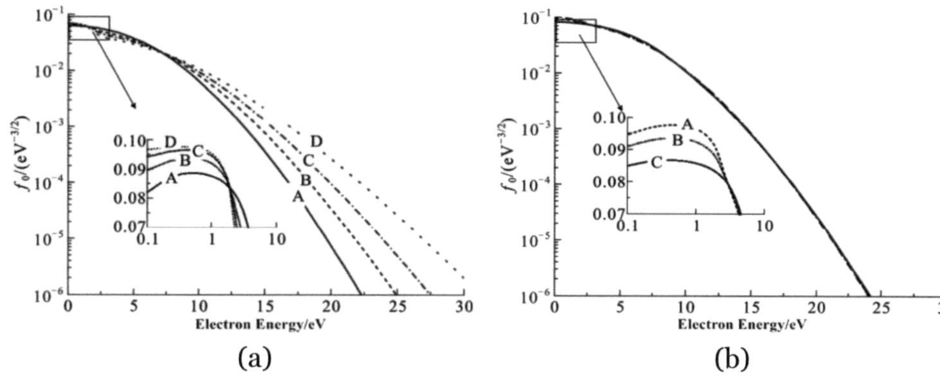

Figure 17.
EEDF of CF_3I ternary gas mixtures. (a) $CF_3I/SF_6/N_2$, (b) $CF_3I/N_2/CF_4$.

does not continue to analyze the reason. In fact, this phenomenon may be related to the abnormal breakdown of the gas mixture containing CO_2.

In addition to binary CF_3I gas mixture, some scholars have studied ternary gas mixture containing CF_3I gas. Xu et al. calculated the electron energy distribution function (EEDF) of CF_3I ternary gas mixtures by solving the Boltzmann equation and propose a new equation to analyze synergistic effect. The results show that if buffer gases such as N_2 or CO_2 are concluded in the ternary mixture, the distribution of low-energy electrons in EEDF increases, leading to the synergistic effect, but there is no synergistic effect with CF4. The mixture of two strongly electronegative gases, CF_3I/SF_6, shows weak negative synergistic effect, while the addition of N_2 or CO_2 can reduce the negative synergistic effect [28]. The EEDF of CF_3I ternary gas mixtures is shown in **Figure 17**.

The computational formula of synergistic effect coefficient proposed by Xu is as follows:

$$\xi = \frac{(E/N)_{\text{lim,mix}}}{\sum x_i (E/N)_{\text{lim,i}}} - 1 \tag{4}$$

where $(E/N)_{\text{lim, mix}}$ is the critical reduced field intensity of gas mixture and $(E/N)_{\text{lim, i}}$ is the critical reduced field intensity of single gas. x_i is the molar fraction of the gas component. Fix the buffer mole fraction of gas to 0, 10, and 30%, and change the mixing ratio of CF_3I and SF_6 gas; the $(E/N)_{\text{lim}}$ and synergistic effect coefficient of CF_3I ternary gas mixtures are shown in **Figure 18**.

3.3 C_4F_7N

C_4F_7N is an SF_6 alternative gas jointly developed by Alstom of France and 3 M of the United States. The commodity name of this gas is Novec 4710, which is an organic compound containing four C atoms and seven F atoms. Its molecular structure is shown in **Figure 19**. The chemical characteristics of the gas are similar to those of fluoro organic gas, and the chemical characteristics are relatively stable, which can achieve good compatibility with other materials in electrical equipment. The relative molecular weight of C_4F_7N is 195.0, and it also has a high liquefaction temperature, which is −4.7°C [29]. Therefore, it also needs to be mixed with buffer gas.

C_4F_7N gas mixture is a popular alternative to SF_6. At present, most researches on SF_6 alternative gases are concentrated on this gas. In order to obtain the optimal buffer gas type and mixing ratio in C_4F_7N gas mixture, Hu et al. studied the power frequency breakdown performance and synergistic characteristics of C_4F_7N/CO_2 and C_4F_7N/N_2 gas mixture with uniform electric field [30]. The gas pressure is

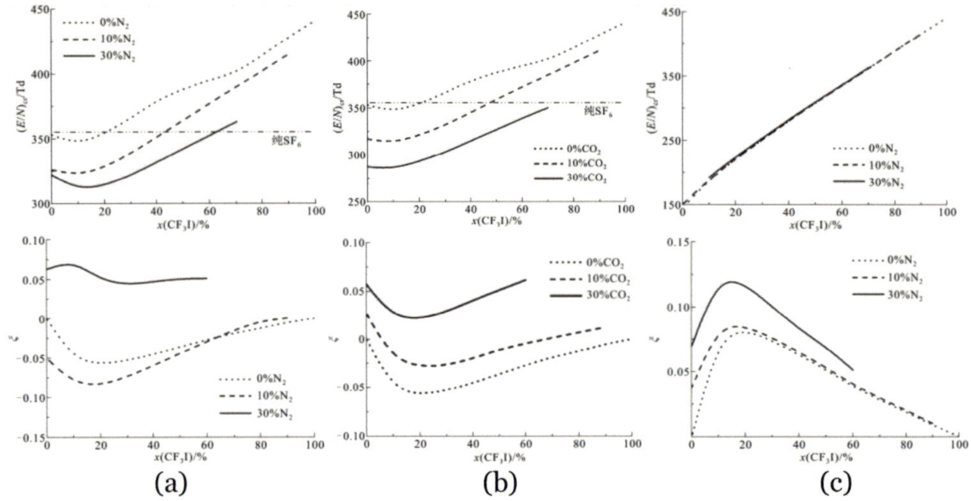

Figure 18.
The (E/N)$_{lim}$ *and synergistic coefficient of* CF_3I *gas mixtures. (a)* $CF_3I/SF_6/N_2$*, (b)* $CF_3I/SF_6/CO_2$*, (c)* $CF_3I/CF_4/N_2$*.*

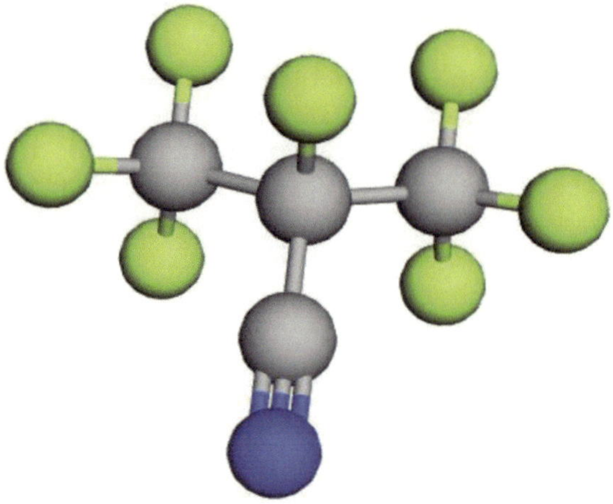

Figure 19.
Molecular structure of C_4F_7N*.*

0.1–0.7 MPa, and the C_4F_7N content ratio in gas mixture is 5–20%. The experimental and calculation results show that the C_4F_7N/CO_2 and C_4F_7N/N_2 gas mixture both have synergistic effects, and the synergistic effect of C_4F_7N/CO_2 is stronger than C_4F_7N/N_2. The interaction between C_4F_7N and CO_2 bimolecular is stronger than that of C_4F_7N and N_2, and the research indicates that there is a certain correlation between the synergistic effect of C_4F_7N gas mixture and the intermolecular interaction. This result presents the theoretical calculation and analysis direction for the qualitative judgment of the synergistic effect of C_4F_7N gas mixture. The synergistic coefficient h of C_4F_7N/CO_2 and C_4F_7N/N_2 calculated by Guo Can's equation is shown in **Table 3**.

Zheng et al. also studied the synergistic effect and insulation performance of C_4F_7N/CO_2 gas mixture and found that the synergistic coefficient was related to avalanche parameters of pure gas. They proposed a graphical method based on the Wieland approximation to calculate the critical electric field of gas mixture.

Gas type	Pressure /MPa	Coefficient *h*			
		5% C_4F_7N	7% C_4F_7N	9% C_4F_7N	13% C_4F_7N
C_4F_7N/CO_2	0.1	0.57	0.63	0.69	0.77
	0.2	0.49	0.59	0.63	0.69
	0.3	0.46	0.52	0.57	0.69
	0.4	0.45	0.53	0.57	0.69
	0.5	0.45	0.54	0.59	0.73
	0.6	0.47	0.55	0.62	0.75
	0.7	0.50	0.58	0.64	0.76
C_4F_7N/N_2	0.1	0.40	0.46	0.53	0.62
	0.2	0.44	0.51	0.57	0.63
	0.3	0.34	0.41	0.48	0.56
	0.4	0.35	0.42	0.48	0.56
	0.5	0.37	0.43	0.48	0.56
	0.6	0.39	0.46	0.50	0.64
	0.7	0.44	0.52	0.58	0.64

Table 3.
Synergistic coefficient h *of C_4F_7N/CO_2 and C_4F_7N/N_2 gas mixture.*

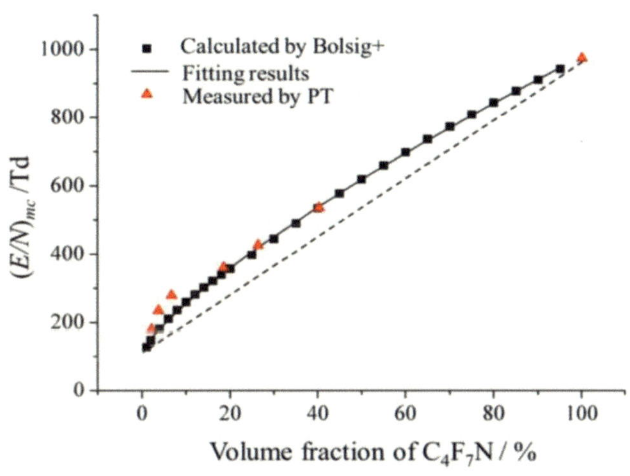

Figure 20.
The relationship of the critical electric field of gas mixture and C_4F_7N content.

Based on the formula, the C_4F_7N content takes 9% in gas mixture tends to be the optimal mixing ratio [31]. **Figure 20** shows the relationship of the critical electric field of gas mixture and C_4F_7N content.

4. Conclusions

With the development of the power system, gas-insulated equipment such as GIS will be more and more widely used. At present, SF_6 gas is still the most used insulating gas in power systems. However, due to the greenhouse effect of SF_6 gas and the consensus of countries on the development of low-carbon clean energy systems, it is imperative to use new environmental friendly insulating gas in power equipment. Looking for SF_6 replacement gas will also be a continuing research hotspot in the field of electrical engineering. According to the current research status, the replacement of SF_6 by single gas and its application in gas insulation equipment is still in the laboratory research stage. The use of an electronegative gas mixed with a buffer gas is a well-feasible solution. This paper reviews the research status of the synergistic effect of insulating gas mixtures including SF_6 gas. It has been found that some achievements have been made in the process and mechanism of synergistic effect, the types of synergistic effects under different conditions, and their influencing factors:

1. SF_6, c-C_4F_8, CF_3I, and C_4F_7N gas all have synergistic effects when mixed with various buffer gases. SF_6/N_2 gas mixture will have negative synergistic effect under the lightning impulse. The synergistic effect of the gas mixture composed of c-C_4F_8, CF_3I and C_4F_7N and CO_2 is more significant than that after mixing with N_2.

2. The synergistic effect of gas mixture is affected by the type of gas and external conditions such as gas pressure, electric field type, voltage type, etc. The synergistic effect of the same gas mixture may change with the change of external conditions, such as from synergistic effect to negative synergistic or from negative synergistic to synergistic effect.

3. The synergistic effect of the gas mixture has a certain relationship with the electronic transport parameters. However, since the assumptions of the parameter calculation differ greatly from the external conditions of the experimental measurements, the interpretation of the synergistic effect is not universal.

4. The synergistic effect of gas mixture has a relationship with the discharge pattern. When synergistic effect occurs, the initial discharge voltage of the streamer discharge rises. When negative synergistic phenomenon occurs, the streamer discharge gradually changes to leader discharge, and the discharge starting voltage decreases.

Author details

Su Zhao and Dengming Xiao*
Department of Electrical Engineering, Shanghai Jiao Tong University, Shanghai, China

*Address all correspondence to: dmxiao@sjtu.edu.cn

IntechOpen

© 2020 The Author(s). Licensee IntechOpen. This chapter is distributed under the terms of the Creative Commons Attribution License (http://creativecommons.org/licenses/by/3.0), which permits unrestricted use, distribution, and reproduction in any medium, provided the original work is properly cited.

References

[1] Koch H, Hopkins M. Overview of gas insulated lines (GIL). In: IEEE Power Engineering Society General Meeting. 2005. pp. 940-944

[2] Yunkun D. Basic Research of the Environmentally Friendly Insulating Gas CF_3I for its Application in Electric Power Apparatus [PhD Thesis]. Shanghai: Shanghai Jiao Tong University; 2016

[3] Hertz W, Motschmann H, Wittel H. Investigations of the properties of SF_6 as an arc quenching medium. Proceedings of the IEEE. 1971;**59**(4):485-492

[4] Malik NH, Qureshi AH, Theophilus GD, et al. Prebreakdown phenomenon in SF6-N2 mixtures at various pressures. Electrical insulation. In: 1978 IEEE International Conference on Electrical Insulation. 1978. pp. 140-145

[5] Abdelsalam M, Abdellah M. Dielectric strength of compressed insulating gases and their mixtures. IEEE Transactions on Industry Applications. 1978;**IA-14**(6):516-522

[6] Wootton RE, Chantry PJ. A critique of methods for calculating the dielectric strength of gas mixtures and a proposed test for, η-synergism. In: Proceedings of the Second International Symposium on Gaseous Dielectrics. Knoxville, USA; 1980. pp. 32-42

[7] Kline LE, Davies D, Chen C, et al. Dielectric properties for SF_6 and SF_6 mixtures predicted from basic data. Journal of Applied Physics. 1979;**50**(11):6789-6796

[8] Zhao S, Xiao D, Xue P, et al. Analysis of insulation performance and polar effect of CF_3I/CO_2 mixtures. IEEE Transactions on Dielectrics and Electrical Insulation. 2018;**25**(4):1364-1370

[9] Ravishankara AR, Solomon S, Turnipseed AA, et al. Atmospheric lifetimes of long-lived halogenated species. Science. 1993;**259**(5092):194-199

[10] Rinsland CP, Brown LR, Farmer CB. Infrared spectroscopic detection of sulfur-hexafluoride (SF_6) in the lower stratosphere and upper troposphere. Journal of Geophysical Research: Atmospheres. 1990;**95**(D5):5577-5585

[11] Reilly J, Prinn R, Harnisch J, et al. Multi-gas assessment of the Kyoto protocol. Nature. 1999;**401**(6753):549

[12] Chantry PJ, Wootton RE. A critique of methods for calculating the dielectric strength of gas mixtures. Journal of Applied Physics. 1981;**52**(4):2731-2739

[13] Christophorou LG, James DR, Mathis RA. Dielectric gas mixtures with polar components. Journal of Physics D: Applied Physics. 1981;**14**(4):675

[14] Okubo H, Hayakawa N. Dielectric characteristics and electrical insulation design techniques of gases and gas mixtures as alternatives to SF_6. In: Gaseous Dielectrics X. Springer; 2004. pp. 243-252

[15] Yamada T, Ishida T, Hayakawa N, et al. Partial discharge and breakdown mechanisms in ultra-dilute SF_6 and PFC gases with N_2 gas. Journal of Physics D: Applied Physics. 2002;**35**(21):2760-2765

[16] Osmokrovic P, Stojkanovic M, Stankovic K, et al. Synergistic effect of SF_6 and N_2 gas mixtures on the dynamics of electrical breakdown. IEEE Transactions on Dielectrics and Electrical Insulation. 2012;**19**(2):677-688

[17] Hayakawa N, Yoshitake Y, Koshino N, et al. Impulse partial discharge characteristics and their mechanisms under non-uniform

electric field in N_2/SF_6 gas mixtures. IEEE Transactions on Dielectrics and Electrical Insulation. 2005;**12**(5):1035-1042

[18] Chen Q, Xiao D, Qiu Y. Discharge characteristics of SF_6/N_2 gas mixture. Journal of Xi'an Jiaotong University. 2001;**35**(4):338-342

[19] Nakanishi K. New gaseous insulation. IEEE Transactions on Electrical Insulation. 1986;**EI-21**(6): 933-937

[20] Takuma T, Watanabe T, Kita K. Breakdown characteristics of compressed-gas mixtures in nearly uniform fields. Proceedings of the Institution of Electrical Engineers. 1972;**119**(7):927-928

[21] Guo C, Zhang Q, Ma J, et al. Negative synergistic effect of SF_6/N_2 gas mixtures in non-uniform electric field under lightning impulse. IEEE Transactions on Dielectrics and Electrical Insulation. 2018;**25**(4):1393-1402

[22] Zhao S, Jiao J, Zhao X, et al. Synergistic effect of c-C4F8/N2 gas mixtures in slightly non-uniform electric field under lightning impulse. In: IEEE Electrical Insulation Conference. IEEE; 2016. pp. 531-534

[23] Xing W, Zhang G, Li K, et al. Experimental study of partial discharge characteristics of C_4F_8/N_2 mixtures. Proceeding of the CSEE. 2011;**31**(7):119-124

[24] Yamamoto O, Takuma T, Hamada H, et al. Applying a gas mixture containing c-C_4F_8 as an insulation medium. IEEE Transactions on Dielectrics and Electrical Insulation. 2001;**8**(6):1075-1081

[25] Zhao S, Xiao D, Xue P, et al. Experimental research on polarity effect of CF_3I/N_2 mixtures under lightning impulse. IEEE Transactions on Dielectrics and Electrical Insulation. 2018;**25**(4):1357-1362

[26] De Urquijo J, Juárez AM, Basurto E, et al. Electron impact ionization and attachment, drift velocities and longitudinal diffusion in CF_3I and CF_3I–N_2 mixtures. Journal of Physics D: Applied Physics. 2007;**40**(7):2205

[27] Jiao J, Xiao D, Zhao S, et al. Insulation characteristics for CF_3I-CO_2 gas mixtures with low proportion of CF_3I in extreme non-uniform electric field. High Voltage Engineering. 2017;**43**(3):772-779

[28] Xu L, Cheng L, Du W, et al. Analysis of the insulation characteristic of CF_3I ternary gas mixture using Boltzmann equation. High Voltage Engineering. 2017;**43**(3):721-726

[29] Li Y, Zhang X, Xiao S, et al. Decomposition properties of C_4F_7N/N_2 gas mixture: An environmentally friendly gas to replace SF_6. Industrial & Engineering Chemistry Research. 2018;**57**(14):5173-5182

[30] Hu S, Zhou W, Zheng Y, et al. Power Frequency Breakdown Experiments and Synergistic Effect Analysis of C_4F_7N/CO_2 and C_4F_7N/N_2 Mixtures. High Voltage Engineering. In press

[31] Zheng Y, Yan X, Chen W, et al. Calculation of electrical insulation of C_4F_7N/CO_2 mixed gas by avalanche characteristics of pure gas. Plasma Research Express. 2019:025013